ROUTLEDGE LIBRARY EDITIONS:
POLLUTION, CLIMATE AND CHANGE

Volume 8

THE VALUE OF THE WEATHER

THE VALUE OF THE WEATHER

W. J. MAUNDER

Routledge
Taylor & Francis Group

LONDON AND NEW YORK

First published in 1970 by Methuen & Co. Ltd

This edition first published in 2020
by Routledge
2 Park Square, Milton Park, Abingdon, Oxon OX14 4RN

and by Routledge
52 Vanderbilt Avenue, New York, NY 10017

Routledge is an imprint of the Taylor & Francis Group, an informa business

© 1970 W. J. Maunder

British Library Cataloguing in Publication Data
A catalogue record for this book is available from the British Library

ISBN: 978-0-367-34494-8 (Set)
ISBN: 978-0-429-34741-2 (Set) (ebk)
ISBN: 978-0-367-36255-3 (Volume 8) (hbk)
ISBN: 978-0-367-36259-1 (Volume 8) (pbk)
ISBN: 978-0-429-34491-6 (Volume 8) (ebk)

Publisher's Note
The publisher has gone to great lengths to ensure the quality of this reprint but
points out that some imperfections in the original copies may be apparent.

Disclaimer
The publisher has made every effort to trace copyright holders and would welcome
correspondence from those they have been unable to trace.

The Value of the Weather

W. J. MAUNDER

METHUEN & CO LTD
11 NEW FETTER LANE LONDON EC4

First published 1970 by
Methuen & Co Ltd
11 New Fetter Lane, London EC4
© *1970 W. J. Maunder*
Printed in Great Britain by
Butler & Tanner Ltd, Frome and London

SBN 416 16060 3   Hardback
SBN 416 18200 3   Paperback

Distributed in the USA by
Barnes & Noble Inc.

Contents

*

Plates

Figures

Tables

Foreword

Man has always been aware that weather affects his life and his activities. Some of the earliest known literature contains mention of storms, floods and droughts as important events in the history or folklore of a group of people. Modern man finds himself living in a very complex environment, with access to tools, facilities and resources that enable him to accomplish things ancient man never imagined. Moreover, modern man finds many of his activities *more* sensitive to weather events than was the case with his ancestors. In addition, some of man's activities are beginning to have substantial influence on the atmosphere.

One of the tools that science and technology have produced gives man the ability to generate and process enormous amounts of weather information. A century or more ago, man rarely knew much about weather events occurring beyond his visual horizon. Modern man can have (if he wants it badly enough to pay for it) a continuous flow of weather information on a global scale. He also has at his disposal the tools to help him analyse and use large quantities of information for his potential benefit.

This book is important because it clearly emphasizes two important ideas:

> Man is affected both by the atmosphere *and* by information about the atmosphere.
>
> Man is not just a passive object, subject to weather events without recourse, but he can and does react to the atmosphere through his ability to make decisions.

In *The Value of the Weather*, Dr Maunder has clearly described (and thoroughly documented) the effect of particular weather events on many of man's activities. This book should prove to be interesting and useful to meteorologists, economists, geographers, political scientists and their students as well as to many readers who are not in any of the above professions but who are concerned about man and his environment.   JAMES D. McQUIGG

E.S.S.A. State Climatologist for Missouri, Columbia, Missouri, U.S.A.

Preface

During the 1960's several notable events took place concerning man and his atmosphere. These included the launching of the first weather satellite, the establishment of the World Weather Watch, routine numerical weather forecasting and the growing awareness of atmospheric pollution and weather modification.

There has also been increasing interest in the benefits and costs of the weather and the climate, brought about by a number of things: notably, man's increasing ability to modify his atmospheric environment, the ever-improving forecasts of tomorrow's weather conditions, the increasing losses of property and income resulting from extreme weather events, as well as the distinct possibility of accurate monthly and seasonal weather forecasts. In addition, the World Weather Watch and Global Atmospheric Research Programmes (both of which are expected to become fully operational in the early 1970's) have focused attention on the need for a critical analysis of the utilization and development of our atmospheric resources.

Most textbooks on the 'atmosphere', whether they be primarily meteorological or climatological in outlook, are chiefly concerned with physical, dynamic or descriptive aspects. One or two books discuss some aspects of the use of the atmosphere as a resource, but with few exceptions no attempt is made to place a value on these resources; nor is any attempt made to assess the economic importance of weather forecasting, weather modification or even the more basic resource – the natural weather and climate – to man.

The purpose of this book is to bring together, for the first time, the most significant and pertinent associations between man's economic and social activities, and the variations in his atmospheric environment. Basically this involves an understanding of, and appreciation for, the resources of our atmospheric environment. To do this, specific attention is focused on economic activities and the weather, the economic analysis of weather, and the benefits and costs of weather information, including weather forecasting and weather modification. In addition some of the many sociological, physiological, political, planning and legal aspects of 'man's' atmospheric resources are discussed.

Specifically, the book has been written in the hope that it will bridge the gap between the physical, dynamical and descriptive aspects of the atmosphere, and the economic, social, political and legal aspects of man and his environment. As such it should appeal to students at various levels in the atmospheric sciences, geography, economics, sociology, agriculture, marketing and forestry. In addition the topics covered in the book should be of interest to various government departments and associated agencies and corporations concerned with the benefits and costs of weather and climate, government and private meteorological services, the commercial and business world, the informed politician, as well as 'John Citizen' who daily is affected by the weather.

Extensive use has been made of the relevant literature in the field of economic climatology and economic meteorology, and reports and references to studies in many parts of the world are discussed. More than 400 of the 750 references have appeared since 1965.

Tomorrow's weather is of importance to many people, but perhaps of greatest concern is understanding why it is of importance, a factor related directly to the recognition that the atmosphere is a resource. *The Value of the Weather* will enable those interested in the atmospheric environment to appreciate fully the necessity for a careful evaluation of these resources. At the same time it will focus attention on an aspect of our environment which for so long has been very much neglected.

ACKNOWLEDGMENTS

It is a pleasure to acknowledge the substantial help I have had from many friends and associates in the preparation of this book. Specifically I wish to thank the New Zealand Meteorological Service, and the Department of Geography at the University of Otago, Dunedin, New Zealand, for their initial encouragement, and the Departments of Geography and Atmospheric Science at the University of Victoria, Victoria, Canada, and the University of Missouri, Columbia, U.S.A., respectively, for their assistance and forbearance while the manuscript was being written. A special word of appreciation is extended to Dr W. R. D. Sewell and Mr A. D. Whitmore of the University of Victoria, and Dr J. D. McQuigg of the University of Missouri for their helpful suggestions, and to

Dr W. L. Decker and Dr J. D. McQuigg of the University of Missouri, for the provision of time and facilities.

Particular thanks are also expressed to the following people who made valuable contributions: Miss Diane Draper, Miss Dawn McMahon and Mr D. J. Murphy of the University of Victoria, Mr P. Prasad of the University of Hawaii, and many of my students, for their constructive comments on preliminary versions of the manuscript; Messrs O. Lange, M. Brown and G. A. Anaka of the University of Victoria, and Mrs Sandra Shore of the University of Missouri, for their assistance with bibliographic details; Miss René Stovold, Mrs Gladys Howard and Mrs Donna Schwarz of the University of Victoria, Mrs R. D. Johnson of Columbia, Missouri, and my wife Melva, for their secretarial assistance; and Mr J. F. Bryant of the University of Victoria, for his cartographic expertise. Photographs were kindly supplied through the courtesy of the Environmental Science Services Administration (E.S.S.A), and *The Columbia Missourian*.

Acknowledgment is also made to several granting agencies who directly or indirectly have supported my research in 'atmospheric resource studies' during the 1960's. Aspects of the several published and unpublished reports of this research are included, where appropriate, in the book. Specific acknowledgment is made to the University Grants Committee, Wellington, New Zealand; the National Advisory Committee on Geographical Research, Ottawa, Canada; the Meteorological Branch, Department of Transport, Toronto, Canada; the University of Victoria, Canada; and the National Science Foundation, Washington, D.C., U.S.A., for their assistance.

The author would also like to thank the following learned societies, editors, publishers, organizations, and individuals for permission to quote material. Specific acknowledgment is given in the text, and in the references.

(1) *Learned Societies:* American Meteorological Society; Association of American Geographers; New Zealand Geographical Society.

(2) *Editors and/or Publishers: Advertising Age*; George Allen and Unwin, Ltd, London; *Business Week*; *Current Affairs Bulletin* (University of Sydney); *The Daily Colonist*, Victoria, British Columbia; *The Economist*; Elsevier Publishing Company, Amsterdam; Gray Printing Company, Du Bois, Pennsylvania; Her Majesty's Stationery Office, London; *International Journal*

of Biometeorology; McGraw-Hill, New York; Prentice-Hall, Englewood Cliffs, New Jersey; Queen's Printer, Ottawa; *Saturday Review*; *Scientific American*; Time Inc.; University of Oklahoma Press, Norman, Oklahoma; University of Washington Press, Seattle; *The Wall Street Journal*.

(3) *Organizations:* Bureau of Agricultural Economics, Canberra; Bureau of Meteorology, Melbourne; Environmental Science Services Administration, Washington, D.C.; Federal Reserve Bank of Chicago; Legislative Reference Service, Library of Congress, Washington, D.C.; Motor Vehicle Branch, Victoria, British Columbia; The National Industrial Conference Board, New York; National Science Foundation, Washington, D.C.; The RAND Corporation, Santa Monica, California; Radio CKLG, Vancouver; The Travellers Research Corporation, Hartford, Connecticut; United Research Company, Cambridge, Massachusetts; Weather Consultants of Canada, Toronto; World Meteorological Organization, Geneva.

(4) *Individuals:* W. S. Barry; W. B. Beckwith; J. C. Beebe; C. von E. Bickert; L. E. Borgman; A. Boyer; T. D. Browne; I. Burton; S. A. Changnon, Jr; Marion Clawson; G. W. Cloos, H. J. Critchfield; L. W. Crow; J. H. Curtiss; F. Danzig; J. C. Day; J. P. Doll; J. H. Duloy; D. L. Eberly; R. D. Fleagle; M. H. Freeman; D. G. Friedman; J. E. Haas; N. L. Hallanger; R. L. Hendrick; J. R. Hibbs; P. V. Hobbs; M. M. Hufschmidt; R. E. Huschke; L. B. Johnson; S. A. Johnson; A. M. Kahan; C. G. Knudsen; H. E. Landsberg; T. W. Langford, Jr; F. P. Linaweaver; F. Linden; F. J. Lyden; W. McDermott; J. E. McDonald; J. W. McLeary; J. D. McQuigg; D. E. Mann; W. H. Miller; T. H. Moorer; E. A. Morris; J. Namias; H. L. Penman; Mary T. Petty; R. R. Rapp; W. O. Roberts; T. P. Rothrock; J. A. Russo, Jr; M. Scarpa; F. Schrivan; W. R. D. Sewell; G. W. Shak; G. A. Shipman; M. Shor; J. S. Smith; A. Spilhaus; G. E. Stout; A. L. Sugg; W. H. Terjung; J. C. Thompson; R. G. Thompson; R. L. Tobin; C. Toebes; S. W. Tromp; R. M. White; O. Wilson; A. D. Woodland; F. H. Zegel.

September 1969   W. J. MAUNDER,*
                                   Department of Atmospheric Science,
                                   University of Missouri
       * *Present affiliation: New Zealand Meteorological Service, Wellington, New Zealand.*

I. Introduction

> The past two or three decades have witnessed great changes in the world – the emergence of many new nations with fragile economies, an explosion in the world's population, growing urbanization, increasing industrialization, greater complexity in the forms of economic and social life and greater interdependence among these forms, new and more sophisticated technology, and such new activities as the exploration of outer space. With these changes has come an increasing awareness of how intimately all the things we do are bound up with the weather. (White, 1967)

These words, spoken on the occasion of the centenary celebration of the Norwegian Meteorological Services, clearly illustrate the importance of an evaluation of our atmospheric resources, for despite the enormous advances that have been made in the atmospheric sciences and weather forecasting, many people justifiably entertain a scepticism about the validity of tomorrow's forecast.

There are many reasons for this lack of confidence, but probably one of the most significant is the fact that the atmosphere has seldom been considered a resource. But the atmosphere *is* a resource and as Miller (1956) states: 'We have to learn how to live within our climatic income.' The growing awareness of the problem of air pollution in most of the world's cities is evidence of man's growing appreciation of a limited climatic income, and air pollution illustrates only one case in which the atmosphere may be considered a resource. Indeed, in an increasingly crowded and fast-moving industrial world, weather as an input, impinges on nearly every economic pursuit from agriculture to supersonic flight. In the United States alone, for example, the costs of extreme weather conditions are estimated to exceed 1,000 lives and $10 billion a year. Accordingly, it is often suggested that if weather could be predicted more reliably and for a longer period of time, and brought under some measure of control, the benefits could be great. But, what are the benefits and are they calculable? What, for example, is the 'value' of one inch of rain to California, Japan or New Zealand? What does a 'wet' spring mean to these areas? Further, what does a *forecast* wet spring mean to these areas, if the forecast proves to be correct, or incorrect?

It is well known that many sectors of the economy are significantly affected by weather conditions. Little is known, however, about the importance of weather-sensitive activities in any regional or national output. A factor of more immediate concern is to determine the extent to which a value can be placed on meteorological information that is made available to, and used by, all entrepreneurs of weather-sensitive activities. There are also many social and psychological benefits of accurate weather information such as the saving of lives and/or an increased sense of security, which cannot be expressed in monetary terms.

The recognition and evaluation of the atmosphere as a resource raises many complex economic, social and institutional problems, and as far as is known, no published comprehensive study on the effect of weather on any regional or national economy is as yet available. During the 1960's, however, there were increasing demands for improved techniques for measuring the impacts of weather and climate on economic activities (see for example: Australian Bureau of Meteorology (1965); U.S. Dept. of Commerce, Weather Bureau (1964); Sewell (1966); McQuigg (1965); Maunder and Whitmore (1969); Sewell, Kates and Maunder (1968); and Thompson (1966)). This rising demand was due to several factors including the increasing use of the atmosphere as a transportation route, increased losses of property resulting from catastrophic weather conditions, increasing awareness of atmospheric pollution, attempts to modify the atmosphere and the increasing ability to forecast the weather accurately, and over longer periods of time.

Serious attempts to evaluate the atmosphere have, unfortunately, evolved only after equally serious attempts at modification of that same environment. It is abundantly clear, however, that we must *first* find out the value of our 'natural' atmospheric resources, for then, and only then, will we be in a position to decide whether these resources are worth forecasting, modifying or controlling.

This book attempts to bring together the most significant and pertinent associations between man's economic and social activities, and the variations in his atmospheric environment. In doing so it is hoped to provide a starting point for econoclimatic studies on a national and international scale, and a greater appreciation of the value of the atmospheric environment.

BIBLIOGRAPHY

AUSTRALIAN BUREAU OF METEOROLOGY, 1965: *What is Weather Worth?* Papers presented to the Productivity Conference, Melbourne, Australia, 31 August–4 September, pp. 1–16.

MAUNDER, W. J. and WHITMORE, A. D., 1969: The value of weather: challenge of assessment. *Aust. Geog.*, 11: 22–8.

MILLER, A. A., 1956: The use and misuse of climatic resources. *Adv. of Sci.*, 13: 56–66.

SEWELL, W. R. D. (Editor), 1966: *Human Dimensions of Weather Modification.* University of Chicago, Dept. of Geography, Research Paper No. 105, 423 pp.

SEWELL, W. R. D., KATES, R. W. and MAUNDER, W. J., 1968: Measuring the economic impact of weather and weather modification: A review of techniques of analysis. In: SEWELL, W. R. D., *et al.*, 1968: *Human Dimensions of the Atmosphere.* National Science Foundation, Washington, D.C., pp. 103–12.

THOMPSON, J. C., 1966: The potential economic and associated values of the World Weather Watch. *World Weather Watch Planning Report*, No. 4, 35 pp.

U.S. DEPARTMENT OF COMMERCE WEATHER BUREAU, 1964: *The National Research Effort on Improved Weather Description and Prediction for Social and Economic Purposes.* Prepared by the U.S. Dept. Commerce, Weather Bureau, for the Federal Council for Science and Technology, Interdepartmental Committee on Atmospheric Sciences, 84 pp.

WHITE, R. M., 1967: Meteorology on a new threshold. *Bull. Amer. Met. Soc.*, 48: 250–7.

ADDITIONAL REFERENCES

ATKINSON, B. W., 1968: *The Weather Business.* Aldus Books, London, 192 pp.

CURRY, L., 1952: Climate and economic life. A new approach. *Geog. Rev.*, 42: 367–83.

DUNCAN, J. F., 1968: Some possible contributions of science to the national economy. *New Zealand Sci. Rev.*, 26: 44–51.

HARE, F. K., 1966: The concept of climate. *Geography*, 51: 99–110.

HORNIG, D. F., 1965: The atmosphere and the nation's future. *Bull. Amer. Met. Soc.*, 46: 438–42.

JACOBS, W. C., 1959: Meteorology applied is a social affair. *Bull. Amer. Met. Soc.*, 40: 179–81.

McGEE, D. A., 1969: Treasures of energy: natural resources of the Ninth and Tenth Federal Reserve Districts. *Monthly Review* (Federal Reserve Bank of Kansas City), February, pp. 3–9.

McMILLAN, J. R. A., 1965: Water, agricultural production and world population. *Aust. Jour. Sci.*, 28: 135–41.

POPKIN, R., 1967: *The Environmental Science Services Administration.* Praeger, New York, 288 pp.

SEWELL, W. R. D., KATES, W. and PHILLIPS, L. E., 1968: Human response to weather and climate – geographical contributions. *Geog. Rev.*, 68: 262–80.

STAGG, J. M., 1961: Meteorology and the community. *Quart. Jour. Roy. Met. Soc.*, 87: 465–71.

WHITE, G. F., 1966: Approaches to the study of human dimensions of weather modification. In: SEWELL, W. R. D., ed., 1966: *Human Dimensions of Weather Modification*. University of Chicago, Dept. of Research Paper No. 105, pp. 19–23.

W.M.O., 1964: *Weather and Man – The Role of Meteorology in Economic Development*. World Meteorological Organization, Geneva, Switzerland, 143.TP.67.

II. Weather variations

A. THE SETTING

Violent weather events, such as hurricanes, tornadoes, major floods and droughts, often result in catastrophic losses of property and income, and on many occasions losses of life. Hurricanes, for example, cause more than $250 million damage in the United States in an average year. In addition, hurricanes have caused more than 20,000 deaths in the U.S. since 1900. Tornadoes in the United States cause further losses in excess of $200 million a year, and result in hundreds of deaths as well. Lightning and hail also cause significant losses.

Longer-term weather variations are also important. During the first half of the twentieth century, for example, much of the Northern Hemisphere experienced a period of 'unusually warm' weather. The fifties and sixties have shown significant signs of a change to a wetter and colder climate, however, and the winter of 1962/63 in Europe, and 1968/69 in Anglo-America, are reminders of the importance of 'climatic' as well as 'weather' changes.

One of the motivations for weather modification is the recognition of the value of certain weather variations as inputs into the production process. Increased precipitation at the right place and time, for example, can improve the operating efficiency of a hydro-electric power system. Recreational opportunity might also be changed by weather modification, such as by increasing the snowfall to prolong the skiing season. In addition, if a substantial modification of weather were possible, shifts in the timing and location of agricultural activities would probably result. Several studies of the immediate impact of weather modification have been undertaken. Many of these are designed to show that an increase in precipitation (or some other weather factor) would increase the output of a particular enterprise such as a farm, or group of farms. Generally, the values are expressed in gross terms. No attempt is made to trace effects beyond the enterprise in question, to discern adverse impacts such as the effect on prices if production is increased, or to weight weather modification with other possible adjustments, such as hazard insurance or irrigation. Such aspects are of course significant, and the

Report of the Task Group on the Human Dimensions of the Atmosphere (Sewell *et al.*, 1968) is of importance in this respect.

Air pollution is also a form of weather modification which affects the 'value of the atmosphere'. Spilhaus (1966) summarized the position admirably.

> In the past we have kicked around our wastes till they got lost – but they don't get lost. Carbon dioxide accumulates in the atmosphere, detergents froth up in our ground water, radioactivity and pesticides are concentrated by the plants and animals we eat. Trail of our jets cause stratus clouds, the fumes of our automobiles and industries cause smog. The growth and concentration of our cities cause climatic changes by waste heat.

It is readily apparent that we are modifying our weather by pouring our wastes into the atmosphere. In some areas we are also modifying the water balance and the heat balance by radically changing the 'landscape', and in most cases we are doing such things inadvertently. We must therefore re-examine present policies and in many cases, develop new policies which will outline sound objectives for the utilization and development of our atmospheric resources.

B. THE NATURAL ATMOSPHERE

1. Violent storms

Violent disturbances in the natural atmosphere result in tornadoes, tropical storms (hurricanes, typhoons, tropical cyclones), hailstorms, lightning, blizzards, extreme icing conditions and severe dust storms. The resulting damage, often severe, may also result in loss of life and inevitably has a weighty price tag attached. The 'damage' caused by a violent storm is generally a function of the severity of the storm, the specificity of the warning, the time of the warning, the population of the area, the construction of the settlement and the organization of the population. For example, *Time* (October 13, 1967) reports that one hailstorm destroyed $2,000,000 worth of tobacco in North Carolina, and that in a single year, hailstorms 'cost' U.S. insurance companies tens of millions of dollars. The cost and damage created by hurricanes is usually on a larger scale. In addition, coastal storms are often severe in some areas, such as the $200,000,000 storm on the eastern seaboard of the United States on March 5–9, 1962 (Mather, Adams and Yoshioka, 1965).

Few calamities can change the appearance of a community as

suddenly and disastrously as a tornado. Homes are reduced almost instantly to a mass of wreckage or carried away entirely. Sometimes a whole town is practically wiped out, as in the case of Coatesville, Indiana, in the tornado of March 26, 1948, when 80% of the town was destroyed (Flora, 1954). The death list is sometimes in excess of 100 and the number of injured is many times more. Property losses have been known to total millions of dollars in a single city.

Less severe, but more prolonged periods of adverse weather are also of consequence, the severe cold period in Europe in the winter of 1962/63 providing a good example of the 'cost' involved. *The Economist* (February 16, 1963), commenting on the effect in Britain, states:

> It may be a year before the full cost of the past seven weeks to Britain is known: after 48 days officially described as 'very cold' . . . most Britons, as individuals, companies, corporations or local governments, are still too busy keeping warm and mobile to have worried about the cost of what will surely be the most expensive two months since the war. . . . Local authorities . . . must expect their tight budgets to burst irreparably at the seams: they have poured our money (but not enough) to clear the streets, keep their water supplies fluid, and give some continuity to the services which, it now appears, they have been running on the far limits of safety for years in hope and expectation that nothing like what has happened could happen. . . .
>
> . . . the householder is in for the leanest spring since the war, faced with fuel bills (when fuel has been available) twice the size of those which adequately bought his snugness in the winters past, . . . arguing with his garage over the staggering demands for the repair of motor cars wrecked in accidents or damaged by frost and precautions (such as salt) against frost, grudgingly giving in to demands for more housekeeping money from his burdened wife (in turn dismayed by the price of what miserable vegetables her greengrocer has managed to find), and perhaps willingly, trying to settle with the plumber. . . . people are needlessly deterred from normal behaviour by alarmist descriptions of what is, although exceptional for Britain, no uncommon thing in the average European, Canadian or American state; thawed plumbing is left unlagged in the vague belief that statistically there can't be another such year soon. . . . £200 million? £1,000 million? It is hoped that somebody with access to a computer will work it all out. Then in the shock that would greet his results, some good may come from the ghastly ordeal of the past months.

a. Tropical storms

In economic terms tropical revolving storms are by far the most destructive weather phenomenon. Hurricanes, in fact, have caused more damage in the United States than any other type of natural catastrophe, with 17,000 lives lost and property damage of $5,000 million in the first 60 years of this century alone (Lane, 1965). An example of the destructive and tragic consequences of hurricanes occurred on June 27, 1957, when Hurricane Audrey moved inland over Louisiana. In less than two days more than 500 people lost their lives, 40,000 to 50,000 cattle died (mostly by drowning), and property damage was estimated at between $150 million and $200 million. By comparison, 57 years earlier, in September 1900, a hurricane sent immense waves surging into Galveston, Texas, destroying 3,600 houses and killing 6,000 people, about one-sixth of the population.

In some other parts of the earth the loss of life has been much greater, particularly in areas subject to Pacific typhoons, and Indian cyclones; the Bay of Bengal experiencing the world's most destructive tropical storms. One of the most severe storms in the last 250 years occurred on October 7, 1737, at the mouth of the Hooghly River near Calcutta, when a violent tropical cyclone and its accompanying 40-foot storm surge destroyed 20,000 boats and other vessels and killed 300,000 people. Typhoon Vera, another example, ravaged Central Japan in September 1959, the wind, sea and rain leaving over 5,000 dead, and more than 40,000 injured or missing. In addition, Vera cut Japan's railway system in 827 places, and destroyed over 40,000 houses (Lane, 1965).

High tides associated with tropical cyclones have considerable economic consequences. Hurricane Betsy, for example, which struck the south-eastern Louisiana coast on the night of September 9, 1965, with winds of 125 mph was, economically, the second most destructive ever to hit the United States.* In this hurricane the total damage was estimated to be over $1,400 million but, as Goudeau and Conner (1968) report, the greatest loss of life and property was realized only after the strong winds had subsided. In this case, the principal causes of losses were floods which had inundated a large section of New Orleans, and most of the parishes in the Delta to the south-east of the city, the floods resulting from the storm surge which moved across the low delta land with the hurricane.

The beneficial aspects of hurricanes, are often overlooked (see

* Hurricane Camille (Aug. 1969) caused an estimated $1,420·75 million damage (A. L. Sugg, per. com.).

however the important contribution of Sugg, 1968), but hurricanes or systems associated with them sometimes bring needed rain to drought-parched agriculture areas. Thus, losses in coastal areas may be offset (on a national scale) by benefits to agricultural land. The hurricane of June 1945, for example, relieved one of the worst droughts in the history of Florida and Hurricane Ginny was equally beneficial in 1963; further, of the hurricanes which struck Puerto Rico during 1899–1928, 60% were considered beneficial, 30% were locally destructive but beneficial in other areas and only 10% were overwhelmingly destructive (Lane, 1965). Accordingly, monetary losses to the economy caused by hurricanes vary considerably (Table II.1). Figures for one storm may range upwards to hundreds of millions of dollars or may be so low as to be of no consequence.

TABLE II.I: *Damage in the United States and Canada (in millions of dollars) from a few selected hurricanes (from Sugg, 1967; and Sugg and Carrodus, 1969)*

Hurricane†		Damage	Hurricane		Damage
Betsy	1965	1,419·8	Audrey	1957	150·0
Diane	1955	800·0	Cleo	1964	128·5
Carol	1954	450·0	Hilda	1964	125·0
Carla	1961	400·0	Florida	1926	111·8
New England	1938	387·1	Isobell	1964	10·0
Donna	1960	386·5	Alma	1966	10·0
Hazel	1954	251·6	Keys	1935	6·0
Dora	1964	250·0	Inez	1966	5·0
Beulah	1967	200·0	Ginny	1963	0·4*

* Loss more than offset by beneficial rains.
† Camille (1969): $1,420·75 million.

The average annual damage from hurricanes in the United States is about $300 million. In comparison, as shown in Table II.2, 'service costs' are estimated to be less than $10 million, such costs including $2·5 million for the cost of aircraft reconnaissance (based upon 150 flights during the 1964 season at $15,000 per flight), and $0·2 million per year for communications.

The cost of protection of property (houses and businesses) is very difficult to assess but it can be estimated from population figures. For example, a cost analysis team of the U.S. Weather Bureau has concluded that it takes $5,000 for protective measures for 1,000 people. Accordingly, if a population of 1·9 million is included in a hurricane warning system, and only 20% of the population makes

B

TABLE 11.2: *Average annual hurricane costs, United States and Canada (in millions of dollars) (from Sugg, 1967)*

Hurricane damage	300·0
Aircraft reconnaissance	2·5
Communications	0·2
Protection (home and business)	2·85
Evacuation	2·0
Special interests	2·0
Satellite	—
	309·55

investment in protective measures (Sugg, 1967), the total protective costs would amount to $1·9 million for the average warning or $2·85 million for the average season. The same analysis team conservatively estimated evacuation costs at $50 to $65 per family or about $15,000 per 1,000 people. Since there were 150,000 evacuees during Hurricane Hilda, and about 350,000 in Carla, evacuation costs of $2·25 million and $5·25 million for these two storms could be assessed, together with another $2·5 million for evacuation during Dora and other hurricanes during the five years 1960 to 1964. This accounts for a total of $10 million, or a yearly average of $2 million as shown in Table II.2.

Estimates for the current extremes of hurricane costs in the United States and Canada are shown in Table II.3. The maximum cost likely is estimated to be $2 billion, Sugg commenting that the $2 billion damage figure seems reasonable since it is only $320 million more than was attributed to the 1955 season and $580 million more than attributed to Betsy in 1965.

TABLE 11.3: *Hurricane costs (millions of dollars) – current extremes for a season (United States and Canada) (from Sugg, 1967)*

	Damage	Recon-naissance	Communi-cations	Protec-tion	Evacua-tion	Special	Total
Min.	0	2·0	0·15	0·0	0·0	0·2	2·4
Avg.	300	2·5	0·2	2·85	2·0	2·0	309·55
Max.	2,000	3·0	0·3	4·0	10·0	3·0	2,020·3

Research on hurricanes has in the main been focused on how to forecast them, and how to adapt and adjust to them. In forecasting, there has been an improvement in prediction, from virtually no warning during the first and second decades of this century to a current forecast 24-hour displacement error of about 100 miles. Nevertheless, the best forecast cannot possibly prevent the structural devastation brought about by a severe hurricane like Camille, Hazel or Betsy, for even with excellent warnings, villages and resort areas have been virtually destroyed by the severe hurricanes. In the case of minimal or moderate storms, however, a good forecast can help to minimize the losses. On the other hand, a poor forecast issued too late, or to too few people, or to too many, can increase costs.

In addition to forecasting their movement, modification of hurricanes is a possibility. Man has modified weather on a small scale (for example, hail suppression and rainmaking), but he does not yet have the power to 'steer' a hurricane. Whether or not he has the power to 'trigger' the occurrence of weather patterns is another matter. In the 1960's the United States has been working on 'Project Stormfury' (Simpson, 1967), which is a programme of attempted modification of full-scale hurricanes. It is realized that man does not at present have the power to disperse hurricanes, but it is hoped that he may be able to reduce their damaging effects while retaining or increasing their beneficial effects. This is one aim of 'Stormfury'. The high cost of this project has often been questioned (for example, fifteen closely co-ordinated planes are required, and each plane costs $15,000 per flight to operate) but when the stakes are considered – a possible billion dollars loss from one hurricane – the expense is trifling, for as *Time* (Copyright Time Inc. 1965) (September 17, 1965) so vividly and tragically reports:

> Betsy was one of the fiercest hellions of them all. Screaming in from the Atlantic, she feinted at the Bahamas and Cuba, veered towards the Carolinas, doubled back again – and stopped teasing. When she did, she exacted a death toll that was expected to go as high as 200.
>
> The hurricane, after hurling 140 m.p.h. winds and massive tides into the Bahamas, blew into the southern tip of Florida, where thousands of tourists and residents fled inland. . . . In the Delta lowlands, where Audrey in 1957 took 518 lives in one Louisiana parish, 250,000 refugees sought shelter in schools and churches. The Delta was even more seriously hurt than the Miami area. Overall property damage was estimated at $500 million.
>
> At week's end, as the hurricane knocked itself out over the Mississippi Valley, the Weather Bureau announced that the name Betsy, used once before as a storm designation, will be retired for at least ten years 'because of this hurricane's infamy'.

Nevertheless, as pointed out earlier, hurricanes do have beneficial effects, the heavy rains accompanying them often reviving drought-affected crops and replenishing supplies of storage water. Further, the benefits are received over a wider area than is ravaged by the hurricane's winds. For this reason all the benefits and all the costs of the tropical storms should be carefully evaluated before modification is allowed to proceed too freely.

b. Tornadoes

Tornadoes are amongst the most awesome of weather anomalies, the speed of a tornado usually allowing a very limited time for precautions to be taken. This is clearly illustrated by the example of an average-sized tornado which cut through the centre of Topeka, Kansas, on June 8, 1966 (Galway, 1966). At 7.02 p.m. this tornado was located by local radar and visual spotting, the city of Topeka was then alerted, and at 7.04 p.m. civil defence sirens sounded. At 7.15 p.m. the tornado entered the city, causing almost total destruction along an eight-mile-long, four-block-wide area. In its wake were 17 deaths, 550 injuries and $100 million property damage (including 800 dwellings destroyed, 810 dwellings with major damage and 400 dwellings with minor damage).

An earlier investigation into tornado damage was made by Flora (1954), who showed that during the period from 1916 to 1952 in the United States, an average of 230 people were killed annually by tornadoes. The yearly death toll varied from 36 in 1931 to 794 in 1925. Property damage in the same period ranged from a low of $2 million in 1916 to a high of over $40 million in 1927 and 1948. In 1927 it was as high as $25,000,000 in Missouri alone, and each year as more people move into these 'tornado-ridden areas' the property loss has increased. In the 36-year period (1916–1952) the total tornado damage amounted to over $500,000,000. For every tornado reported, Flora estimated that 1·26 people die, 16 are injured and property valued at $122,750 is destroyed.

Tornadoes can occur singly, or may also occur in families. One such 'family' of 47 was investigated by Hardy (1966). These tornadoes, which developed on April 11, 1965, tore their way across heavily populated portions of Illinois, Iowa, Michigan, Ohio and Wisconsin during the late afternoon. The resulting toll was 257 killed, over 3,000 injured and property loss of $500,000.

In recent years an improved tornado warning system has been in effect in the United States. Its value can be seen by considering the destruction caused by the Tri-state tornado of 1925, which according to Changnon and Semonin (1966) killed 695, injured 2,027 and caused damage amounting to $40,000,000 (in 1964 dollars). The cause of the high death-rate was attributed to: (1) the size of the storm, (2) lack of tornado forecast, (3) lack of immediate warning due to lack of communication caused by destruction of phone lines, (4) high speed of the tornado, (5) lack of adequate shelter and (6) lack of recognition of the approaching storm as a tornado. Four of the six factors (i.e. 2, 3, 5, 6) have changed since 1925, which generally means a lower tornado fatality rate today. The factors which determine tornado intensity and velocity still evade us, however, and as the population and development of tornado areas increases, so probably will the tornado dollar damage.

Tornado modification is the remaining alternative to reduce property damage, but modification is hampered by the rapid speed of development of the tornado and the difficulty of pinpointing where it will develop. In addition, until the genesis of tornadoes is better understood, actual modification is unlikely to be very successful, and it could possibly generate unknown side effects.

c. Hailstorms

The comparatively common hailstorm usually causes losses in the United States exceeding $50 million per year, damage to crops comprising about 80% of all hail losses reported. Table II.4 gives the losses from selected destructive hailstorms.

TABLE II.4: *Losses from the most destructive hailstorms in the United States* (from 'Hailstorms of the United States', by Snowden D. Flora, Copyright 1956 by University of Oklahoma Press)*

Date		Location	Damage
1953	June 21	Wichita, Kansas	$9,180,000
1953	July 2	Kimball, Banner and Cheyenne Counties, Neb.	$6,000,000
1951	May 29–30	Wallace County to Kearny County, Kansas	$6,215,000
1951	June 21–22	Sherman County to Reno County, Kansas	$5,716,000
1951	June 23	Sedgwick County to Allan County, Kansas	$14,340,000

* To 1953

An examination of hailstorms in Illinois and their effects has been made by Changnon (1967). He established that the occurrence of crop hail damage was closely associated with the high-speed surface winds. These data also indicated that crop damage, in general, was related to the occurrence of relatively large stones ($\frac{3}{4}$ inch or larger in diameter), and long durations of hailfall. In many instances, however, heavy crop damages occurred with hailstones of half inch or less in diameter.

In other areas of the world, hailstorms are equally destructive, and in a few localities loss of life results. This is particularly true of India, where there are records of more people being killed by hailstorms than in any other country. Lack of adequate shelter and congestion are two of the reasons for this. The greatest loss of life from a single hailstorm occurred on April 30, 1888, in the Moradabad and Beheri districts, when 246 persons perished from hailstone injuries. On another occasion (June 19, 1932), in Honan Province of China, 200 persons were killed and thousands injured in a severe hailstorm (Flora, 1956).

Until recently the only protection farmers had against heavy crop losses from hail has been insurance. Some growers still prefer to carry their own risks, but crops in the United States are now insured annually for over $1,000 million. A basic problem of crop hail insurance is the rate structure to be charged by the insurance company. These must of necessity be higher than corresponding rates on property such as buildings and automobiles. The most important factor in determining the rates is the 'loss ratio', this being the expression in dollars of an insurance company's loss per $100 of insurance written in a particular region. This is sometimes called the 'loss cost' and is computed from the experience of insurance companies for each county or township. The loss ratio may vary greatly depending upon the kind of crop insured, the average frequency and severity of hailstorms and the ground elevation. Flora (1956) indicates that in some counties in Kansas, where elevations are below 1,000 feet, the loss ratio is as little as or even less than $1 per $100, whereas in some of the high elevations in California, loss ratios exceed $12 per $100.

For centuries attempts have been made to reduce or to prevent hail falling on vulnerable areas. Lane (1965) mentions that the ancient Greeks used to sacrifice a lamb or pullet when hail threatened, and in the Middle Ages peasants erected tall poles bearing strips of parchment inscribed with incantations against the hail god. More recently, other preventive measures have been used, and most of these measures depend upon the same principle, namely to supply more nuclei similar in shape to ice crystals, on which hailstones can develop, on the assumption that the more nuclei in the cloud, the smaller the size of the individual stones. But in practice the theory rarely worked, mainly because before a potential hail cloud could be identified and seeded, its hailstones had already grown large enough to cause damage. United States and Soviet Union meteorologists among others, however, are continuing research into hail modification, Soviet achievements in this field being impressive. *Time* (October 13, 1967) reports, for example, that the Soviet solution to this problem was to use radar to identify the cloud as soon as it began to form hailstones, and to deliver a load of silver iodide to the cloud by means of a non-splintering anti-aircraft shell. During the 1965 application of this technique, according to the report, crop loss from hail was reduced to 3% in the protected areas, compared with a 19% loss in adjacent unprotected fields. In some areas the loss was cut to a tenth of normal.

d. Lightning

Lightning, in the United States, accounts for more *direct* deaths than any other weather phenomenon. Zegel (1967), in a study of such deaths during the period 1959 to 1965, showed that the average annual death-rate was approximately 140, compared with 85 as a result of tornadoes. He also found that about 75% to 85% of all lightning casualties (combined deaths and injuries) were male, and that about 70% of all injuries and fatalities occurred in the afternoon.

Fatalities and injuries attributed to lightning do not necessarily result from being struck by lightning. Some of the most disastrous losses of life have occurred when aircraft have been struck by lightning, one notable example being the explosion of a passenger jet liner over Elkton, Maryland, in December 1963, killing all 81 on board. In this case lightning is believed to have caused the explosion of residual fuel vapour in one of the aeroplane's outboard wing tanks as it passed through an unseasonable and particularly violent thunderstorm (Zegel, 1967). Other catastrophes have occurred when houses have been set on fire by a flash of lightning.

Lightning also causes a substantial amount of property damage, either from fires, explosions or falling debris. In addition, millions of dollars' worth of timber is burned each year in fires caused by lightning, this loss in some areas representing more than 70% of all forest fire losses. Because of these losses, weather modification in the form of lightning suppression is carried out (with some degree of success) in many forested areas. In the United States, property damage caused by lightning is reported in *Storm Data*, and Zegel has re-estimated these dollar values as shown in Table II.5.

TABLE II.5 : *Estimated property damage caused by lightning* (*from Zegel, 1967*)

1959	$15,025,000
1960	$18,110,000
1961	$14,540,000
1962	$18,910,000
1963	$19,340,000
1964	$11,300,000
1965	$17,400,000
Average	$16,375,000

e. Blizzards

'City Blizzard Warning Out' cried the front-page headline of the *Edmonton Journal* for December 14, 1964. The story below the headline gave the details of how the storm was expected to sweep over Alberta that night, and warned both urban and rural residents to take all possible precautions.

The blizzard which developed that night raged all day on the 15th and continued through the 16th in some areas. The combination of very low temperatures and very strong winds made it the most severe blizzard in the memory of even the oldest 'old-timers'.

These comments by Storr (1965) in a report on the 'Great Prairie Blizzard of 15 December 1964' point to one of the winter's most unwelcome events. In this case although no human lives were lost by direct exposure, livestock and wildlife losses were high, some ranchers reporting half their herds lost. Approximately 1,000 head were lost in southern Alberta during the storm, the four-week cold spell which followed reminding 'old-timers' of the winter of 1906/7 when half of Alberta's cattle perished. In addition, wildlife officials estimated the destruction of 50 to 70% of the pheasant population in southern Alberta.

A comparable 'blizzard' occurred in Chicago when 23 inches of snow fell on January 26–27, 1967. By afternoon on the first day the Chicago metropolitan area of over seven million people was all but paralysed, and as J. S. Smith (1967) reports:

> Although the storm was not technically a blizzard, it mattered little to the Chicagoans who endured wind gusts of more than

60 m.p.h. combined with temperatures in the upper 20s. Snow drifts of 4 to 8 feet were common; some mounted to 10 to 12 feet. . . .

There were the inevitable tragedies that occur with snow-storms, especially the thickly populated areas. More than 45 deaths were attributed to the storm. Most resulted from heart attacks brought about by exhaustion from walking through the deep snow or shovelling it. A few persons were found frozen near their stalled cars and one woman was frozen in a city bus. . . . By Sunday, January 29th, major highways were partially open but traffic was confined to one lane in most areas. They were opened to two-way traffic during Monday, but side streets in the area were impassable well into February.

The thousands of stalled vehicles in the streets was perhaps the biggest hindrance keeping Chicago from returning to normal, the clearing of some streets taking ten times longer than normal. Delaying the task even longer, according to Smith, were the numerous breakdowns of heavy equipment, much of which was 15 years old or more.

The cost of the Chicago blizzard was very high with both Smith (1967) and Ward (1967) commenting that losses were well over $100 million. Smith, for example, states that the total cost of business losses during the storm was estimated at $150 million by a representative of the Chicago Association of Commerce and Industry. This figure, it should be noted, represents only the amount of goods and services which went unsold, other business losses being 'incalculable'.

2. Precipitation extremes

Any evaluation of atmospheric resources requires answers to at least two questions: first, what is the 'value' of the resource; second, how much of the resource is 'required'. The answers to both questions vary considerably both in time and space, but with respect to water resources, some kind of water resource book-keeping is necessary if a better appreciation of the value of precipitated water is to be obtained.

The place of water in any nation's economy is however measurable. In a Presidential Address to the New Zealand Hydrological Society, Toebes (1967) said that in surveying water problems one must first look at how much water there is and then how we utilize it.

As a starting point, Toebes suggests that the average annual rainfall of New Zealand is approximately 65 inches, about 40% returning to the atmosphere by evaporation and transpiration. The remainder returns to the oceans by means of stream flow, together with a small quantity by ground-water flow. In considering how much water New Zealand uses, estimates based on 1966 data give an approximate picture of annual consumption (Table II.6).

TABLE II.6: *Water use in New Zealand* (from Toebes, 1967)*

Used by economic vegetation in transpiration		15·00	
Used by hydro-electric development		0·50	
Used by industrial water supply	0·007 ⎫		
Used by urban water supply	0·010 ⎬	0·04	
Used by rural water supply	0·021 ⎭		
Used for irrigation		0·14	
Total economic use			15·70
Returned to atmosphere without beneficial use		11·00	
Returned to oceans without beneficial use (not considering fisheries, wet-lands biology or water sport)		38·30	
Total non-economic use			49·30
Total precipitation			65·00

* Assuming that use of ground water and re-use of water is minimal at present. Values are in inches.

These data show what looks to be an 'exceedingly rosy picture' for the whole of New Zealand, but as Toebes says, the rainfall is unequally distributed and the major water user, the soil, is certainly not in an optimum water condition throughout the year. Further, he says that 'We do not yet know how much we are short, and much hydrological research has to be done before even first figures are available.'

The control or lack of control of water, and the associated erosion, are major problems in many areas of the world. In some areas of New Zealand, for example, up to 80% of the land has been seriously affected by erosion. This is, according to Toebes, 'water action which is not beneficial, but without hydrological studies we cannot tell to what extent'. To be able to 'tell to what extent' requires considerable hydrological research; however, New Zealand had an expenditure of only $0·5 million in 1966, which is an insignificant amount compared with the huge amounts spent on water use and water control (Table II.7).

Such a disparity in the allocation of expenditures on water resources is not of course limited to New Zealand. Indeed, a great deal of any economy (progressive or backward) is based in the final

TABLE II.7: *Expenditures on water use, water control and hydrological research in New Zealand, 1966 (adapted from Toebes, 1967)*

Hydro-electric development	$47·7 million
Industrial water supply	$ 2·1 ,,
Urban water supply	$ 6·3 ,,
Rural water supply	$ 3·1 ,,
Bridges and culverts	$ 6·0 ,,
Flood control	$ 5·0 ,,
Erosion control	$ 0·6 ,,
Irrigation	$ 0·5 ,,
Pollution control	$ 2·0 ,,
Flood damage to highways	$ 1·4 ,,
Urban storm-water control*	$ 5·4 ,,
Hydrological research	$ 0·5 ,,

* Including sewerage, excluding private expenditure.

analysis on water. It is clear therefore that the evaluation of water resources – which are fundamentally linked to variations in precipitation and the water balance – cannot properly be neglected.

The increasing urbanization of society and the growth of the modern city is a theme subjected to frequent comment by geographers, sociologists, planners and politicians. Few, however, have commented on the relationships of urban growth to water demand, this lack of comment, in a sense, paralleling a seeming lack of public interest. As Tweedie (1967) comments, in most cities it is the problem of transport rather than that of water supply which seems to be in the most personal and most pressing issues, and it is traffic flow, rather than water flow, which most absorbs the attention of planners.

In the associated field of cultural hydrology, other related comments are made by Miller (1966) concerning the application of studies on how flood plains are used, what measures might reduce flood damages, and how man and water might co-exist on sites that streams have created (White, 1964; Kates, 1964; Murphy, 1958). The work of this programme by the Department of Geography at the University of Chicago is a major contribution to an understanding of the man–land–water system in a particular site. In another study, Brater (1968) suggests that as population density and land values increase, especially in urban areas, the detrimental effects of floods become a tremendous economic burden as well as a serious threat to the health and well-being of the community and its individual citizens. The damage from floods can, of course, be alleviated if the magnitude and frequency of future flood events can be predicted so that economically feasible relief measures can be planned and warning systems developed.

The economic consequences of reduced precipitation in the form of drought are also of considerable concern, and in many areas the effects of drought last much longer than the effects of excessive precipitation. Droughts have always been of particular significance to the marginal farming areas of the world, but lack of precipitation can also have serious consequences in optimum agricultural areas, as well as in urban and industrialized areas. This is particularly the case in many parts of the United States and the Soviet Union. Moreover, in many key agricultural producing areas, even relatively small negative departures from the normal rainfall can have significant economic effects.

a. Floods

Several properties of floods can be measured and classified. These include magnitude, seasonality, frequency, velocity and load. All these factors are associated with the 'value' of flooding.

The magnitude or dimension of a flood, measured in cubic feet per second or an elevation in feet, determines the area that will be inundated. The seasonality of a flood is also important. A spring or autumn flood, for example, may result in a higher degree of destruction to agricultural areas than a corresponding winter flood. The frequency of floods is usually computed by the application of the statistics of extreme values, and a knowledge of flood frequency is of considerable value, particularly in relation to costs and benefits of flood insurance, and the design of structures for flood prevention.

The velocity of flood waters, together with the magnitude of the flood, generally determines the kinds of structural damage. They are also relevant factors with regard to stream load and channel scouring. A further property of a flood is the load, which refers to the sediment carried by a flood. The deposition of sediment on agricultural land is generally beneficial, but the deposition of sediment in urban areas is usually destructive.

The severity of these factors varies considerably both in time and in space, but evidence from many areas of the world indicates that floods cause a greater loss of life and property than any other national disaster. Losses through flooding are reported almost daily in our newspapers, and various accounts of some of the great floods are available. Lane (1965) in his book *The Elements Rage* provides many examples.

The floods in England and Holland on January 31 and February 1, 1953, for example, are described in detail, the 'cost' of the invading sea in Holland being reported as 1,835 killed, 72,000 evacuated, more than 3,000 houses destroyed and 40,000 damaged, and 625 square miles flooded. Nine per cent of all agricultural land and 3% of the dairy land were covered by the sea, and over a quarter of a million livestock and poultry were lost, the total damage in Holland being estimated at £150 million.

Although very destructive, sea floods are much less frequent than those resulting from the overflow of inland waters. With the occupancy of more and more land near rivers, such flooding has become more costly, and in the United States alone flood damage averages over $300 million a year. In some years the total damage exceeds one billion dollars.

Anglo-America suffers greatly from floods, some of the most severe flooding resulting from hurricanes which can deposit rains of the order of 24 or more inches in 24 hours. Factors other than the rainfall are also important. These include the amount of melted snow, the accidental damming of rivers, the build-up of river bottoms through accumulation of silt and the bursting of river banks. The Mississippi river system has seen some of the greatest floods in the world, the most costly flood in the United States being the June/July 1951 floods of Kansas–Missouri, which Lane reports had an estimated cost of $935 million.

Other very costly floods have occurred in other parts of the world, particularly in China. One notable flood (the world's most lethal flood according to Lane) was the overflowing of the Hwang-Ho and Yellow Rivers in September and October 1887 when the river rose 70 feet. The flood inundated 50,000 square miles, and in addition, 2,000,000 people were made homeless, 300 villages were swept away, and 900,000 people were killed.

Most of the above reports are concerned with the damages resulting from floods. A number of papers, however, have suggested methods for reducing flood losses. Burton (1965) discusses in some detail flood-damage reduction in Canada. He says that although in many communities floods remain a constant threat to life and property, the threat is not always recognized and is sometimes ignored. Nevertheless, when a major flood occurs as in the lower Fraser valley in 1948, the Red River valley in 1950, and in Toronto in 1954, everyone is made aware of the power of this natural hazard and its great capacity for destruction. Disasters such as these, involving major cities (Vancouver, Winnipeg and Toronto), usually lead to action, and attempts are made to prevent the occurrence of similar events in the future. At such times also, major shifts may take place in government policy (see for example Sewell, 1965). However, as Burton points out, a common approach to flood damage has been corrective rather than preventive. He further states:

> The continued occurrence of destructive floods, both large and small, provides a measure of the nation's adjustment to this environmental hazard. In spite of highly advanced engineering and communications technology, floods continue to plague us, and there are no signs of a progressive reduction of flood damages. Intelligent and informed appraisals of the problem range from the view that individual managers of flood-plain property make adjustments consistent with their own best interests and that hence human use of flood plains tends toward an economic optimum; to the notion that flood protection provided at public expense serves only to subsidize and support uses of land that are already unduly expensive.*

In the United States, a major element in the flood-control pro-gramme has been heavy reliance on engineering works such as dams and levees, other possible adjustments such as land-use regula-tion, encroachment statutes and building codes being used in only a few areas. Burton indicates that in spite of an average annual expenditure of between $300 million and $400 million on flood-control works in the United States, the rising toll of flood losses has not been arrested, and a major reason for this is the continuing human invasion of flood-plain land.

Four case studies are given in Burton's analysis, and they illus-trate the range and variety of flood problems, and the kinds of adjustments that are being made. The case of metropolitan Toronto and its flood-control project provides an example of a way in which methods of flood control can be considered together with other goals into a multi-purpose project. Burton gives some background into the Toronto Plan, in which the Don Valley Conservation Authority of Toronto in 1950 suggested that flooding is not a very serious problem (on the Don) and does not call for costly measures of control. However, in 1954, Hurricane Hazel caused considerable property damage in the Toronto area (including an estimated $3·5 million in the Don Valley alone) and resulted in the loss of 81 lives. Subsequently, a plan for flood control and water con-servation for the metropolitan area of Toronto was formulated, and

* Quotation and Table II.8 on page 23 reproduced with permission of the Queen's Printer for Canada.

as Burton says 'It is widely acknowledged that Hurricane Hazel changed views of flood problems in Toronto and southern Ontario.

The tangible benefits of the Toronto plan are shown in Table II.8 and it is clear that in this case a natural disaster, in the form of Hurricane Hazel and the associated flooding, had the effect of generating activity and concern with respect to both the positive and negative aspects of a water resource.

Several alternatives to flood control have been suggested; a convenient and effective way of coping with risks being flood insurance. Lind (1967) discusses some of the advantages of flood insurance and says that in practice any programme of flood insurance will probably not provide complete protection, and the premium will not bear exactly the desired relation to the expected value of losses. If, for example, it is desired to cover all the costs of providing the insurance, including a return to the company which underwrites the insurance, then the premium will be set at a level clearly above the expected value of losses. This is essentially the procedure used by private insurance companies. On the other hand, Lind shows that if the government administers the programme, the premium can be set equal to the most reliable estimate of expected losses, the government absorbing any difference between this estimate and the true value of expected losses.

TABLE II.8: *Tangible benefits of the Metropolitan Toronto plan**
(from Burton, 1965)

Type of benefit	Value of benefits†	Per cent
Flood-damage reduction	$12,447,000	22·4
Recreational use of reservoirs	$21,203,000	37·6
Indirect benefits	$ 9,335,000	16·5
Others (including water supply, fishing, outside reservoirs, pollution abatement, and recreational use of acquired flood-plain lands)	$13,240,600	23·5
Total	$56,225,600	100

* Derived from *Plan for Flood Control and Water Conservation*, Metropolitan Toronto and Region Conservation Authority, Woodbridge, Ontario, 1959.
† Figures expressed in present value, assuming a 100-year life of construction works and an interest rate of 4%.

b. Droughts

Drought is generally considered to be a period of moisture deficiency.

The required size of the deficiency is not, however, agreed upon, and drought quite justifiably means different things to different people. Definitions of droughts (absolute, partial, agricultural) are available, and they all endeavour to describe or measure the availability of moisture. But whatever the definition, and whatever the kind of drought, there remains the problem of too little 'water' required by too many factors (people, animals, acres of agricultural land, trees).

Droughts have affected most areas of the earth at one time or the other, but perhaps the continent of Australia can be cited as a prime example of the overriding 'control' which water has on an economy. In some years in Australia, rainfall is much above average and the continent benefits by generally increased agricultural production. On the other hand, rainfall is usually less than average in approximately four years out of five. Stock losses and hazardous forest fires, associated with the 'dry years', are therefore an almost constant threat to at least one or two areas of the continent.

The costs of drought in Australia, as elsewhere, are difficult to estimate. The Australian *Current Affairs Bulletin* 'Drought' suggested that fluctuations of two or more times the average in production were evidence of the effects of drought, but in reality it is not that simple. For example, it is often very difficult to establish real cause and effect relationships between drought and total economic production. The *Bulletin* (Anonymous, 1966) continues:

> The mind may well boggle at estimates of the cost of a drought calculated to include wool shorn from sheep never conceived and wheat not harvested from crops never sown. Even if it can be admitted that one can in some way lose something which never came into being, the calculation of values of the items becomes very involved.

The effects of drought on the Australian economy have been discussed by a number of authors, and the survey by Heathcote (1967) provides a critical insight into many of the problems relating to the evaluation of drought conditions. He notes in particular the negative effects of droughts. These include spasmodic and incessant effects. The former are the relatively short-run and immediate results of the period of drought, and they are separated by the sequence of their occurrence:

(1) Disaster – the immediate reduction of surface and soil moisture resources. This has an immediate and disastrous effect upon a range of economic activities from crop and fodder growth to water navigation.

(2) Emergency – the immediate day-by-day reactions to the disaster from the emergency effects of drought.

(3) Rehabilitation – the follow-up measures to restore losses and restart production.

Incessant effects of drought, on the other hand, are those effects which result from the annual costs of preparing against the recurrence of drought. Four subdivisions (related to costs) are recognized by Heathcote in his study:

(1) Abnormal service capacities to meet abnormal demands for water, such as a storage or flow capacity well in excess of normal use. Where such capacity is held, its annual maintenance is an incessant cost of drought.

(2) The maintenance of an overseas credit reserve to 'cover the fluctuations of Australian production'. This is in part a drought security measure which in recent years has varied from $800 million to $1,400 million.

(3) The cost of drought research.

(4) The indirect result of the general attitude to resource development in the drought-prone areas. This has meant that in most areas there has been only limited investment and attempts at intensification have had only qualified success.

As with most weather variations – extreme or otherwise – the effects of drought are not all negative. For example, in certain ecological contexts the effect of droughts are often beneficial. Perry (1962) found that the drought of 1958–1961 in Central Australia had caused little long-term damage to the vegetation and that the perennial grasses, at least, were probably in a healthier condition than after a more normal succession of seasons, because the death of livestock had reduced the grazing pressure.

In addition to these ecological benefits there are the direct economic benefits. To many people, drought means increased business, such as the cartage of water and feed supplies. Nevertheless, it is as well to bear in mind that in some cases those receiving the benefits are in another part of the world. A specific case, for example, would occur if drought in Australia enables the wheat farmer of the Canadian prairies to export additional wheat to those areas unable to be supplied by Australia (or vice versa). Such transactions do, of course, have much wider international implications than the 'simple' export of wheat from one country to another.

The influences of drought are often long lasting. Cumming (1966), reporting on the 1966/67 drought in Australia, suggests that the 'recovery period' for sheep numbers in New South Wales would be about nine years, and for Queensland about sixteen years. He indicated that shorter periods of recovery are possible in both states, but it is clear that it normally takes several years, and in some cases a decade or more, for the sheep population to return to its pre-drought size.

Another Australian study has been made by Duloy and Woodland (1967), who use an econometric model to assess the effects of a 20% across-the-board drought year reduction in farm output. The 1944/1945 drought is given as an example of a severe drought (Table II.9), and then using various assumptions, the authors compute the effect of a drought on farm income, and the resultant effects upon the gross national product, imports and the balance of trade (Table II.10). These results show that the depressing effect on the G.N.P. was exhausted after about three years, and that the damage in the balance of trade is substantial only in the first year.

TABLE II.9: *Effect of the 1944/45 drought in Australia upon the output of some rural products (from Duloy and Woodland, 1967)*

Product	Unit	1943/44	1944/45	Percentage change
Wool	'ooo lb greasy	1,169	1,016	−13·1
Wheat	million bushels	110	53	−48·2
Sugar	'ooo tons	524	670	+27·9
Meat	'ooo tons carcass	1,043	983	− 5·8
Milk for all purposes	million gallons	1,067	1,012	− 5·2

TABLE II.10: *Effects of a drought upon gross national product and the balance of trade ($ million)* (from Duloy and Woodland, 1967)*

Year	Change in farm income†	Change in G.N.P.	Change in imports	Change in balance of trade
1	−266	−285	− 17	−249
2	−133	−263	−116	− 17
3	− 67	− 85	− 17	− 50
4	− 34	+ 11	+ 39	− 72
5	− 17	+ 11	+ 25	− 68
6	− 9	− 9	0	− 9

* The figures are rounded to the closest $ million. The marginal propensity to import is taken as 0·471. 1959/60 prices are given.
† Also assumed to be equal to the change in exports.

As in all such studies, many assumptions and simplifications are made, and, as the authors state:

> A more satisfactory study would look closely at the different effects of droughts on different products to obtain better estimates of output changes, and how these are distributed over time. It would enquire into the likely effects on receipts of these output changes for different commodities, and would entail a detailed study of farm costs.

In terms of livestock numbers, the effects of droughts are very evident, a study of the Queensland cattle industry showing, for example, that substantial reductions in cattle numbers through death have occurred in some drought years. An example given is for 1946/47 (see Table II.11). The average annual number of deaths in the four 'normal' years (i.e. 1945/46, 1947/48, 1948/49, 1949/50)

TABLE II.11: '*Losses*' *of cattle in Queensland, 1945/46–1949/50**
(*from* Quart. Rev. Agric. Econ., *6: 19*)

	Deaths	Slaughterings	Live exports (overland)	Total
1945/46	362,000	1,007,000	389,000	1,758,000
1946/47	881,000	804,000	381,000	2,066,000
1947/48	277,000	1,157,000	397,000	1,831,000
1948/49	328,000	1,149,000	365,000	1,842,000
1949/50	229,000	1,101,000	361,000	1,691,000
'Normal' Average	299,000	1,103,000	378,000	1,780,000
per cent	16·8	62·0	21·2	100·0

* Deaths and slaughterings from *Queensland Year Book*, 1950. Slaughterings for calendar years, starting with 1945; they include cattle slaughtered on stations and holdings. Figure for live exports supplied by the Queensland Government Statistician.

were 299,000. In 1946/47 deaths were 881,000 giving an excess of deaths in that year of 582,000. Not all of these would have ultimately come to market, and as shown in Table II.11, 'normally' about 16·8% of cattle 'losses' in Queensland are the result of death from natural causes. This suggests that, of the 582,000 cattle which died 'prematurely' from drought in 1946/47, 98,000 would have died from natural causes in the succeeding years, leaving 484,000 which, but for the drought, would have been available for slaughter or export.

During the mid-1960's, a relatively severe drought affected many properties in Australia, and a comprehensive survey by the Australian Bureau of Agricultural Economics (Anonymous, 1969) attempted to assess its financial implications. The report indicates that although the effects of a drought vary between properties according to their individual characteristics, the severity of the drought, its geographical incidence, and the specific drought and pre-drought policies adopted, it is almost inevitable that the 'total returns' of a pastoral property will fall as a result of both a decrease in wool produced and reduced livestock trading, depending on how many stock are sold or die and any adjustments made to breeding. On the other hand, if a selling policy is adopted, the survey points out that it is possible that *cash* returns may be quite large in a drought year although this will often be largely offset by reduced returns from wool. The financial impact of the 1964–1966 drought on properties in the pastoral zone of New South Wales and Queensland is given in Table II.12, the decline in total returns and the increase in total costs as compared with the base period (1960/61–1962/63) resulting in a substantial fall in net farm income in 1964/1965, and in 1965/66. The average net farm income in 1965/66 was in fact negative, and as shown in the table, the greatest absolute decline occurred in the pastoral zone of New South Wales, where net farm income dropped from $A8,834 in 1964/65 to − $A10,311 in

TABLE II.12: *Financial impact of drought on properties in two areas of Australia (from Wool Economics Research Report, No. 15: Anon., 1969)*

Area	1960/61 to 1962/63	1964/65	1965/66	1966/67
New South Wales Pastoral Zone	($A)			
Net farm income	15,629	8,834	−10,311	10,544
Cash balance	7,193	12,592	−1,792	536
Interest paid	803	729	932	1,090
Cash investment (incl. land purchases)	8,049	4,539	1,690	5,077
Queensland Pastoral Zone	($A)			
Net farm income	9,139	2,570	−7,902	9,134
Cash balance	7,933	4,292	1,735	4,275
Interest paid	1,050	1,401	1,863	1,522
Cash investment (incl. land purchases)	4,291	4,397	1,371	4,168

1965/1966, a decrease of $A19,145 or 217%, whereas the largest proportional fall occurred in the pastoral zone of Queensland, where net farm income declined from $A2,570 to —A7,902 or more than 400%.

Australia is, of course, not the only country to suffer from drought, the mid-1960's drought in the north-east of the United States giving emphasis to the fact that drought affects urban dwellers as well as sheep and cattle. In this drought, a major problem was the insufficient supply of domestic water. *Time* (July 16, 1965), reporting on the effects of the drought, stated that in normal times New York City industries and her 8,000,000 inhabitants use a daily *per capita* average of 154 gallons. At one stage during the drought, however, New York City's 476·5 billion gallon capacity reservoirs held only 240·7 billion gallons, forcing the city to impose restriction to reduce the *per capita* consumption to 125 gallons.

Such an example can be considered trivial when compared with the droughts that have occurred in China, the Soviet Union and South America. Furthermore, the failure of crops in India and Pakistan, which may cause millions to starve, occurs in part as a result of the non-arrival of the monsoonal rains. Nevertheless, drought and its associated factors, lack of drinking water, lack of water for industry and the ever-present problem of forest fires, are important facets of urban living in industrialized nations.

C. THE MODIFIED ATMOSPHERE

1. Air pollution

Environmental pollution is probably best described as 'the wrong thing in the wrong place at the wrong time'. Such a definition includes many things, including air pollution.

Air pollution is indeed part of our twentieth-century environment, and one may ask why? There is no simple answer, but even a most cursory survey of the development of our society points clearly to one fact. Consciously or unconsciously man has taken his natural environment for granted, often using its ingredients as limitless resources. But man is learning the hard way that these are finite resources (the water, the land and the atmosphere), and are capable of being over-taxed and made incapable of performing the functions demanded of them.*

* A critical appraisal of how man is using these resources is available in *Pollution: what it is . . . what it does . . . what can be done about it* (Maunder, 1969a), a publication resulting from a series of lectures given at the University of Victoria, B.C.

Scientific and technological developments have brought us a long way from the time when man was compelled by environmental stresses of temperature extremes, high winds, flooding, drought, snow and various types of severe storms to remain in only those areas of the world where tolerable environmental limits occurred naturally. But as we construct cities, transportation systems and facilities for production we find that the surface characteristics are modified and that the air environment (particularly the heat and water balance, and the composition of the lower troposphere) is changed.

What is the nature of such changes? Bryson (1967) has some interesting observations. He indicates, for example, that the observed dustiness of the air over the United States, *outside* the cities, doubled from 1957 to 1963. European cities are also producing big clouds of smoke and dust, as are the relatively unindustrialized countries of Asia and Africa. The effect on the radiation balance is probably even more noticeable, for since 1950 the earth in places has cooled off as much as it had warmed in the previous 150 years, and the amount it has cooled is easily within the range of effect of the increase in dust in the atmosphere. The cause-and-effect relationship of 'dust' in the atmosphere is, nevertheless, subject to considerable debate, and Bryson (1968) considers some of the difficulties encountered.

a. Components

Basically there are two types of air pollution: natural and man-made. Examples of natural air pollution include particulates entering the atmosphere from volcanoes, deserts and other sandy areas such as beaches, or from naturally started forest fires. Pollen from various forms of vegetation may also constitute air pollution. Man-made pollutants on the other hand, which in many cases are more detrimental to health, emanate from automobile exhausts, burning of fuel for power and heat, burning wastes, manufacturing processes, demolition and construction operations, dirty and littered streets, atomic fallout and jet aircraft.

Polluted city atmospheres may be classified into two major types according to their chemical composition: the 'London' type, composed principally of sulphur compounds from the burning of coal,

The comments which follow, concerning air pollution, are in part from two chapters of this publication, Chapter 1: 'What is Pollution' (Maunder, 1969b), and Chapter 3: 'Weather, Weather Modification, and Air Pollution' (Maunder, 1969c). Additional viewpoints in regard to the legal, medical and engineering aspects of pollution are given in Chapter VI, Section E of this text.

and the 'Los Angeles' type, composed principally of petroleum products. In most urban areas, however, a combination of both types usually prevails.

In large urban areas air pollution is often of real significance. In New York City, for example, one finds a daytime population of 9,000,000 people who are subjected to discomfort, hazard and damage by air pollution. These pollutants are made up of deposited soot and dust at the rate of 60 tons per square mile per month, suspended particulate matter of 280 microgrammes per cubic metre, sulphur dioxide of 0·2 ppm and other pollutants in proportion (Benline, 1965).

The automobile is a very important source of air pollution basically due to the inefficient combustion of fuels. The hydrocarbons and nitrous oxide given off by the combustion of petroleum are at first neither visible nor irritating; however, when these substances are exposed to sunlight for an hour or so, they undergo an important chemical change, yielding ozone and other reactive compounds. Hydrocarbons also escape both through the exhaust and as vapours from the fuel tank and carburettor vents. In addition air pollution in the form of highly pulverized rubber and asphalt is generated by abrasion of tyres upon streets. The daily output of every 1,000 operating automobiles in an urban community has been estimated to burden the air with 3·2 tons of carbon monoxide, 400 to 800 pounds of organic vapours and 100 to 300 pounds of nitrous oxides, plus smaller amounts of sulphur and other chemicals (McDermott, 1961).

b. Weather aspects

Many variables affect the intensity of air pollution. In most metropolitan regions, municipal installations, households, industrial plants and automobiles give off approximately the same combination and relative volume of chemicals to the air, but whether these contaminants become a community problem, at any one time, depends on the interactions between population density, topography and the weather.

In this respect, three significant meteorological variables are wind direction, wind speed and the vertical temperature profile. Generally, the air is warmer at the ground and colder above; however, a temporary ceiling to the atmosphere (the thermal inversion) often occurs, which prevents the rise of pollutants and forces the diffusion to be mainly horizontal. Consequently, when an inversion occurs in the lower troposphere of a heavily populated region, the same air must accumulate a much higher concentration of pollutants, a fact which can, on occasion, give rise to chronic bronchitis–emphysema (see McDermott, 1961, for the case of Donora, Pennsylvania, in 1948).

Ground-level concentrations of pollutants at a distance from a specified source are highly dependent on meteorological conditions. For example, the ability of the atmosphere to disperse pollutants can vary according to conditions by a factor of perhaps a million. In particular, the areas to be affected at any time depend on the direction of drift of air at appropriate levels and on the temperature stratification at the time. In breezy daytime conditions the dispersal is generally rapid, since the irregular motions cause a rapid spread of gases through a wide cone. By contrast, in typical night or early morning conditions, the usual temperature stratification with relatively cool air near the ground results in the suppression of eddy motions, an absence of mixing, and at times the complete stagnation of the air at low levels. In these circumstances pollutants tend to remain concentrated at some level that is partly dependent on the chimney height and the temperature of the emergent gases.

Specific applications of pollution accumulation have been made. For example, Hilst (1967) says that under normal conditions of wind and mixing, accumulation of pollutants increases as the square root of distance. On this basis for a continuous city of uniform pollution sources, the concentration of a single, conservative pollutant is $Y = K(X)^{\frac{1}{2}}.Q$, where Y is the street level of concentration, K is a proportionality constant dependent upon wind speed and atmospheric stability, X is the downward distance from the edge of the city and Q is the source strength per unit area and time.

If, therefore, there is a critical concentration (Y_c) which is not to be exceeded, we can deduce that there is a limiting size of the city X_c, which depends upon the meteorological conditions K, and the source strength Q. This distance is given by $X_c = Y_c^2/K^2Q^2$, where the value of K is deduced for the meteorological and topographical conditions encountered in any particular urban site. Hilst says that for purposes of illustration we may assign a typical value for K associated with minimal ventilation and vertical mixing, of the order of 50. Taking Q as 10^6 tons per year over a 300 square mile area (or 5×10^{-5} g/m²/sec), and Y_c as 1 ppm:

$$X_c = 1/(2500 \times 25 \times 10^{-10}) \text{ m}$$
$$= 160 \text{ km} = 98 \text{ miles}$$

Accordingly, for this particular set of circumstances, the limiting length of the city is about 100 miles, and the world's larger metropolitan areas (for which the above estimates of SO_2 emissions are conservative) are already approaching these critical dimensions.

More important perhaps is the use of such equations as a guide to design criteria. Specifically, for instance, halving the source strength (Q) uniformly over the city increases Y_c by a factor of four, whereas doubling the wind speed has the same effect. An alternative design for air resource management therefore is the deliberate spacing of source and non-source areas.

The importance of sunlight on smog formation is significant. In Los Angeles, for example, the air is often strongly oxidizing, the oxidant being mostly ozone. A considerable amount of ozone is formed in the upper atmosphere by the action of ultraviolet radiation, but this ozone does not descend to earth during smog conditions because of the very temperature inversion that intensifies smog. However, since high oxidant or ozone values are found only during daylight hours, a photo-chemical reaction appears to take place when one or more ingredients of smog are exposed to sunlight – which is of course abundant in the Los Angeles area. The answer to the puzzle of the oxidizing smog of the Los Angeles area thus lies in the combination of heavy automobile traffic and copious sunlight (Haagen-Smit, 1964). Similar photo-chemical reactions do of course occur in other cities, and the Los Angeles type of smog appears to be becoming more widespread.

c. Urban and rural atmospheres

Concentrations of population give rise to many problems. Among these are problems associated with changes that occur in the atmospheric environment of most urban areas, some of these changes being of considerable magnitude and of concern to our health and well-being.

Two major changes have been introduced by man. One is the radical alteration of the surface, the other is the continuous addition of a wide variety of substances (gaseous, liquid and solid) to the air (Landsberg, 1961). The change of surface particularly involves the radiation balance, and in some locations where a new effective surface is created, the runoff and evaporation character is considerably altered. The air motion is also basically affected by changes in the roughness parameters, and on a microscale, wind channelling occurs. Accordingly, even a city without heating devices, industry or motorized traffic would create a climate different from that of the surroundings.

Cities differ from the countryside in most aspects of the climate. The city itself is the cause of these differences: its compact mass of buildings and pavements obviously constitutes a profound alteration of the natural landscape, and the activities of its inhabitants are a considerable source of heat. Together these factors account for five basic influences that set a city's climate apart from that of the surrounding area (Lowry, 1967). First, there is a marked difference between surface materials in the city and in the countryside, the predominantly rocklike materials of the city's buildings and streets conducting heat about three times as fast as it is conducted by wet, sandy soil. Second, the city's structures have a far greater variety of shapes and orientations than the features of the natural landscape. Third, the city is a prodigious generator of heat, particularly in the winter, when heating systems are in operation. Fourth, the city has distinctive ways of disposing of precipitation. In addition, the air in the city is different in that it carries a heavy load of solid, liquid and gaseous contaminants.

d. Economic aspects

Of all the factors considered when assessing atmospheric pollution economic aspects tend to predominate. But in order to adequately evaluate costs, we must consider *all* the costs and *all* the benefits. Employment opportunities, for example, are closely associated with industry, and many industries under present economic conditions could be said to necessitate some air pollution.

The financial cost of air pollution is immense. In New York City, for example, air pollution adds several hundred dollars per year in washing, cleaning, repairing and repainting bills to the budget of the average family. And, in the United States, property damage due to airborne pollutants is estimated to be more than $10 billion a year, this representing a cost per person in the U.S. of over $50.

The estimation of damages resulting from air pollution requires knowledge of many factors, including the source of emission, the initial pollutants emitted, the rate of emission, the direction in which the pollutants will move and the concentration of pollutants. These factors are important in both urban and rural areas, since pollutants carried aloft are often deposited many miles from the source, causing considerable damage to crops and natural vegetation. Fruit-producing orchards, for example, suffer under the effects of air pollution, and it has been shown that fruit trees in polluted areas are 10% smaller and produce 10% less fruit with diminished Vitamin C content than trees in clean air (Bagdikan, 1966).

The major damaging pollutants emitted into the atmosphere are nitric oxide, sulphur dioxide and various hydrocarbons, but an extremely difficult and not fully understood problem is the formation of new substances in the atmosphere, which in many cases are potentially more harmful than the initial pollutants. This process of transformation is accelerated by solar radiation, and results in the formation of ozone, peroxyacyl nitrate, formaldehyde, nitrogen dioxide and other compounds.

Air pollution represents an external cost imposed upon the community. Such external costs include damages to health and impairment of physical efficiency, reduction of crop and orchard yields, reductions in the output of dairy herds and shortening the useful life of the herd, increases in agriculture costs for such items as fertilizers, acceleration of the deterioration of paint as well as the oxidation of metals, greater hazards for air transportation, increases in the soiling of merchandise and the creation of the veil of pollution which hides the natural beauty of the environment. Many of these external costs can be measured and an economic price assigned to them (Ogden, 1966). For example, the reduction of agricultural output and the increased cost due to air pollution is a tangible effect and amenable to measurement. It appears, however, that markets for clean air are few; consequently there exists an air pollution problem in most urban areas today, which shows few signs of any rapid abatement.

e. Decision-making

The recognition that local pollution occurs, *and* the removal of its cause, may not in itself be a solution to the *total* pollution of the environment. Urban areas, for example, may 'solve' their own major pollution problem by the high-level dispersal of smoke. This may, however, simply transfer pollutants downwind to a much larger area, thereby affecting visibility, fog formation and crop production in this area.

Communication between scientists, engineers and decision-makers is therefore essential. The meteorologist, the climatologist, the engineer, the biologist, the economist, the physician, the planner, the geographer and other associated specialists, must be able to identify, measure and interpret the air pollution problem where it occurs, and to transmit this information in terms understandable to the decision-makers – whether they be municipal councils, provincial legislators, state or federal representatives. They in turn must be aware of public opinion, so as to interpret the scientists' findings for the well-being of the majority.

2. Atmospheric modification

Weather and climate affect economic and human activities in pervasive ways. Studies have shown that the profitability in certain economic activities depends very much on fluctuations in the weather, obvious examples being agriculture and air transportation. Other investigations have indicated that some human behaviour patterns are influenced by weather variations. It is natural, therefore, that man should try to find ways of adjusting to, and even altering, the weather. Such 'adjustments' usually involve one of two things: man can either accept weather and climate as given and alter his patterns of activity to accommodate them, or he can try to alter the processes in the atmosphere which produce the weather and the climate.

In the first case he can either do nothing, reschedule activities (providing he has faith in the weather that is forecast) or improve 'the technology of weather protection' such as by developing drought-resistant crops, providing more economical air conditioning or even building 'hurricane-proof' cities. Alternatively, he can alter the amount or the temporal distribution of particular weather elements, by modifying or changing the atmospheric circulation in relation to a small area (an airport), a medium area (an urban area), a large area (a river basin or a state) or even possibly an extensive region such as a continent.

For the most part man has concentrated his attention on the first alternative. That is, he has moved to those locations where the climate is most acceptable for personal comfort and where economic activities can be pursued at reasonable cost. He has also tried to develop ways of reducing the impact of variations in the weather. However, there now seems to be agreement that in certain areas, under certain circumstances and at certain times, weather modification in the form of an increased precipitation of the order of 10 to 15% can be achieved, and that the dispersal of cold fog and the suppression of hail and lightning are feasible. Further, it is clear that on a microscale, weather modification *is* a reality. Nevertheless, there is still considerable doubt as to just how 'successful' the weather modifiers have been considering the *overall* effects in an area.

a. Types of atmospheric modification

Atmospheric modification means many things to many people. The term is often considered synonymous with weather modification, and weather modification usually means cloud seeding by dry ice, silver iodide or similar agents which are designed to increase rain, suppress hail or dissipate supercooled fogs. This meaning of weather modification usually implies vast public benefit, if it is successful. Atmospheric modification embraces goals far more than 'simple' cloud seeding, however. Attempts to influence the general circulation by initiating triggering impulses (such as inducing a change in planetary wave number during the progression of the seasons) are suggested by Roberts (1965) to be a form of weather modification, as are the prospects for initiating self-propagating climatic changes.

Atmospheric modification also includes 'inadvertent' atmospheric modification, such as the modification of the atmospheric environment of an urban area by pollutants, as well as the alteration of the water and heat balance brought about by urban growth. We can also include the possible changes in the radiative equilibrium that might be produced by the stimulation of cirrus cloud formation by jet aircraft contrails, and the modification of atmospheric circulation by surface features such as crops, parking lots, ploughed fields and trees. In addition, Roberts indicates that inadvertent atmospheric modification also includes the effects of rocket exhaust effluent on the higher levels of the atmosphere, and the effects of an injection of charged particles into the Van Allen radiation belts.

The results of modifying the atmosphere (whether intentional or unintentional) show that before we can understand the limits of man's capability to undertake such a modification we must know what effect the unmodified atmosphere would have had on various activities. Such modifications produce serious fears in the minds of many thoughtful people, and as W. C. Roberts (per. com.) notes:

> I have been intrigued by the tendency . . . for cirrus clouds to grow from a few wisps of contrail clouds left by a conventional jet airplane. . . . If we fly larger numbers of even larger jet aircraft, will the effect of such clouds be enhanced to the point where they alter significantly the large-scale atmospheric radiation budget? Will weather or climate patterns be altered, perhaps even irreversibly, in ways that with our present knowledge we cannot even detect?

Possibly even more relevant, in the light of the frequent occurrence of growing cirrus in the jet trail patterns at lower levels, have we perhaps *already* done this? Further, has inadvertent modification already altered large-scale rainfall or temperature regimes?

b. Urban modification

Since a large proportion of the world's population, particularly in North America, Europe and Australia, live in cities, the climate of cities is of importance to many people. Some of the most significant characteristics of the man-made urban climate were discussed at a recent symposium (McCormick, 1961).

The effect of pollution on the radiation balance was reviewed by Sheppard (1958). He showed that there is a reduction in solar energy reaching the ground in polluted urban areas. Mateer (1961) also found that solar radiation in Toronto averaged 3% higher on Sundays than on weekdays and that during the heating season (October to April inclusive) the increase was 6%. Meetham (1945) compared ultraviolet radiation in the centre and on the outskirts of Leicester, England, over $2\frac{1}{2}$ years. He concluded that in winter the removal of all smoke would have increased ultraviolet radiation by 30% in central Leicester, whereas in the summer the change would have been only 3%. The urban energy released by combustion processes are also important. Kratzer (1956) has summarized the position, and says that for the built-up area of Vienna (without parks) the artificial heat supply per annum is one-sixth to one-fourth that provided by direct solar radiation, and for Berlin the ratio is one-third.

The characteristic warmth of a city is called the urban heat island. It is at a maximum at night when skies are clear and winds are light, and its seasonal maximum is summer to early autumn (Munn, 1966). A beneficial effect of the night-time heat island is to lengthen the frost-free period of the year. Daytime city temperatures are usually about the same as, or even slightly lower than, those in the suburbs; however, Chandler (1962) has presented a case in which there was a daytime heat island in London. Of interest is the effect of the size of the city on the intensity of the heat island. Chandler suggests that the magnitude of the heat island is not a linear function of area or population. It appears, in fact, that a city is a collection of microclimates, the character of the built-up area immediately surrounding the station probably being more important than the size or form of the city. In addition to the heat island effect, downtown areas of cities also have fewer natural sources of evaporation and transpiration. The water vapour content of city air is therefore less in summer, as evidenced by lower frequencies of dense fog than in nearby countryside. By contrast, in winter, the situation is often reversed because of the emission of water by combustion processes.

The city's effect on its own climate is, therefore, complex and far-reaching. Lowry (1967) quotes a study by H. E. Landsberg, who concluded that cities in the middle latitudes received 15% less sunshine on horizontal surfaces than is received in surrounding rural areas, and that they received 5% less ultraviolet radiation in summer and 30% less in winter. Landsberg's figures also show that the city, compared with the countryside, has a 6% lower annual mean relative humidity, 5 to 10% more precipitation, 5 to 10% more cloudiness, 25% lower mean annual wind speed, 30% more fog in summer and 100% more fog in winter (see Table II.13).

The advantages and the disadvantages of city climate testify to the fact that the city's climate is distinctly different from that of the countryside, and a fuller understanding of the climatic changes created by a city may make it possible to manage city growth in such a way that the effect of troublesome atmospheric changes will be minimal. A knowledge of the climate of cities has many important applications, notably in bioclimatology, industrial climatology and economic climatology. In particular, in cities where pollution levels are becoming alarmingly high, it will become increasingly necessary to select new areas for industrial zoning on the basis of meteorological factors, as well as taking into account variations in the modified urban weather in transportation planning.

c. 'Natural' atmospheric modification

The previous section has considered the changes which occur in the atmosphere as a result of 'inadvertent' man-made changes; what follows is a brief examination of the status of intentional weather modification, which has been admirably summarized by Fleagle* (1968).

The basis for artificial modification of clouds was established in 1933 and 1938, when Bergeron and Findeisen advanced a theory of precipitation based on the co-existence of ice crystals and super-cooled droplets in clouds at the same temperature, and on the rapid

* The quotations on pages 42, 43, 44 are from *Weather Modification: Science and Public Policy*, edited by R. G. Fleagle (Copyright 1968 by University of Washington Press).

TABLE II.13: *Climatic changes produced by cities (from Landsberg, 1960, pp. 325, 326)*

Element	Comparison with rural environs
Contaminants:	
dust particles	10 times more
sulphur dioxide	5 times more
carbon dioxide	10 times more
carbon monoxide	25 times more
Radiation:	
total on horizontal surface	15 to 20% less
ultraviolet, winter	30% less
ultraviolet, summer	5% less
Cloudiness:	
clouds	5 to 10% more
fog, winter	100% more
fog, summer	30% more
Precipitation:	
amounts	5 to 10% more
days with 0·2 in.	10% more
Temperature:	
annual mean	1 to 1·5°F more
winter minima	2 to 3°F more
Relative humidity:	
annual mean	6% less
winter	2% less
summer	8% less
Wind speed:	
annual mean	20 to 30% less
extreme gusts	10 to 20% less
calms	5 to 20% more

transfer of vapour from the droplets to the crystals. Later, in 1946 Schaefer demonstrated that solid CO_2 (dry ice) dropped into super-cooled clouds rapidly transforms the droplets into ice crystals, and in 1947 Vonnegut demonstrated a similar effect using silver iodide (AgI) particles. These experiments were carried out under the general direction of Irving Langmuir, and as Fleagle comments they 'received wide and enthusiastic acclaim and influenced the federal government through the Defense Department to support Project Cirrus, a five-year series of field experiments, conducted by the General Electric Company'.

Field experiments were conducted in Ohio, the Gulf states and California, and these clearly demonstrated the capability for cloud modification, but led to generally conservative assessments by the U.S. Weather Bureau of the economic importance of cloud seeding. From 1951 to 1953 a further series of experiments were carried out with support of the U.S. Defense Department, and at about this time a thriving business in cloud seeding sprang up. Research groups in cloud physics were also established in England, Australia, France, Canada, Switzerland, Japan and the U.S.S.R. In the United States the President's Advisory Committee on Weather Control was formed in 1953 to evaluate experiments in weather control; the committee reported in 1958 that seeding was effective in producing an average 10 to 15% increase in precipitation in mountainous areas of the western United States in winter. Not unexpectedly, the report was vigorously attacked by some statisticians on the grounds that proper randomization of the cloud-seeding trials had not been achieved, and controversy regarding this problem continues.

Commercial cloud seeding was initiated in the United States in the late 1940's in response to public interest in the General Electric experiments and the enthusiastic claims that these experiments stimulated (Fleagle, 1968). By the early 1950's, 10% of the land area of the United States was under commercial seeding operations, three to five million dollars being spent annually by ranchers, towns, orchardists, public utilities and resort operators in the hope of economic benefits. However, some of the effects were not so obvious, and the size of commercial operations receded sharply. Fleagle reports that accumulating experience and results of research have brought about a second phase of growth in operations, more stable than the earlier one, and in 1965 the National Science Foundation reported that commercial operations in the United States covered 58,000 square miles in twenty-six states.

Field experiments in cloud modification have also occurred in other countries, notably Australia, Japan, Israel, Kenya, Spain, Italy, the Soviet Union, Canada and South Africa. In some cases impressive successes in hail prevention have been claimed.

Continuing in his review, Fleagle says that in the United States, the RAND Corporation in 1962 (in a study of weather modification commissioned and supported by N.S.F.) emphasized that progress in weather modification was inextricably linked to progress in understanding atmospheric processes. They recommended increased attention to observational techniques and instrumentation, the dynamics of convection, the transfer mechanism between various scales of motion, cloud particle growth, long-period climatic

c

changes, inadvertent climate modification and mathematical and laboratory studies of the large-scale circulation.

In the following year a panel of the National Academy of Sciences Committee on Atmospheric Science was appointed, under the chairmanship of Gordon J. F. MacDonald, to study the potential and the limitations of weather modification. An interim report, issued in 1964, concluded in part that it had *not* been demonstrated that precipitation from winter orographic storms can be increased significantly by seeding (National Academy of Sciences – National Research Council, 1964). This statement, according to Fleagle, attracted the vigorous dissent of commercial operators, and resulted in a thorough analysis of extensive commercial cloud-seeding records by James E. McDonald, a University of Arizona cloud physicist. Fleagle comments:

> McDonald, working with a single-minded purpose through the spring and early summer of 1965 with great quantities of newly available unevaluated data, reached the conclusion that these data contained convincing statistical evidence of increased precipitation resulting from seeding. In one of the dramatic high points in the history of weather modification, McDonald presented his evidence to the MacDonald panel and later to the academy committee. In January of 1966 the committee [National Academy of Sciences – National Research Council, 1966] reported that 'there is increasing but still somewhat ambiguous statistical evidence that precipitation from some types of clouds by storm systems can be modestly increased or redistributed by seeding techniques. The implications are manifold and of immediate national concern.'

The report to the National Academy of Sciences fell into what Fleagle called 'a sea already strongly agitated by cross currents'. However, two further reports, one by the Special Commission on Weather Modification of the National Science Foundation (National Science Foundation, 1966), and the other by the U.S. Weather Bureau (Gilman *et al.*, 1965), advocated weather modification experiments. At the same time increasing attention was being focused on to the 'non-scientific' aspects of weather modification, and two significant publications concerned with the human, economic, legal and political aspects of weather and weather modification (Sewell, 1966; Sewell *et al.*, 1968) were subsequently published.

Two methods are generally suggested whereby precipitation may be triggered artificially. They are the introduction into a cloud of large hygroscopic particles or liquid water droplets in order to initiate the coalescence mechanism of rainfall production, and triggering of the Bergeron–Findeisen mechanism by producing ice crystals in a cold cloud by introducing artificial ice nuclei or some other suitable material (Hobbs, 1968).

In a cloud that does not reach the freezing level, there is no possibility of ice crystals. Therefore, the only way in which precipitation particles can be produced by artificial means is the coalescence mechanism. Further, since the growth of precipitation-size droplets in some warm clouds may be restricted by the lack of sufficient numbers of large condensation nuclei, it seems feasible that certain warm clouds might be induced to precipitate by introducing into the cloud large hygroscopic particles. Hobbs, in looking at the question states:

> The first experiments of this type were carried out in 1950 by the French cloud physicist Henri Dessens. Dessens seeded a number of cumulus and stratocumulus clouds with small particles of sodium chloride. On five occasions the seeding was followed by a fall of rain upwind from the generators. In England in the summer of 1952 isolated cumulus clouds were seeded with finely ground sodium chloride which was dispersed from an aircraft. In these experiments there was no observable effect on four occasions while on one occasion a shower developed. A more extensive series of experiments was carried out in East Africa in 1952 when bombs impregnated with sodium chloride were exploded just above the bases of cumulus clouds. . . . during a seeding period of about one month the total rainfall in an area 6 to 12 miles downwind was about 6 inches in excess of that which fell in a similar period of nonseeded days, but the rainfall over an area 5 miles upwind was also greater during the seeded period by 2 to 3 inches.

A more direct method of modifying the droplet spectrum of a natural cloud is to introduce water droplets directly into the cloud. Hobbs (1968) reports that Bowen carried out experiments along these lines in Australia in 1952. In these experiments situations were chosen in which several similar non-precipitating clouds were present, the behaviour of a seeded cloud being compared with the non-seeded clouds. In the majority of the cases the seeded cloud behaved differently from the neighbouring clouds, and on many occasions the seeded cloud was observed to precipitate. More recently, reports Hobbs, the Cloud Physics Group at the University of Chicago has described experiments in which water was released into the bases of warm cumulus clouds in the Caribbean. In these experiments two similar clouds were chosen and one was subsequently seeded. At a seeding rate of 450 gallons per mile, 22 out of 46 seeded clouds developed precipitation compared with only 11 of the unseeded clouds.

Experiments along these lines are continuing in many areas, and most results are inconclusive. Nevertheless, as Hobbs indicates, the results do show that under certain circumstances the seeding of warm clouds with hygroscopic nuclei or water droplets 'appears to produce rain'.

The second method of artificially producing precipitation, that of introducing artificial ice nuclei into a cold cloud, has been experimented with to a considerable extent. In this method, the basic principle is that in a cloud that extends above the freezing level, there coexists supercooled water droplets and ice particles. However, the concentration of natural ice crystals is sometimes less than that required for the efficient initiation of the Bergeron–Findeisen precipitation process. Consequently, it appeared reasonable to suppose that it would be possible to induce a cold cloud to precipitate by introducing into the cloud artificial ice nuclei or some other material that would cause ice crystals to appear.

Field experiments using the ice crystal technique have been carried out in more than a dozen countries, and as with the modification of warm clouds, many of the trials give conflicting evidence. It is generally accepted, however, that in cold clouds a 10 to 20% increase in precipitation can occur in certain circumstances.

Not surprisingly, many problems are raised in the evaluation of weather modification field tests. As indicated by McDonald (1968), one of the problems is that the atmosphere is a physical system characterized by many degrees of freedom and exhibiting enormous variability. Attempts at execution of the type of 'controlled experiments' that are possible in many of the physical and engineering sciences is therefore almost impossible. As McDonald states:

> With so broad a spectrum of natural variability, evaluation efforts will forever be plagued by the question so succinctly put in the famous *New Yorker* cartoon about rainmakers, which showed two robed clerics quizzically gazing out of a rain-spattered church window asking each other, 'Is it theirs or ours?'
> To ask whether it's 'theirs or ours' is to ask whether an observed weather process occurring after a modification test is merely one that would have occurred in the absence of the modification effort or whether it actually constitutes a significant departure from what was going to happen before we carried out our modification effort. To answer this question involves looking for an

artificially stimulated and probably small change against a confusing background of natural fluctuations of wide amplitude.

The many difficulties relating to these problems are reviewed by Chapman (1968). He suggests that the whole problem of statistical evaluation of cloud seeding is whether we can accept the observations as being random and independent, and whether a simple model can be established relating the control and target areas. The same problem also arises in evaluating other kinds of weather modification, such as hail suppression and hurricane modification. Nevertheless, Chapman says that of the many projects reviewed, significantly more of these showed positive results than negative ones. Further, while a number of the projects studied showed statistically significant positive results, few if any showed significant negative results. Accordingly, in viewing the total set of data, 'there is indeed "increasing evidence" of some effect' (Chapman, 1968).

In addition to cloud seeding, weather modification in the form of hail suppression, lightning suppression and hurricane modification has been carried out.

In hail suppression, the main objective is to increase the number of hailstones, but to decrease their size. Experiments have been conducted in several countries, notably the United States, France and the Soviet Union. The Russians have reported the more promising results. Hobbs (1968) reports that in experiments carried out in the Caucasus and Transcaucasus, silver iodide or lead iodide is shot directly into the region of the cloud containing a high concentration of liquid water. During the years since 1964 damage caused by hail in the target area has decreased to one-third to one-fifth of that occurring in unseeded areas. Hobbs states further that 'for the year 1967 the Russians estimated that the decrease in hail damage to crops that resulted from their hail-prevention program amounted to 30 million rubles, and that the cost of the program was 1·5 million rubles'. The results are possibly inconclusive, however, for Hobbs points out that the Russian experiments have not been carried out in a manner that permits statistical evaluation.

Lightning suppression has been carried out by the United States Forest Service (Project Skyfire) for over a decade. The results are in general inconclusive, although Hobbs (1968) reports that a two-year field programme (1960/61) in Montana showed that the number of cloud-to-ground discharges on seeded days was 38% less than on unseeded days.

Hurricane modification is also an important part of man's attempt to modify natural atmospheric processes. The principle involved is that if a hurricane is seeded with sufficient artificial ice nuclei, the added heat of fusion could cause a change in the hydro-static pressure field. Such a change would then lead to an outward movement of the rotating winds and a corresponding reduction in their intensity. Experiments through Project Storm-fury have been used to seed at least three hurricanes (Esther in 1961, Beulah in 1963, and Debbie in 1969). In commenting on the two earlier hurricanes Hobbs (1968) states that 'Although the changes that were observed in these two hurricanes following seeding were not inconsistent with the theoretical predictions, these changes were not so large that they fall outside the range which would be attributed to natural fluctuations.'

Many other aspects of weather modification have been com-mented upon, and the reader is referred to the extensive literature on the subject including the following appraisals of 'national' weather modification activities: Soviet Union (Fedorov, 1967, Battan, 1965); Australia (Smith, Adderley and Bethwaite, 1965; Warner, 1968); the United States (Droessler, 1968; Simpson and Simpson, 1966; Sewell, 1966); Canada (Godson, Crozier and Holland, 1966); Kenya (Sansom, 1966); Peru (Howell, 1965); and Israel (Gabriel et al., 1967).

d. Human aspects

It is clear that modification of the atmosphere – whether intentional or otherwise – can be done. However, little thought has gone into some fundamental questions regarding weather modification. The basic scientific question may well be – can you modify the weather? The more fundamental questions are – should you do so, where should you do so, and what safeguards are there either in the form of laws or compensation to provide for the errors which occur? (Maunder and Sewell, 1968).

What is also important is the possibility of ascertaining the potential of extensive urbanization for causing large-scale changes of climate over entire continents. The evidence is not yet substantial enough to show that urbanization does cause such changes, but it is sufficient to indicate that the possibility cannot be ignored, and the acquisition of more knowledge about the climate of cities may in the long run be one of the keys to man's survival (Lowry, 1967).

We can invent the future, not merely predict it. We can mould our future environment and not merely tolerate it. But, as Spilhaus (1966) comments:

> Can anyone doubt that we are modifying our weather by the wastes we are pouring into the atmosphere? That we are modifying our ground water by excessive pumping which on the coast allows the salt water to seep in? If we are doing such things inadvertently, surely we must have ways to do the things we need on purpose.

Although some remedial action has been taken, in most areas modification of the atmosphere is still largely uncontrolled, due probably in no small measure to man's general disregard for the resources of his atmospheric environment. It is essential, therefore, that we achieve a much greater appreciation and understanding of the natural atmosphere, for then and only then will we be able to use, and if necessary modify, the atmosphere for the good of all mankind.*

BIBLIOGRAPHY

ANONYMOUS, 1966: Drought. *Current Affairs Bulletin* (Dept. Adult Education, Univ. of Sydney), 38(4): 51–64.

ANONYMOUS, 1969: An economic survey of drought affected properties – New South Wales and Queensland 1964–65 to 1965–66. *Wool Economic Research Report*. No. 15, Bureau of Agricultural Economics, Canberra.

BAGDIKAN, B. H., 1966: Death in our air. *Saturday Evening Post*, October 8.

BATTAN, L. J., 1965: A view of cloud physics and weather modification in the Soviet Union. *Bull. Amer. Met. Soc.*, 46(6): 309–16.

BENLINE, A. J., 1965: Air pollution control problems in the city of New York. *Trans. N.Y. Acad. Sci.*, 27(8): 916–22.

BRATER, E. F., 1968: Steps toward a better understanding of urban runoff processes. *Water Resources Research*, 4(2): 335–48.

BRYSON, R. A., 1967: Where is science taking us? *Sat. Rev.* (N.Y.), April 1, pp. 52–5.

BRYSON, R. A., 1968: 'All other factors being constant . . .' *Weatherwise*, 21(2): 56–61.

BURTON, I., 1965: Flood-damage reductions in Canada. *Geog. Bull.*, 7(3/4): 161–85.

CHANDLER, T. J., 1962: London's urban climate, *Geog. Jour.*, 127: 279–302.

CHANGNON, S. A., 1967: Areal–temporal variations of hail intensity in Illinois, *Jour. App. Met.*, 6(3): 536–41.

CHANGNON, S. A. and SEMONIN, R. G., 1966: Tornado disaster in retrospect. *Weatherwise*, 19(2): 56–65.

* Further comment on the human aspects of atmospheric modification including the legal, economic, political and ecological aspects, are given in Chapter VI, Section D.

CHAPMAN, D. G., 1968: Statistical aspects of weather and climate modification. In: FLEAGLE, R. D., ed., 1968: *Weather Modification Science and Public Policy*. University of Washington Press, Seattle, pp. 56–68.

CUMMING, J. N., 1966: The effects of the 1965/66 drought on sheep numbers and the expected rate of recovery. *Quart. Rev. Agric. Econ.*, 19(4): 169–76.

DROESSLER, E. G., 1968: First national conference on weather modification – conference summary. *Bull. Amer. Met. Soc.*, 49: 982–6.

DULOY, J. H. and WOODLAND, A. D., 1967: Drought and the multiplier. *Aust. Jour. Agric. Econ.*, 11(1): 82–6.

FEDOROV, E. K., 1967: Weather modifications. *Bull.*, 16(3): 122–30.

FLEAGLE, R. D. (Editor), 1968: *Weather Modification: Science and Public Policy*. University of Washington Press, Seattle, 147 pp.

FLORA, S. D., 1954: *Tornadoes of the United States*. University of Oklahoma Press, Norman, Oklahoma, 221 pp.

FLORA, S. D., 1956: *Hailstorms of the United States*. University of Oklahoma Press, Norman, Oklahoma, 201 pp.

GABRIEL, K. R., AVICHAI, Y. and STEINBERG, R., 1967: A statistical investigation of persistence in the Israeli artificial rainfall stimulation experiment. *Jour. App. Met.*, 2: 323–5.

GALWAY, J. G., 1966: The Topeka tornado of 8 June 1966. *Weatherwise*, 19(4): 144–9.

GILMAN, D. L. et al., 1865: *Weather and Climate Modification*. Report to the Chief of the Weather Bureau, Department of Commerce, Washington, D.C.

GODSON, W. L., CROZIER, C. L. and HOLLAND, J. D., 1966: An evaluation of silver iodide cloud seeding by aircraft in Western Quebec, Canada, 1960–63. *Jour. App. Met.*, 5(4): 500–12.

GOUDEAU, D. A. and CONNER, W. C., 1968: Storm surge over the Mississippi river delta accompanying hurricane Betsy, 1965. *Monthly Weather Rev.*, 96(2): 118–24.

HAAGEN-SMIT, A. J., 1964: The control of air pollution. *Scientific American*, 210(1): 24–31.

HARDY, W. E., 1966: Tornadoes during 1965. *Weatherwise*, 19(2): 18–22.

HEATHCOTE, R. L., 1967: *The Effects of Past Droughts on the National Economy*. Paper presented to A.N.Z.A.A.S. Conference, Melbourne, January.

HILST, G. R., 1967: What can we do to clear the air? *Bull. Amer. Met. Soc.*, 48: 710–12.

HOBBS, P. V., 1968: The scientific basis, techniques, and results of cloud modification. In: FLEAGLE, R. G., ed., 1968: *Weather Modification: Science and Public Policy*. University of Washington Press, Seattle, pp. 30–42.

HOWELL, W. E., 1965: Twelve years of cloud seeding in the Andes of Northern Peru. *Jour. App. Met.*, 4(6): 693–700.

KATES, R. W., 1964: *Hazard and Choice Perception in Flood Plain Management*. University of Chicago, Dept. of Geography, Research Paper No. 78, 157 pp.

KRATZER, P. A., 1956: Das Stadtklima. *Wissenschaft (Braunschwerg)*, 90: 184 pp. (quoted from McCormick, 1961).

LANDSBERG, H. E., 1960: *Physical Climatology*. (Second edition) Gray Printing Co., Dubois, Pennsylvania, 446 pp.

LANDSBERG, H. E., 1961: City air – better or worse? In: MCCORMICK, R. A., ed., 1961: *Symposium: Air Over Cities*. Tech. Rep. No. A62–5, U.S. Public Health Service, Cincinnati, pp. 1–22.

LANE, F. W., 1965: *The Elements Rage*. (Revised edition) Chilton Books, Philadelphia, Penn., 346 pp.

LIND, R. G., 1967: Flood control alternatives and the economics of flood protection. *Water Resources Research*, 3(2): 345–58.

LOWRY, W. P., 1967: The climate of cities. *Scientific American*, 217(2): 15–23.

McCORMICK, R. A. (Editor), 1961: *Symposium: Air Over Cities*. Tech. Rep. No. A62–5, U.S. Public Health Service, Cincinnati, 240 pp.

McDERMOTT, W., 1961: Air pollution and public health. *Scientific American*, 205: 49–57.

McDONALD, J. E., 1968: Evaluation of weather modification field tests. In: FLEAGLE, R. G., ed., 1968: *Weather Modification: Science and Public Policy*. University of Washington Press, Seattle, pp. 43–5.

MATEER, C. L., 1961: Note on the effect of the weekly cycle of air pollution or solar radiation at Toronto. *Inter. Jour. Air Water Poll.*, 4: 52–4.

MATHER, J. R., ADAMS, H. and YOSHIOKA, G. A., 1965: Coastal Storms off the eastern United States. *Jour. App. Met.*, 3(6): 693–706.

MAUNDER, W. J. (Editor), 1969a: *Pollution: What it is, what it does, what can be done about it*. A publication of the Evening Division, University of Victoria, B.C., 115 pp.

MAUNDER, W. J., 1969b: What is pollution? In: MAUNDER, W. J., ed., 1969: *Pollution: What it is, what it does, what can be done about it*. A publication of the Evening Division, University of Victoria, B.C., pp. 1–10.

MAUNDER, W. J., 1969c: Weather, weather modification and air pollution. In: MAUNDER, W. J., ed., 1969: *Pollution: What it is, what it does, what can be done about it*. A publication of the Evening Division, University of Victoria, B.C., pp. 21–30.

MAUNDER, W. J. and SEWELL, W. R. D., 1968: Adjustments to the weather – choice or chance? *Atmosphere*, 6: 93–6, 105–8.

MEETHAM, A. R., 1945: Atmospheric pollution in Leicester. *D.S.I.R. Tech. Paper*, No. 1, H.M.S.O., London, 161 pp.

MILLER, D. H., 1966: Cultural hydrology: A review. *Econ. Geog.*, 42(1): 85–9.

MUNN, R. E., 1966: *Descriptive Micrometeorology*. Academic Press, New York, 245 pp.

MURPHY, F. C., 1958: *Regulating Flood-Plain Development*. University of Chicago, Dept. of Geography, Research Paper No. 56, 216 pp.

NATIONAL ACADEMY OF SCIENCES – NATIONAL RESEARCH COUNCIL, 1964: *Scientific Problems of Weather Modification*. National Academy of Sciences – National Research Council, Publication No. 1236, Washington, D.C.

NATIONAL ACADEMY OF SCIENCES – NATIONAL RESEARCH COUNCIL, 1966: *Weather and Climate Modification: Problems and Prospects*. National Academy of Sciences – National Research Council, Pub. No. 1350, Vols. I and II, Washington, D.C.

NATIONAL SCIENCE FOUNDATION, 1966: *Weather and Climate Modification.* Report of the Special Commission on Weather Modification, National Science Foundation, Report No. 66–3, Washington, D.C.

OGDEN, D. C., 1966: Economic analysis of air pollution. *Land Econ.,* 42: 137–47.

PERRY, R. A., 1962: Notes on the Alice Springs area following rain in early 1962. *Arid Zone Newsletter,* pp. 85–91.

ROBERTS, W. O., 1965: Atmospheric modification. *Bull. Amer. Met. Soc.,* 46: 775–8.

SANSOM, H. W., 1966: The use of explosive rockets to suppress hail in Kenya. *Weather,* 21(3): 86–91.

SEWELL, W. R. D., 1965: *Water Management and Floods in the Fraser River Basin.* University of Chicago, Department of Geography, Research Paper No. 100, 163 pp.

SEWELL, W. R. D. (Editor), 1966: *Human Dimensions of Weather Modification.* University of Chicago, Dept. of Geography, Research Paper No. 105, 423 pp.

SEWELL, W. R. D. et al., 1968: *Human Dimensions of the Atmosphere.* National Science Foundation, Washington, D.C., 174 pp.

SHEPPARD, P. A., 1958: The effect of pollution on radiation in the atmosphere. *Inter. Jour. Air Water Poll.,* 1: 31–43.

SIMPSON, J., 1967: An experimental approach to cumulus clouds and hurricanes. *Weather,* 22(3): 95–114.

SIMPSON, R. H. and SIMPSON, J., 1966: Why experiment on tropical hurricanes? *Trans. N.Y. Acad. Sci.,* 28(8): 1045–62.

SMITH, E. J., ADDERLEY, E. E. and BETHWAITE, F. D., 1965: A cloud seeding experiment in New England, Australia. *Jour. App. Met.,* 4(4): 433–41.

SMITH, J. S., 1967: The great Chicago snowstorm of '67. *Weatherwise,* 20(6): 248–53.

SPILHAUS, A., 1966: Goals in geotechnology. *Bull. Amer. Met. Soc.,* 47: 358–64.

STORR, D., 1965: The great Prairie blizzard of 15 Dec. 1964. *Weather,* 20(12): 370–2.

SUGG, A. L., 1967: Economic aspects of hurricanes. *Monthly Weather Rev.,* 95(3): 143–6.

SUGG, A. L., 1968: Beneficial aspects of the tropical cyclone. *Jour. App. Met.,* 7: 39–45.

SUGG, A. L. and CARRODUS, R. L., 1969: Memorable hurricanes of the United States since 1873. *ESSA Technical Memorandum, WBTM–SR–42,* U.S. Dept. of Commerce, Weather Bureau.

TOEBES, C., 1967: Presidential address: The place of hydrology in New Zealand's economy. *Jour. Hydrol.* (N.Z.), 6: 57–8.

TWEEDIE, A. D., 1967: Water and the city: prospects and problems. *Australian Geog. Studies,* 5(1): 1–14.

WARD, R. A., 1967: The great midwest snowstorm of January 1967 in the Chicago–Calumet area. *Weatherwise,* 20(2): 68–70.

WARNER, J., 1968: A reduction in rainfall associated with smoke from sugar-cane fires – An inadvertent weather modification. *Jour. App. Met.,* 7: 247–51.

WHITE, G. F., 1964: *Choice of Adjustment to Floods*. University of Chicago, Dept. of Geography, Research Paper No. 93, 164 pp.

ZEGEL, F. H., 1967: Lightning deaths in the United States: A seven year survey from 1959 to 1965. *Weatherwise*, 20(4): 167–73, 179.

——, 1963: Uncommon cold spell, a national economic disaster. *The Economist*, 206: 603–4, Feb. 16, 1963.

——, 1965: Weather. *Time*, July 16, 1965, p. 24.

——, 1965: Weather. *Time*, Sept. 17, 1965, p. 21.

——, 1967: Meteorology – Firing back at hail. *Time*, Oct. 13, 1967, p. 73.

ADDITIONAL REFERENCES

BARRY, R. G. and CHORLEY. R. J., 1968: *Atmosphere, Weather and Climate*. Methuen, London, 319 pp.

BATTAN, L. J., 1969. *Harvesting the Clouds – Advances in Weather Modification*. Doubleday & Co., Inc., Garden City, 148 pp. (American Meteorological Society: Science Study Series).

BOOTH, A. W. and VOELLER, D., 1967: *Social Impact of Meteorological Drought in Illinois*. Water Resources Center (Washington, D.C.), Rep. No. 9, 31 pp.

C.C.W., 1969: Exceptional weather events in 1968: *WMO Bull.*, 18: 72–80.

CHANGNON, S. A., JR, 1966: Disastrous hailstorms on June 19–20, 1964, in Illinois. *Crop-Hail Insurance Actuarial Assoc., Research Report* (Chicago), No. 31, 37 pp.

CHANGNON, S. A., JR, and SCHICKEDANZ, P. T., 1969: Utilization of hail–day data in designing and evaluation of hail suppression projects. *Monthly Weather Rev.*, 97: 95–102.

COULTER, J. D., 1965: Flood and drought in northern New Zealand. *New Zealand Geog.*, 22: 22–34.

DIGHTMAN, R. A., 1968: Central Montana rainstorms and floods. June 6–15, 1967. *Monthly Weather Rev.*, 96: 813–23.

DUNN, G. E. and MILLER, B. I., 1964: *Atlantic Hurricanes*. (Revised edition) Louisiana State University Press, Baton Rouge, 377 pp.

FETERIS, P. J., 1955: £330,000 hail damage in fifteen minutes, analysis of a devastating hailstorm. *Weather*, 10: 223–32.

HARTMAN, L. M., HOLLAND, D. and GIDDINGS, M., 1968: Effects of hurricane storms on agriculture. *Water Resources Res.*, 5: 555–62.

HEATHCOTE, D. L., 1969: Drought in Australia: a problem of perception. *Geog. Rev.*, 69: 175–94.

HUFF, F. A., 1969: Climatological assessment of natural precipitation characteristics for use in weather modification. *Jour. App. Met.*, 8: 401–10.

JANZ, B. and TREFFRY, E. L., 1968. Southern Alberta's paralyzing snowstorms in April, 1967. *Weatherwise*, 21: 70–5, 94.

KAHAN, A. M., STINSON, J. R. and EDDY, R. L., 1969: Progress in precipitation modification. *Bull. Amer. Met. Soc.*, 50: 208–14.

McQUIRE, K., 1962: Economic effects of drought on 12 properties in the West Australian pastoral zone. *Quart. Rev. Agric. Econ.*, 15: 87–94.

MATHER, J. R., FIELD, R. T. and YOSHIOKA, G. A., 1967: Storm damage hazard along the east coast of the United States. *Jour. App. Met.*, 6: 20–30.

MOHNEN, V. and VONNEGUT, B., 1968: Weather modification and air pollution. *Proc. 1st Natl. Conf. Weather Modification*, Albany, New York, pp. 228–40.

RAYNER, J. N. and SOONS, JANE M., 1965: The storm in Canterbury of July 12–17, 1963. *New Zealand Geog.*, 21: 12–25.

ROONEY, J. F., JR, 1967: The urban snow hazard in the United States. An appraisal of disruption. *Geog. Rev.*, 57: 538–59.

RUSSELL, J. A., 1967: Weather modification – then what? *Jour. Geog.*, 66: 100–1.

SAUER, J. D., 1962: Effects of recent tropical cyclones on the coastal vegetation of Mauritius. *Jour. Ecology*, 50: 275–90.

SCHAEFER, V. J., 1969: The inadvertent modification of the atmosphere by air pollution. *Bull. Amer. Met. Soc.*, 50: 199–206.

STOW, C. D., 1969: On the prevention of lightning. *Bull. Amer. Met. Soc.*, 50: 514–20.

YAO, A. Y. M., 1969: Climatic hazards to the agricultural potential in the North China Plain. *Agric. Met.*, 6: 33–48.

III Economic activities

A. THE SETTING

Large monetary losses are frequently caused by adverse weather conditions. For example, losses in the farming industry are closely associated with the frequency of hail, drought, excessive rain, snow and frost; the construction industry may be inactive for several weeks as a result of heavy snow or continued rain; and the economic fortunes of the business world are often associated with unseasonal weather conditions. In addition, fog may disrupt airline traffic causing a substantial loss of revenue to the airlines, as well as losses of income to those who are affected directly or indirectly by shipping and passenger delay. However, despite these 'everyday' weather-affected events, there have been relatively few systematic studies of the 'dollar value' of weather variations as they are related to economic activities.

The many agroclimatological studies which have been made, for example, are mainly restricted to showing the relationship between variations in temperature, precipitation, evapotranspiration and sunshine (or a combination of these) to productivity. During the last two decades, however, there have been important advances in methodologies for measuring the dollar impact of weather on agricultural activity. Nevertheless, considerable research remains to be done before the total effects of weather on the agricultural sector of an economy can be properly assessed.

Other sectors of the economy are also sensitive to weather variations. The transport industry, for instance, is especially sensitive to weather conditions such as fog, snow and visibility, and the several studies completed in the aviation sector of the transport industry in the United States have sought to identify the losses associated with specific weather elements. The impacts of weather on the most efficient use of highways, seaways and railways have also been investigated to a limited extent, as has the effect of weather changes on the sale of particular products, and on specific types of business outlets in the retail trade sector of the economy.

Econoclimatic studies have, therefore, been undertaken in limited

53

fields; nevertheless, no study appears to have been made of the total impact of weather on a regional economy, nor has there been a large-scale attempt to study the importance of the weather to the economic activity in a nation as a whole. One approach to the problem of a regional econoclimatic analysis is proposed by the World Meteorological Organization (1967) (*World Weather Watch Planning Report* No. 17). The suggestion is to draw up a schedule of weather-sensitive activities by sectors and branches of the national economy. Thus, agriculture would be divided into crop and live-stock farming, forestry, and intra-structural activities such as irrigation and land reclamation. Each of these branches would then be further subdivided into specific commodities such as cereals, fruits, vegetables and dairy products. Fisheries which are particularly weather-sensitive, whether in terms of inland water-resource control or of ocean-going trawler fleets, would have a separate listing in the schedule. The same procedure could be adopted for each of the other main economic sectors – construction, power generation, commerce, transport, communications, mining, manu-facturing and tourism, as well as public utilities such as urban drainage systems and urban water supply.

The subdivision of the various economic sectors into branches, however, raises problems, and according to the Report, one problem is the degree of detail to be aimed at, and another is the danger of double counting. The first problem involves the full nomenclature of industrial activities, some of which are particularly weather-sensi-tive, while others are relatively insensitive to weather conditions. Similarly, with the second problem, some branches of the food-processing industry, including beverages, may experience changes in both supply and demand, and both may cause additional costs. Thus, double counting must be avoided to ascertain the true effects of weather variations. Any attempt to 'isolate' the effect of weather variations on a specific sector of the economy is, therefore, difficult. The kind of difficulties faced by economists is seen by the comments of Cloos (1967) on the 1967 winter in Chicago:

> Economic conditions in January and February often provide a poor basis for evaluating future developments. Many activities are at seasonally low levels in these months and abnormally mild or severe weather can strongly influence year-to-year comparisons. This effect is doubly important when opposite conditions pre-vail in consecutive years. . . . In early 1966, relatively pleasant weather aided many types of activities, especially residential con-struction and auto sales. This year, however, storms and record

snowfalls tended to keep potential shoppers at home, hampered excavations and concrete work and movements of goods in and out of plants. The Chicago area – with a population of more than 7 million was particularly hard hit. Weather influences, although not subject to precise measurement, should be kept in mind when recent economic data is analyzed.

Statements similar to the above are to be found in dozens of periodicals, newspapers and reports. They are interesting, even newsworthy, but not especially useful, for the information given is not at all helpful for making decisions. Thus, although it would be very desirable to report on some regional econoclimatic studies (see, however, Chapter V), the subsequent chapter is by necessity essentially a review of some of the studies that have been made linking the dollar value or the potential dollar value of weather variations to specific activities in the primary, secondary and tertiary sectors of the economy. These include such varied 'activities' as wool production, tree growth, fishing, construction, electric power consumption, ship routing, the commodities markets and consulting meteorologists. It should be noted, however, that although the influence of weather on economic activities is not at present subject to precise measurement, it is clear that such influences are in many cases significant. Furthermore, it can be expected that future research on the measurement and understanding of the impact of weather on man's economy will pay handsome dividends.

B. PRIMARY ACTIVITIES

1. Agriculture

The effects of weather variations on agriculture are often far-reaching. Generally, only the direct effect of weather on the yield and/or quality of the specific crops is considered, but any analysis of the total crop industry shows that the influence of the weather does not end on the fields of the farm. As *Time* (November 10, 1967) relates:

> The effects of the bad growing season have rippled all through California's multimillion agricultural empire. Late fruits reached canneries at the same time as on-schedule tomatoes, causing so much of a jam that a great deal of fruit spoiled while it was waiting to be canned. Farmers, with their orchards maturing late, had no school children available to help harvest the crops, and the supply of temporary Mexican labor has been reduced since the law covering the use of *braceros* was tightened three years ago. Governor Reagan angered unions by permitting convicts to help with the harvest. (Copyright Time Inc. 1967.)

California, and the world, is in fact dependent to a very large extent on agricultural production, and agricultural production in turn is closely related to climate. The 'normal' climate is, however, rarely experienced, and what occurs most of the time are variations from the normal, variations which, as Thompson (1964) points out, while not random, are not rhythmic enough to produce a prediction equation. Nevertheless, significant departures of one or more consecutive seasons from the normal climate do occur, and they are especially important if they occur in the world's major food-producing area, for the world's food supply is far from solved. Indeed, the meteorologist, like any other scientist, owes the world a living, and it is 'his fundamental duty to put his abilities at the disposal of the common good' (Smith, 1967b).

The importance of the atmospheric environment in the agricultural potential of various areas of the world has of course attracted the attention of several meteorologists and climatologists, but, as Chang (1968) pertinently comments, the present situation too often seems to ignore the atmospheric elements. Chang points out, however, that under a given set of economic conditions the best agricultural land use scheme is rigidly determined by the physical environment. Several other examples are given to demonstrate the 'deterministic' effect of climate on agriculture, and Chang notes further that in agricultural land use planning it is not sufficient to know *only* that a crop can be grown in a certain area, but that it is also important that an estimate of the yields of these crops be made.

The estimation of the precise economic value of climate and weather in their influence on crop and livestock production has not, however, received as much support from researchers as has that of the physical and biological relationships (Maunder, 1968b). It is, of course, a particularly difficult problem and, as Curry (1958) points out, although farmers all over the world assess weather as a resource, the formal expression of this evaluation is far from easy. Despite these difficulties, solutions to this problem can be found, however, even if only very approximately. It is therefore essential that more research along these lines be done, in view of the fact, as stressed by Watson (1963), that 'climate determines those crops the farmers can grow, weather influences the annual yield, and hence the farmers' profit, and more important, especially in underdeveloped countries, how much food there is to eat'.

The rapid increase in productivity in the United States during the past 35 years has raised the question as to the relative contributions of technology and weather. Bean (1967) in discussing this question asks: 'How much of this lift in per acre yields is due to the doings of man, – with what he does with his land, his seeds, his fertilizer, his pest and disease control, his row spacings, and how much may be the result of a fundamental change in weather?' Commenting further, Thompson (1964) says that there has been a growing tendency to believe that technology has reduced the influence of weather on grain production so that we no longer need to fear shortages due to unfavourable weather. Thompson notes, however, that '. . . there is increasing evidence . . . that a period of favourable weather interacted with technology to produce our [U.S.] recent high yields, and that perhaps half of the increase in yield per acre since 1950 has been due to a change to more favourable weather for grain crops'.

Irrespective of the importance of the overall trends in the long-term seasonal weather, there remain the day-to-day, week-to-week and season-to-season weather variations (see, for example, Decker, 1967), and it is these 'short-term' variations that are of fundamental importance. The significance of agroclimatological studies is therefore not lessened by advances in technology. Indeed, increasing pressure of populations on world food supplies suggests that such studies will be of even more value in the future than they are today.

a. Climatic variations and agricultural production

In recent years there have been a considerable number of agroclimatological studies linking climatic variations with variations in agricultural production. The analysis in these and many other similar studies have generally been of a statistical nature associating agricultural production in a selected small area (such as a county, shire or experimental farm) with the climatic variations as reported at a nearby climatological station. An attempt has been made in some cases to collate such statistical analyses for a number of adjacent areas and obtain what could be classified as an agroclimatological model for a relatively large area. There have, however, been few if any studies relating variations in agricultural production for a nation as a whole and the variations in the nation's climate.

To this end, an attempt was made to express in simple terms the relationship between agricultural and climatic variations as they apply to the Australian continent (Maunder, 1968d). Australia stands out as a prime example of the importance of the climatic environment (see, for example, Tweedie, 1967), for in no other area of continental proportions do weather and climatic variations – particularly rainfall – play such an important national role. It is true that many other areas, such as the eastern part of the United States, experience extreme weather variations, but the variations in rainfall in Australia present a constant reminder to the Australian farmer, forest fire-fighter, businessman and politician that water is essential to the well-being of every Australian.

The basic climatological information used was obtained from the maps of Australia (published annually by the Director of Meteorology, Melbourne), showing the areas of the continent with rainfall above or below the average. An analysis of the rainfall patterns as revealed by the maps for the 25 years 1941 to 1965 was made, and Figs. III.1 & III.2 show in graphical form the percentage of Australia with above average rainfall expressed as two time series, in conjunction with small inset maps showing the extent of the areas with above average rainfall. It may be noted that, over the 25-year period, the *average* percentage of Australia with rainfall above average was only 40%. Indeed in only five of the 25 years (1947, 1950, 1955, 1956, 1960) did more than *half* of Australia receive more than the *average* rainfall, while in five other years (1943, 1945, 1957, 1961, 1965) less than 20% of the continent received more than the average rainfall.

The associations between the 'time trends' of agricultural production for Australia and the corresponding percentage of the country with above average rainfall, for 19 agricultural factors were considered, and this revealed that in many cases there appears to be a relatively 'strong' association between fluctuations in some aspects of total Australian agricultural production and the variations in the proportion of Australia having an above average rainfall. On the other hand, there are many instances where there does not seem to be any real association. This, of course, is not unexpected since the data considered for both rainfall and agricultural production were deliberately chosen for the whole of Australia. Nevertheless, the analysis indicated that the proportion of Australia receiving above average rainfall is in many cases a good index of the Australian agricultural production for that season or in some cases for the following season.

Associated with the supply of water for agriculture are the properties of evaporation and transpiration, and the concept of potential evapotranspiration, all of these factors being especially

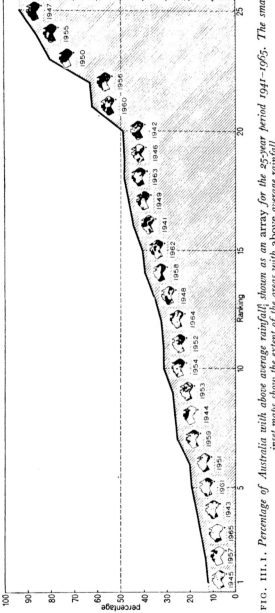

FIG. III.1. *Percentage of Australia with above average rainfall, shown as an array for the 25-year period 1941–1965. The small inset maps show the extent of the areas with above average rainfall.*

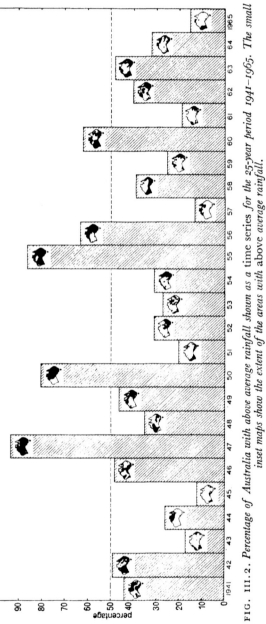

FIG. III.2. Percentage of Australia with above average rainfall shown as a time series for the 25-year period 1941–1965. The small inset maps show the extent of the areas with above average rainfall.

important in the assessment of the water needs of agriculture. For example, Curry (1962, 1963) considered the climatic resources of livestock farms in New Zealand from the point of view of potential evapotranspiration. In a later paper, the seasonal programming of atmospheric resources is considered, and Curry (1966) suggests that four alternatives exist for improving production, each of which should be considered: first, the manipulation of storage devices; determination of the relevant period for which rain probabilities are to be assessed; third, seasonal forecasting on a probabilistic basis; and fourth, rainmaking.

Actual evapotranspiration, which is directly related to plant growth (Arkley and Ulrich, 1962), is an important consideration in all kinds of agriculture. Knowledge of these and other biological relationships allowed Flinn and Musgrave (1967) to develop a plant–soil moisture simulation model, inputs to the simulation model being parameters reflecting the water-holding capacity of the soil in question, the type of crop grown and weather data. The authors aggregated the daily results over the irrigation season to derive an index of plant growth, and the model highlighted the importance of the time application of irrigated water. Irrigation as such is not discussed in this book,* but it is pertinent to point out that the benefits from a programme of scientific irrigation (see, for example, Thornthwaite and Mather, 1954) will accrue not only to the irrigation farmer through increased yields and profits but also to his neighbour who turns to irrigation, and to those who provide various services to the farmer, such as the manufacturers and dealers of irrigation equipment, the electric power companies, the fertilizer industry and the general business community.

In general, there are three ways of establishing the importance of climate–agriculture relationships. The first is the study of the fundamentals of plant–climate relationships, namely, the radiation and moisture balance for various crops in various climatic environments. Monteith (1965) indicates, in fact, that the analysis of the radiation climate is a central problem of agricultural meteorology since the rates of photosynthesis depend on the receipt of visible light, and the rates of transpiration depend on the net exchange of radiation by a crop canopy. Second, the study of agricultural data and climatic data for a number of places within a given area, for as

* The reader is, however, referred to the many references on the subject, including Ehlers (1960), Hargreaves (1956), Headley and Ruttan (1964) and Hagan et al. (1967).

long a period as consistent records of both agriculture and climate allow, may permit the deducing of agroclimatological relationships from analyses of the data. However, despite the many useful agroclimatological relationships obtained using such methods, there are many associated problems, and in particular it is very difficult to isolate the contribution made by any one of the various components of the climate such as temperature, day length, light intensity or precipitation. However, the third method, controlled environments, does permit in many cases an analysis of the importance of the individual climatic factors, since in a controlled environment the climatic factors can be varied one at a time. Nevertheless, Evans (1963) and McWilliam (1966) sound a note of warning about the extrapolation of results derived in a controlled environment to field situations. One inherent problem of this approach is the virtual impossibility of simulating natural environments under controlled conditions. It is not possible, therefore, to predict the entire range of plant performance under field conditions on the basis of controlled studies.

The 'significance' of climatic variations in their associations with agricultural production is, however, difficult to determine, and Smith (1961) says that '. . . if . . . we examine recent climatic variations through the eyes of a statistician, we may be forced to conclude that none of them are "significant". . . . This may be acceptable to the pure scientist, but it could tend to infuriate the man who has experienced what *appears to him* [italics mine] to be a radical change in the circumstances in which he has to live and work.' Thus, although some agroclimatological relationships may not be statistically significant, they may nevertheless be very important to the plant, to the farmer, and to an area's economy.

Agroclimatological investigations are usually designed to show the association between several climatic elements and various agricultural factors. In addition, the prediction of agricultural production based on a statistical analysis of weather and yields is possible, particularly if some kind of multiple curvilinear analysis is made. Thompson (1962), for instance, has suggested that in some cases forecasts of yields may be made. However, he also points out that there are many limitations to the solution of this apparently 'simple' problem. Watson (1963), moreover, indicates that only when more detailed knowledge has been acquired of how and at what growth stages climatic factors influence yield, will it be possible to derive complex variates that give appropriate weight to the different factors for correlation with yield in naturally varying climates, and to use them to predict yields from meteorological records.

The forecasting of agricultural yields on the basis of meteorological conditions is nevertheless a very important aspect of agroclimatological studies, and Bourke (1963) emphasizes that the importance of the agricultural forecasting is so great that '. . . much of our efforts in the next ten years must necessarily be devoted to finding some reliable basis for extrapolating beyond the short-term weather forecasts adequate for agricultural tactics, to estimates of long-term weather trends which will be applicable to the even more important field of agricultural strategy'.

As already pointed out, however, there has within recent years been a considerable number of questions raised regarding the relative effects of weather and technology upon the current high levels of production, and in this regard Shaw and Thompson (1964) point out that in its simplest form one could say that yields are a result of weather, of technology and of a weather–technology inter-action. Moreover, it should be emphasized that favourable weather alone will not produce high yields unless adequate technology is used, and improved technology alone will not produce high yields without adequate weather. Favourable weather conditions could, therefore, be said to allow technology to express itself to its fullest potential. In addition, the general problem of agroclimatological relationships is accentuated since it involves a consideration of biology, physics, meteorology and economics. Penman (1962) states, for example, that physically several divisions of agroclimatological relationships are possible. He comments:

> . . . there is the obvious first-order effect in which we know that crop growth depends on weather, and that seasonal and secular changes in yield are caused by weather changes. . . . Second-order effects are those arising from pests and diseases, nearly all of which have strong associations with weather in their incidence, in their development, or in their spread. . . . Third-order effects, too, have a biological origin – man. . . . The fourth-order effects may be peculiar to a species or even to a variety of plant. They are the effects that involve 'trigger action' or the existence of threshold values; to which for completeness, might be added the effects of meteorological abnormality, sometimes manifest in frost, rain, hail or wind damage.

Sanderson (1954), who made a detailed study of the methods of crop forecasting, also comments that among the various branches of economic activity agriculture is the only one which seems to be destined forever to be subject to wide and irregular fluctuations of output, the year-to-year variations of agricultural output being largely determined by physical factors. However, economic factors are also important, and Sanderson rightly points out that variations in the acreage planted to a given crop in response to fluctuations in price are, as a rule, confined to land which is marginal from the point of view of that crop, and that the price received in the previous year may also be responsible for variations in the use of input factors, such as fertilizers and sprays.

A final comment on the general relationship between climate and agriculture is from Landsberg (1968), who indicates that the problem of feeding the growing world population is likely to become the most important issue within the next decade. The weather, says Landsberg, is the most important variable in crop production, and it enters in two ways into the problem of adequacy of food supplies. First, through the obvious hazards to crop plants, such as freezes, hail, drought or pest-provoking weather conditions, and second, through a land use which is wasteful of climatic potential. This second factor is unfortunately often ignored, and although an analysis of the total energy available for agriculture is difficult, Landsberg suggests that a profitable use may be made of a systems approach. This is shown in Fig. III.3 and indicates the prominent role weather plays in the basic inputs, and also the value of weather information for the control complex where counter-measures to adverse weather conditions are planned and executed.

> In a complete systems analysis, the weather factors and their influences on the system could be readily lifted out and analyzed as a subsystem. A considerable part of the basic work for quantitative analysis of a weather subsystem in agriculture has already been done or data for it are readily available. Many probabilistic expressions for climatic risks have been established. Multiple regressions for the effects of various weather elements on specific crop plants are also available. For most major crop plants the optimal phenological conditions are well known. If all this information is incorporated into the system, it becomes possible to assess the influence of any given variable singly or in combination with others. . . . It is a challenge for agricultural meteorologists to add to their normal micrometeorological orientation a problem of widest scope and implications. (Landsberg, 1968.)

b. Effects of weather on specific crops

The effects of weather on specific crops are many, and space permits

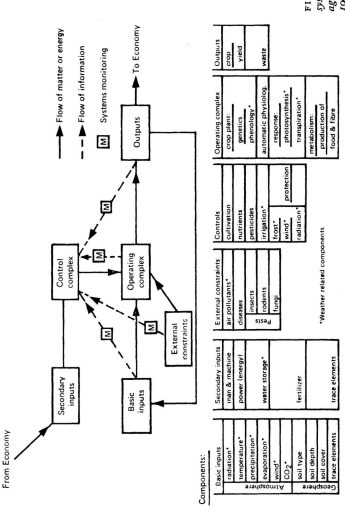

FIG. III.3. A scheme for a systems analysis of land use in agriculture (after Landsberg, 1968).

mention of only a few of the studies completed.* It should be noted, however, that the majority of crop–weather studies are not concerned with 'dollar values', and what follows is an attempt to emphasize only those studies which have a potential dollar orientation. Reference should be made, however, to Chapter V (and in particular Table V.1) in which the 'dollar values' of several agricultural crops in New Zealand are discussed (Maunder, 1965a, 1966b, 1968a).

Wheat. It is clear from the results of many studies that definite climatic factors determine where and to what extent wheat can or cannot be grown economically, and further, that weather conditions have a marked influence on the growth of wheat as well as on the yield and quality of its grain. A study by Nuttonson (1955), for example, makes use of phenological data (such as dates of sowing, emergence, heading and ripening) and of some meteorological records to ascertain wheat–weather and wheat–climate relationships. Such analyses are particularly valuable when the relevant phenological data (and appropriate climatic data) are available. However, in many cases such data are not available; consequently, this brief review of literature is mainly confined to 'conventional monthly studies'.

The main relationship between rainfall and wheat yields per acre, shown by most studies, is the requirement that moisture must be generally adequate and consistent throughout most of the growing season. For example, the importance of moisture during the early-middle growing season is indicated by Davis and Palleson (1940) in the Dakotas, Hopkins (1935) in the Canadian Prairies, Frisby (1951) in the Great Plains, and Tauheed (1948) in investigations on the relationship between rainfall and wheat yields in New Zealand. However, pre-seasonal rainfall is indicated as being important by Pengra (1946) in studies in the Great Plains and Army *et al.* (1959) in the southern High Plains, area of the United States, whereas Davis and Palleson (1940) in North Dakota suggested that while mid-summer rainfall is beneficial, rainfall during the late summer is detrimental. In the semi-arid areas, as would be expected, dry conditions usually have an adverse effect on yield, and this is suggested by Hewes (1958) and Johnson (1959) in their studies in areas of central-southern United States, and by Cornish (1950a, b) in South Australia. Millington (1961), however, in his studies in

* Only studies relating to wheat, oats, barley, corn, potatoes, tea, apples and raisins are considered in this section.

South Australia, states that rainfall in the month following sowing is detrimental to high wheat yields, and he shows that the wheat yield per acre is reduced by nine bushels for every inch of rainfall in the month following sowing.

Temperature conditions are also considered significant, the importance of early summer temperatures being noted by Frisby (1951) in the Great Plains and Maldenhaugher and Westin (1959) in South Dakota. On the other hand, Army and Hanson (1960) in their studies in the Plains Area of Montana indicate that high temperatures were especially harmful after the crop had headed. The added effect of sunshine has not been considered by many researchers, one of the few studies correlating sunshine duration with wheat yields being made by Tippett (1926) at Rothamsted, England.

Four other studies could be mentioned. First, Thompson (1962) used multiple curvilinear regression analyses to differentiate between the influences of weather and technology in the production of wheat in five states in the Great Plains, for the period 1935–1961. Suggested important climate/wheat associations included the benefit of pre-seasonal rainfall in Nebraska and Kansas, and of relatively wet and warm conditions in mid-spring and late spring in the Dakotas. Second, Johnson (1964) showed that the average contribution to yield of an inch increment of stored moisture (in the Great Plains) was estimated to be 2·44 and 2·74 bushels per acre for spring and winter wheat, respectively. Third, Asana and Williams (1965), using controlled environments, indicated that there was a mean reduction in wheat yield of 16% for a 6°C (11°F) rise in 'day' temperature from 25°C (77°F) to 31°C (88°F). Fourth, Hewes (1965), in reviewing wheat failures in the dry farming region of central Great Plains from 1939 to 1957, showed that drought, wind and hail were the chief causes of wheat failure, a notable loss being the 77·2% failure of planted wheat in one year in Stanton County, Kansas, as a result of wind.

Oats. The main relationship between climate and oat yields appears to be in the beneficial effect of precipitation especially during the period of active growth, and of cool temperatures, particularly during the early summer. However, in some cool climates it appears that a warmer than average summer is beneficial for above average oat production.

Zacks (1945) indicates that in southern Ontario a wet early and mid-summer is one of the important climatic factors influencing high oat yields. Similar results were indicated by Coffman and Frey (1961), who suggest that in the United States precipitation from mid-spring to early summer is important, and that in Canada precipitation is desirable from late spring to late summer.

With respect to temperatures, cool growing conditions are the main features stressed in the literature. For example, Hooker (1922) and Martin and Leonard (1949) have shown the importance of a cool summer, although Geddes (1922) concluded that in three north-east counties of Scotland oats require a mean temperature above normal, particularly during the sowing period and for some weeks immediately before harvest. Coffman and Frey (1961) suggest that in the United States oat-growing areas cool conditions are necessary during periods of germination and of greatest growth, and Grainger et al. (1954), working in southern Scotland, indicated that low mid-summer temperatures favoured high grain yields.

Sunshine factors have been discussed briefly in a few papers. In Georgia, with autumn-sown oats, Crowder et al. (1955) found that growth during early and mid-winter was dependent on temperature and solar radiation. Grainger et al. (1954) in Scotland also suggested the benefit of sunny conditions. Maunder (1965b), however, indicated that in southern New Zealand a cloudy mid-summer is beneficial for above average oat yields.

Barley. The benefit of rainfall during the critical jointing period (generally early summer for spring sown barley) and of relatively cool summer conditions are the main features of most barley/climate research.

Relatively dry conditions in the early part of the season are suggested as being important by Wishart and Mackenzie (1930), and Russell and Bishop (1933) from their experiments conducted at Rothamsted, England. The late summer/early winter to mid-spring period is generally cited as not requiring excessive moisture. For the latter part of the season, however, there is fairly general agreement on the beneficial effect of rains (see, for example: Hooker, 1922, Weaver, 1943, and Malcolm, 1947). However, Leonard and Martin (1963) note that barley grows better in moderate rather than excessive rainfall.

High temperatures are generally detrimental to barley production, the preference of relatively cool growing conditions being indicated by Leonard and Martin (1963), Malcolm (1947) and Weaver (1943). The latter paper showed that in the United States, a decrease of $1°F$ in the average temperature for the mid-spring to mid-summer period, 'increased' the barley yield by 1·6 bushels per acre.

Few results of the effect of sunshine on barley production have been published, and the comment by Russell and Bishop (1933), working at Rothamsted, England, that '. . . it seems to matter little (to yield) whether the spring is sunny or not', may still be relevant.

Corn. Many papers have been published on the effects of climate on corn yields, the majority of these relating to conditions in the mid-west of the United States. A review of some of the earlier literature in the field is given by Basile (1954). Rose (1936), for example, assessed the relationship between climate and corn yields in several mid-west states, and Hendricks and Scholl (1943) measured the joint effects of rainfall and temperature on corn yields in Ohio, Indiana and Iowa.

Several additional papers have reported on specific aspects of the corn–climate relationship. Stacey *et al.* (1957), for example, state that in Georgia the value of an inch of rainfall declines for a time after the usual planting date and then increases sharply during the latter part of the growing season, especially when associated with the higher temperatures which are normal for that period. In another study, in South Africa, some aspects of corn production and droughts from glass-house experiments are discussed by Grant (1960). She noted that the corn yield was proportional to the water consumed except at the highest level of drought incidence, when the yield per unit water used was reduced.

A more detailed study was reported by Runge and Odell (1960) on the relation between precipitation, temperature and the yield of corn on a farm in Urbana, Illinois. Odell (1959) further comments that a 64-day period, extending from 50 days before full tassel to 14 days after full tassel, was found to be of utmost importance to the corn crop. He notes that a yield reduction of 0·05 bushels per acre is associated with each 1°F above normal maximum temperatures when the corn is near full tassel, while a rainfall one inch above normal in the week before full tassel will usually increase the yields by about four bushels.

Another climatic factor, soil temperature, was investigated by Willis *et al.* (1957), who state that the most favourable soil temperature at the four-inch depth for corn growth in central Iowa appears to be about 75°F. They comment that the results emphasize the importance of considering soil temperature in tillage and soil management studies. Soil moisture is also a significant factor, Holt *et al.* (1964) indicating the importance of stored soil moisture at planting in western Minnesota and eastern South Dakota. Another report on the effects of soil moisture and temperature on the growth and yield of corn in Kentucky is given by Benoit *et al.* (1965).

Corn–weather studies are continuing to be made, such as the study by Changnon and Neill (1968), who point out that a major problem facing all corn–weather studies has been to ascertain accurately the roles of weather factors, treated individually or collectively, and those of man-made (technology) factors which interact to produce a corn yield. They note, however, that assessment of the current role of weather in corn production is essential because of the rapid increase in national and world food needs and the need to establish proper governmental controls on crop production.

Studies relating to these needs (e.g. Thompson, 1966; Shaw and Durost, 1965) have produced conflicting results concerning the role of weather and that of technology. However, the availability to Changnon and Neill of nine years of recent (1955–1963) data on weather conditions, corn yields and associated agricultural practices on cash-grain farms in a small area of central Illinois offered them a unique opportunity to make a mesoscale study of corn–weather relations during this latest technology phase. In addition, it allowed an assessment of corn–weather relations derived from actual farming operations as opposed to those established using experimental farm data or that for large areas. The data for the study were obtained from 60 farms, 49 recording rain gauges and seven temperature stations located in a 400-square-mile area, the nine years of data at the 60 locations providing a sample size of 540 yield values for correlation with weather and agronomic observations.

The study area is representative of a high-value, cash-grain farming region, and in 1963 the value of the crops produced from this area was $16,500,000. In the study, weekly, monthly and seasonal rainfall and temperature values, plus agronomic data, were correlated with corn yields during 1955–1963. July and August mean temperatures and cumulative degrees above 90°F during July and August had the 'strongest' correlations with corn yield, -0.50, -0.69 and -0.51, respectively. The results suggest, therefore, that the lowering of high temperatures in July and August would be of considerably greater value in increasing corn yields and, as pointed out by Nicodemus and McQuigg (1969), such cooling could very well be produced by the generation of cirrus contrail clouds. The results also indicated that added water from irrigation or man-made rainfall would be beneficial only in late June and July, but that the added water would not produce significant increases in yields because of the relatively poor correlation between rainfall and corn yields in this area.

Potatoes. The absence of climatic extremes appears to be of importance in potato production, an adequate supply of moisture, a temperature of 18–20°C (64–68°F), and an average light intensity, rather than 'extremes' of these elements, being suggested as the optimum climate. However, potatoes are grown in many different climatic regions, and it is clear that the optimum conditions are not the same under all conditions. For example, Ivins and Milthorpe (1963) suggest that relatively low temperatures, high light intensity and short days accelerate the development of potatoes, and accordingly stimulate tuber growth. On the other hand, high temperatures, low light intensities and long days delay tuber formation. They stress, however, that temperature, light intensity and sometimes day length show optimum curves for stem elongation, leaf weight and other components.

In other studies, Bushnell (1925) showed that tubers, after two months of growth at temperatures ranging from 20° to 29°C (68–84°F), decreased in weight with increasing temperature, and that no tubers were found at 29°C (84°F). Low night temperatures are, however, important and Gregory (1954) obtained a much higher yield at 30° day–17° night (86–63°F) than at 23° day–23° night (75–75°F) (365 and 159 g, respectively). The importance of frosts on potato production is also very important, and Dingwall (1953), writing in New Zealand, suggested that a late ground frost of 6°F or more occurring while plants are establishing will check growth severely.

Tea. The influence of climatic factors on the yield of tea has been detailed in a few papers, a recent investigation on the yield of tea in the Assam Valley of India being made by Sen, Biswas and Sanyal (1966).

In a review of previous studies, the authors noted the papers of Portsmouth (1957) on tea production in Ceylon, and Shelford (1953) and Laycock (1958) on tea production in Nyasaland, and they indicate that one of the important findings from Shelford's studies is that the season's crop is largely determined by the performance of the bush in the cold season and the first months of summer.

In the paper by Sen *et al.*, the authors reached a number of conclusions. First, a study of moisture, rate of evaporation and rainfall suggested that the season could be divided into four time groups: the dry, cold weather from January to March; the period of early rains from April to June; the period of main rains from July to September; and the dry period from October to December. Second, of all the climatic variables, rainfall and mean temperature during January to March proved most beneficial to the early crop, the rate of increase in yield with increase in rainfall being inversely proportional to the average rainfall. In addition, an increase in rainfall during January to March proved more beneficial to the crop when the mean temperature was high, this supporting the view that rain during the night followed by sunshine during the day is beneficial to crops. Third, rainfall and mean temperature during January to March were highly correlated with the main tea crop. Fourth, rainfall and mean temperature during October to December were the only climatic variables which were significantly correlated with the late tea crop.

Apples. Many references are made in the literature on the effect of temperature on apple trees and apple production. Chandler (1957), for example, says that killing winter nights, spring frost, lack of winter chilling and the expense of combating diseases and insects, are all factors limiting production. Minimum winter temperatures are also mentioned by Childers (1961), who states that ' . . . probably the most important factor influencing the distribution of the fruit industry is minimum winter temperatures'. Shoemaker and Teskey (1959) indicate, however, that the rate of drop and duration of freezing temperatures may be more serious than the actual degree of cold reached. This point is further emphasized by Childers (1961), who points out that the amount of injury from freezing in dormant tissue is influenced by three factors: the rate at which the temperature falls, the duration of the low temperature and the rate of thawing. The effect of cold temperatures on blossom is reviewed by Shoemaker and Teskey (1959), who say that blossoms are killed at 25–28°F. However, they comment that only about 5% of the blossoms on an apple tree in heavy bloom are required to set fruit in order to produce a commercial crop, and hence a good set of fruit sometimes occurs after earlier examination of the blossoms seemed to indicate a crop failure.

The economics of minimizing losses due to cold temperatures is also particularly important, and Kemp (1956) in one pertinent study discusses a Central Otago orchard in New Zealand and shows that the amount of oil used annually normally varies with the amount of frost. In 1953, one of the worst frost years since 1933, over 6,000 gallons of oil were burnt in the orchard to combat cold temperatures.

Summer temperatures are also important, however, and Shoe-maker and Teskey (1959) state that the optimum temperatures for pollination and fertilization are 70–80°F with moderate relative humidity. They further indicate that temperatures below 40° and above 80° inhibit pollen germination, that some pollen germination takes place at 40–50°, and that temperatures in the range 60–70° give satisfactory germination. Shoemaker and Teskey also mention that summer temperatures and sunshine are important in develop-ing red colour, without which prices may be lowered.

The normally detrimental effects of low temperatures on apple trees and the subsequent fruit production is stressed by many research papers. However, other weather factors are also impor-tant, and in a paper on fruit production at Nelson, New Zealand, Rigg (1959) says that there were three seasons in the period 1940–1957 in which abnormal weather seriously affected the yield of orchard and the quality of the fruit. In the 1942 season a violent gale in early autumn caused the loss of an estimated 800 cases of pears and apples which were ready for picking; in 1957, a hailstorm in the preceding late spring caused serious damage to both foliage and fruit; and in 1949, a very dry summer seriously affected the sizing of the fruit.

These observations point out, therefore, that irrespective of the 'normal' climatic requirement or the 'normally' adverse climatic factors which influence apple production, there are other factors such as gales, hailstorms, droughts and out-of-season frosts, which in some seasons are equally if not more important.

Raisins. The utility of weather forecasts to the raisin industry has been assessed by Kolb and Rapp (1962), and is discussed in Chapter VI.

In another paper, Lave (1963) considers the effect of weather on raisin growing in the lower San Joaquin Valley of California, an area that produces 90% of all U.S. raisins. Under ideal conditions, 3,000 degree days are necessary for raisin grapes to be of drying quality, the degree day being considered an index of the total heat supplied to the vine, and is computed by subtracting 50° from the mean temperature of a day. That is a day with a high of 100° and a low of 50° is measured as 25 degree days, the number of degree days being computed between the time of budding and the first

D

autumn rain over 0·10 inches. In his analysis, Lave shows that in California the weather is the significant uncontrolled factor of production, and he suggests that each additional degree day will lead to the production of 194 additional pounds of raisins. Thus, one additional day with mean temperature of 85° before the first rain would increase production by 6,790 pounds. Moreover, because grapes are severely damaged by rain, the prediction of rainfall is particularly important to the raisin grower, and Lave calculated that the value of being able to forecast rain three weeks in advance is $91 per acre, which if realized for each bearing acre in 1960 would have had a potential value of over $20,000,000.

c. Effects of weather on animal and pastoral production

In reviewing the potential of pastoral and grassland farming in New Zealand, Sears (1961) stated:

> The products of a country depend as much on what its people want to produce, as on its natural advantages. New Zealand has a tremendous climatic advantage for grassland agriculture and over the past century has developed to be one of the world's leading exporters of grassland products. . . . New Zealand's real future, at least on present knowledge, must still depend on a wider and more intensive development of her grassland agriculture. . . .

These comments serve to point out that in some parts of the world agriculture based on grassland farming is extremely important. Moreover, if one considers the pastoral products of milk, butter, cheese, meat and wool, and their world distribution in terms of exports, then several countries, notably Australia, New Zealand, South Africa and Argentina, are especially important, particularly in view of the fact that their climatic year is 'opposite' to that of the major importing areas of the world. The importance of the weather conditions in these areas is illustrated in a review by Maunder (1963), who also points out that about 38% of the butter exports, 94% of the mutton and lamb exports, 57% of the beef and veal exports, and 86% of the wool exports originate in Southern Hemisphere' countries. In view of these high percentages, and the associated fact that pastoral farming in most Southern Hemisphere countries is a particularly significant part of the respective national economies, it is not surprising that a considerable amount of research on the effect of weather on pastoral production has been done in these areas. This is particularly the case in Australia and New Zealand, and what follows is a brief review of some of these studies that are either value or production orientated.

Dairy production. In New Zealand, the effect of climatic variations on dairy production has been reviewed by Maunder (1966a), who notes that about one-third of the gross farm income and about one-quarter of the foreign exchange earnings in New Zealand come from dairy production.

In a study of dairy-farming trends in the Waipa county of New Zealand from 1941 to 1950, Hutton and Wright (1952) stated that '. . . climate is the dominant factor affecting per cow and per acre production'. Such comments on the importance of climatic fluctuations on butterfat productions are fairly common; nevertheless, little work has been published on the specific importance of the various aspects of the measurable climate on butterfat production. It is clear, however, from the many papers on grassland farming in New Zealand, that an adequate supply of moisture – especially during the summer – is essential for above average dairy production. This is emphasized by Johnson (1955) and McMeekan (1960) among others, while du Faur (1962) indicates that areas regularly experiencing hot, dry summers will have lower pasture yields. For example, Johnson, in an analysis of factors affecting butterfat supplied to Waikato factories, states that the marked effect of variations in summer rainfall on dairy production in the Waikato is clearly brought out. He further comments that since one-third of the total cows in milk in New Zealand are in this area, the variation in summer rainfall at dairy farms in the Waikato will have a reasonably strong influence on total butterfat production in New Zealand, and even on aggregate farm output. McMeekan (1960) also comments on the importance of rainfall in the summer, and states that after the peak pasture production is reached in October and November production falls to much lower levels in summer and may even reach zero proportions in February.

A considerable amount of research has also been done into the effects of climate on pasture production, the various papers of Mitchell (1959, 1960) and others giving details regarding the variations in pasture growth from area to area, season to season, and within seasons. For example, Filmer (1960) indicates that the rate of pasture growth in late spring may be twice that in the autumn, three times that in the late summer and ten times as great as the rate in winter. Filmer also indicates that production of pastures in a very good year may be 50% higher than in a very bad year. This is also indicated in a paper by Iverson and Calder (1956), who show that pasture production on 'light land' in the Canterbury district of New Zealand, varied in the two consecutive springs 1954/55 and 1955/56 from 422 lb to 1,938 lb, whereas the corresponding summer variations in the same two seasons were 357 and 16 lb, thus giving a difference of 1,922 lb between the spring and summer of 1955/56.

The relationship of monthly pasture production with monthly temperature, precipitation and computed actual evapotranspiration was examined by Taylor (1954), temperature being closely related to growth during the months from May to November inclusive at most of the eight stations used. In addition, computed actual evapotranspiration proved, according to Taylor, to be 'very significantly' related to pasture growth at all stations. Another investigation into the relationship between climate and livestock in New Zealand was made by Curry (1958), who assessed the association between evapotranspiration and dry matter production of ryegrass–white clover pastures. Curry found that for each inch of evapotranspiration in a month above 0·9 in., a ryegrass–white clover pasture will produce 1,200 lb of dry matter. He further noted that a monthly evapotranspiration of 1·4 in. must take place before milk can be produced when stocking is at one cow to the acre. The effect of droughts on dairy and pasture production is therefore of some importance, and Ward (1951) mentions the dangers of maximum stocking. Davey (1962) also shows that in the Manawatu area of New Zealand, butterfat production per cow dropped from 369 lb in 1960/61 to 301 lb in the drought year of 1961/62.

Variations in rainfall and temperature are therefore of some importance to dairy production in New Zealand, and Maunder (1965a), using a multiple regression model, showed that in general the significant climatic factors associated with above average butterfat yields per cow were relatively wet conditions in each of the months October to February, cool conditions in November and February, warm conditions in January and cloudy conditions in December and February. The effect of such climatic variations on agricultural incomes is reviewed in another paper (Maunder 1968a), and this is discussed in some detail in Chapter V, under Regional Applications.

Agroclimatological studies related to pasture and dairy production are also available for countries other than New Zealand, such as the studies by Smith (1967a, 1968) on British grassland farming. The first paper considers the importance of potential transpiration, which Smith says takes into account radiation, temperature, humidity, wind and rainfall, combined in a way that provides a useful index for the growth of grass. Using this index, a useful parameter in assessing potential grass production and the associated forms of husbandry is formulated. The second paper (Smith, 1968) considers the problem of forecasting annual milk yields, and the author indicates that forecasts of the annual (April–March) milk yield of recorded herds in England and Wales can be made nine months ahead, at the end of June, using April–June milk production data and the June rainfall with a mean percentage error of 0·31.

Meat production. The effect of climatic variations on meat production in New Zealand and Australia is reviewed by Maunder (1967b). Two aspects of the problem appear to be most important: first, the seasonal growth of pastures, and their effect on meat production; and second, the effect of drought on sheep and cattle numbers. The first problem, that of the seasonal growth of pastures, has already been noted in the previous section, but it is appropriate to point out that the variability occurs not only on a seasonal basis, but also from year to year and from area to area, and that all three variations affect production and the value of the meat.

Perhaps the major climatic problem facing annual production, however, is the effect of drought, which is a serious problem to many aspects of successful meat production. In Australia the drought problem is particularly severe, and a report by Carney (1950) outlines some of the problems and discusses the measures employed by graziers in the northern section of the Hay district in New South Wales. He states:

> The most common policy was to reduce stock numbers by selling a portion of the flock, and during the last drought 69% of the properties in the Hay area reduced their flocks in this way. . . . In the early stages of the drought fair prices were obtained, but as the drought progressed and the market deteriorated, good three- and four-year-old ewes realized only 6s. 6d. per head, while lambs brought only 2s. 6d. In some cases around Hay, graziers even offered sheep free to anyone willing to remove them from the property.

Carney comments that losses of stock in the Hay District were severe, the total sheep carried being reduced from a normal figure of 880,000 (10-year average) to 500,000 at December 31, 1945.

This represented a reduction of 43%, which he says can be attributed mainly to deaths caused by drought conditions, to selling surplus stock, and to a great reduction in the number of lambings.

In Queensland the influence of drought conditions on beef and cattle numbers is also of importance, and Patterson (1957) says that the duration and distribution of rainfall, temperature and evaporation have exerted a tremendous influence on the structure of the industry. A five-year moving average of the weighted indexes for Queensland as a whole together with beef cattle numbers from 1894 to 1955 are given in the paper, and illustrate the close relationship of cattle population to seasonal conditions throughout Queensland. Kinsman (1953) also indicates that the six major droughts in Queensland between 1920 and 1950 resulted in an annual average loss, good years and bad together, of about $8\frac{1}{2}\%$ of marketable beef.

In a review of climatic variations and meat production in New Zealand, Maunder (1967b) points out that it is apparent that any adverse change from the 'normal' seasonal growth of pasture is detrimental to animal production, the adverse climatic factors including excessive cold and/or inadequate rain in spring, summer drought, and the late arrival of autumn rains. Many other factors, climatic and non-climatic, however, are also important, and Coop (1955), commenting on the apparent complexity of the climate–lamb growth relationship in Canterbury, poses the question as to whether the reasons for the variations in growth are a matter of dry matter content or intake, or, as he says, '. . . is it purely a matter of sunshine and temperature?'

Wool production. Wool production in New Zealand has a considerable worldwide importance, and the review of climatic variations on wool production in New Zealand (Maunder, 1967a) is pertinent to any placement of a value of weather on world wool production.

An assessment of wool production per sheep in New Zealand was made by Philpott (1956), who analysed wool production and wool production per sheep shorn in each of the seasons 1918/19 to 1954/1955. This analysis showed that the wool shorn per sheep varied from 7·4 lb in 1930/31 to 9·9 lb in 1954/55, giving a rise of about 0·5 lb per sheep every 10 years. Several marked fluctuations occurred, however, such as the variation from 7·4 lb per sheep in 1930/31 to 8·1 lb per sheep in 1931/32, and the 1·0 lb per sheep increase from 8·8 lb in 1943/44 to 9·8 lb per sheep in 1944/45, the difference in total New Zealand wool production in the latter two seasons being

42,000,000 lb, which at a value of $0·25 per lb is worth over $U.S.10 million.

There are many reasons for these differences in wool clips, including the time at which the sheep are shorn, but clearly climatic variations are one of the important aspects to be assessed, and some of the influences of climate on wool growth in several countries are now discussed.

The first unequivocal demonstration that climatic factors influenced wool growth, according to Coop and Hart (1953), was that of Ferguson *et al.* (1949) in Australia, who found that wool grew fastest in summer and slowest in winter, thus giving a close correlation with mean temperature. Subsequent investigations have been made by Stevens (1958), Storey and Ross (1960) and others. Stevens (1958) states, for example, that in New Zealand the period of least growth is June–September, but from then until the peak is reached in March–April there is a rapid and progressive increase. This is followed by a comparable decrease during winter until the lowest point in August–September. The total clip for the year according to the Stevens experiment was about 13·5 lb, March contributing about 1·75 lb but August only 0·75 lb, or less than half as much.

The importance of rainfall on wool growth has been reported by Hoxmark (1928) from studies in Argentina, and by Dowsett (1946) and others in Australia where most work has shown the general benefit of rainfall. For example, Dowsett showed that the summer and autumn rainfall had the greatest influence on fleece weight in the following season, and that winter rainfall was the least influential. Of greater importance is the effect of drought conditions on wool production; in a review Franklin (1962) indicated that climatological records and fluctuations in livestock numbers and in wool production reveal that no other single factor has exacted such a heavy toll on the sheep industry. He further points out that the problems of effective drought control constitute one of the greatest challenges to efficient and increasing primary production. Specific examples are given by Franklin, and he notes that in 1944 the sheep population of Australia was 123 million. Then followed '. . . the disastrous drought of the middle forties when sheep numbers declined to less than 96 million . . . and wool production fell at the same time from 3·6 million bales to 2·9 million bales'. The economic consequence of such a drop in wool production was further investigated by Franklin, who indicates that had it been possible to prevent those losses of 1944/46, the cumulative wool cheques over the succeeding ten years may have totalled an additional £A600 million, equivalent to over $U.S.1,400 million. A more recent study by the Bureau of Agricultural Economics in Australia (Anonymous, 1969) suggests, however, that these estimated losses grossly overstate the position, since they were arrived at by the simple multiplication of the estimated difference in sheep numbers, average wool cut per head and wool prices, none of which is independent.

Most research on the effect of climatic variations on wool production lays emphasis on wool quantity rather than wool quality. However, Moule (1962) notes the apparent beneficial effects of drought on wool quality and he comments that year-to-year variations occur in the qualities predominating in the wool clips of the different states of Australia. These variations, he says, have been greatest in Queensland, where the proportion of the wool clip of 70s or finer quality rose from 5% in 1945/46 to 10·6% in the drought year 1946/47, and from 3·8% in 1950/51 to 10·1% in the drought year of 1951/52.

In the preceding section an attempt has been made to summarize the pertinent comments from a few of the many papers written in recent years on the effects of climate on wool production and sheep numbers. It is clear that the relationship between wool quantity, wool quality and climatic conditions is not a simple one. Nevertheless, Doepel and Turner (1959) do suggest that the most extensive and more readily apparent of the changes in the environment from one season to the next are those that come about as the result of the variation in the weather. Similar observations are also made by Storey and Ross (1959), who investigated the effect of shearing time on wool production. They indicated that some of the observed wool clip variations were due to the time of shearing, but they also pointed out that most of the differences were seasonal in character and depended mainly on the weather and the availability of feed in the winter and early spring months.

d. Hail and frost damage

Hail, frost and freezing temperatures are of special concern to many facets of agriculture, particularly to the sensitive crops such as berry and stone fruits in the spring. Several investigations have shown the adverse physical, biological and economic effects of hail, frost and/or freezing conditions on crops, and the reader is referred to specific papers such as the World Meteorological Organization (1963) publication *Protection Against Frost Damage* and the paper by

Crawford (1964). In the latter paper Crawford developed a mathematical model describing the heat balance of the 'crop' layer, which he says may be readily used by meteorologists, engineers and agriculturalists who are concerned with frost protection.

Of possibly greater economic effect is the effect of hail, and an important 'value' aspect of the hail–crop relationship is hail insurance, which is discussed later in this chapter. Crop losses for most states in the United States are available from the Crop–Hail Insurance Actuarial Association, and this data was utilized by Changnon and Stout (1967) in a study of crop–hail intensities in the northwestern United States. An index called the 'loss cost' is used by the Actuarial Association and it represents the total storm-day losses divided by the liability for the area with loss, the ratio being then multiplied by $100. For example, a storm causing $100,000 in losses in an area with $100 million in liability would have a loss cost of $0·10. For each state and for each month, Changnon and Stout computed a median storm-day loss cost value for either corn or wheat or both. These were defined as intensity indices, the values for months with eight or more storm-days ranging from a low of $0·0001 to a high of $0·0270. To simplify these median loss cost values, the dollar designations were removed and multiplied by 1,000. Thus, the median monthly loss cost of $0·0270 became a monthly hail intensity index of 270. Monthly intensity indices were then computed for the available data on corn and on wheat, the relevant data for 14 states for wheat being shown in Table III.1. Among the conclusions reached by the authors was that the intensity in the peak intensity–crop damage months, June and July, decreases rapidly and steadily eastward from the Rocky Mountains. Hail in Illinois and Indiana, for example, is only 5–10% as intense on the average as that in the mountain states.

Other aspects of the problem of hail in American agriculture are discussed by Lemons (1942), and for specific data on hail–crop insurance, the reports prepared by Changnon (1960, 1966) for the Crop–Hail Insurance Actuarial Association are of interest.

e. Insects and disease

In recent years it has become obvious that many animal populations carry on a very intimate relationship with the 'natural' atmospheric environment. It is also evident that very little is known about how most of the earth's fauna types are specifically affected by climate, a fact which is especially regrettable when viewed against man's

TABLE III.1: *Monthly data on hail damage to wheat, 1957–1964*
(*after Changnon and Stout, 1967*)

State	June a*	June b*	July a*	July b*	August a*	August b*
Texas	25	146	5	5	0	0
New Mexico	10	260	3	15	0	0
Oklahoma	22	36	3	4	0	0
Kansas	29	80	14	6	0	0
Colorado	21	234	20	121	4	6
Nebraska	28	90	24	42	4	2
Wyoming	11	240	14	270	4	77
North Dakota	23	25	29	107	25	20
South Dakota	23	78	28	110	15	15
Minnesota	13	40	20	40	13	10
Montana	20	26	25	161	20	30
Oregon	3	38	4	17	3	14
Washington	4	30	4	10	2	14
Idaho	7	19	8	31	10	30

a*: Average number of storm-days.
b*: Intensity indices (altered median loss cost values; see text for details).

increasing ability to control and manipulate the atmospheric environment. In regard to agricultural production, this lack of bio-climatological knowledge is of special concern, for if climatological control is undertaken without a thorough study of the biological consequences it will cause, it is possible that plagues of pests will arise and cause the probable extinction of many useful species. Devastating population imbalances in the past have resulted from climatological fluctuations and atmospheric pollution. Man must therefore realize the implications of his 'tinkering' with the atmosphere, and resist the desire to enter into a field that outwardly promises quick economic gains.

In most aspects of agriculture, insects and diseases play an important part in both the quantity and quality of the commodity produced, whether this be apples, wool, corn or milk. Such insect populations and plant diseases are often weather-related, and although man can effectively control many of these diseases, the associated loss in the value of the crop produced can in many ways be indirectly attributed to a period of 'adverse' weather conditions.

Facial eczema, a disease of sheep and cattle, for example, is responsible for serious production losses in New Zealand in some years. It is caused by the fungus *Pithomyces chartarum*, the growth of the fungus being strongly influenced by climatic conditions. For example, Lancashire and Keogh (1964) state that the fungus develops mainly in the summer and autumn when earth temperatures at 4 in. are 65°F or higher. They further comment that rain is necessary both for the infection of dead plant material by fungal spores and also for the rapid sporulation of the fungus. In another study, Mitchell *et al.* (1959) said that an examination of factors which could influence the growth of the fungus and the development of facial eczema toxicity in pasture indicated that growth and sporulation on dead grass leaves and stems is most rapid under warm (70–80°F) moist conditions. Moreover, although it is difficult to estimate the loss in wool production as a result of the disease, it is clear that the occurrence of facial eczema can have serious effects.

Most diseases flourish best in certain kinds of climate; some diseases prefer cool regions, others the warmer areas, some require a great deal of moisture, others require much less. Within the range of conditions that permit its existence, a plant disease may therefore be an insignificant or an important factor in crop production. This depends, in general, on how exactly the local or seasonal conditions fit the requirements for its development and spread. On the other hand, climate can constitute an effective barrier against the advance of a plant disease from one region to another.

The connection between disease and weather is often fairly obvious – such as the attacks of potato blight following a cool rainy period (Blair, 1959) or the dependence of the apple scab on rain. However, with some other diseases the critical period is passed long before the attack becomes evident (see Wang, 1963). For example, the temperature of the preceding winter appears to be the decisive factor in the occurrence of bacterial wilt of sweet corn, and the severity of wheat leaf rust often depends on temperature and moisture during late winter and early spring. Consequently, the weather at the time that a plant disease is most conspicuous is not necessarily reliable evidence to the conditions that encourage disease development. Moreover, a change in the intensity of one weather element may bring about changes in the whole disease relationship. A change in temperature, for example, may make existing moisture conditions more or less favourable to attack by the pathogen, or it may increase or decrease the vulnerability of the host. Conversely, a change in moisture supply may require a corresponding change in temperature if the disease is to progress unchecked. The successful forecasting of critical weather conditions associated with the probable outbreak of a plant disease can therefore be of considerable economic value, particularly if suitable fungicides have been developed to counteract the disease.

The application of insecticides to control plant disease and/or insects also has meteorological implications. Naya (1967) points out, for example, that the best insecticide might give bad results if it is applied incorrectly or at the wrong time. It is therefore advantageous that the farmer know the habits of the insects, the properties of the insecticides and the influence of the weather on both the insects and the insecticides.

f. Agriculture and the value of the weather

The weather and climate exert a major influence on many economic activities, but few activities appear to be more profoundly affected than agriculture. Some of the studies cited above have shown, for example, how crop yields vary with the amount and timing of precipitation, and how yields from pasture through livestock vary with weather conditions. The uncertainty of weather also plays a major role in farm operations, and if farmers could be more certain of the occurrence of precipitation at a particular time, it is claimed that farm operations could be more efficient.

Unfortunately the number of agroclimatological studies related to productivity per unit, particularly in terms of value per unit to either the farmer or the community, is small. Nevertheless, as has been demonstrated by the author, it is possible to measure both the probability of occurrence *and* the effects on farming income of significant climatic variations as these in turn affect agricultural production (Maunder, 1968a). For example, it is suggested that in the South Auckland area of New Zealand, which has a dairy cow population of over 750,000, a 'wet' January (occurring once in about six years) will be associated with an 'increase' of about $N.Z.2,000,000 in the income of dairy farmers in the area, compared with a corresponding 'decrease' in incomes once in about eight years as a result of a 'dry' January.

These and other similar results suggest therefore that weather modification, especially where it is designed to augment natural precipitation, would be especially appealing. But weather modification is only one means of combating weather variations and, as pointed out in Chapter VI, it is probable that a better use of weather forecasts, as well as improved weather forecasts, may be equally beneficial. In addition, as Castle and Stoevener (1966) point out, many factors must be considered in estimating benefits of increased precipitation or a reduction in temperatures. Account should be taken, for example, of the losses which may result in one region from the increases in production in another region, and there is also a need to weigh the costs of obtaining the benefits claimed for weather modification against the costs of obtaining the same benefits by some other means, such as by 'importing' water and using it for irrigation. Moreover, weather modification designed to affect this supply of water would not normally occur at the scene of production. This point is made by Kirkbride and Trelogan (1966) and emphasizes that it may not be the precipitation that falls during the crop-growing season that is of importance to the supply of water, since the critical periods for weather modification to supplement irrigation water supplies may occur during periods when few crops are in the ground.

Many problems are associated with the economic evaluation of climate and agricultural production, and before any large-scale weather modification is attempted in agricultural areas it is essential that a much better understanding of the specific relationships between the elements of the atmosphere and unit productivity be obtained. For example, we need to know the specific effects of an additional 10 hours of sunshine, 2 inches of rainfall, or 100 growing degree days on the quantity *and* the quality of production per unit, whether this be cotton production per acre, milk production per cow, wool production per sheep or apple production per tree. Indeed, if we do not know these values, and many other related values – both agricultural and climatic – then we are hardly in a position to decide whether it is economically desirable to modify the natural weather conditions through irrigation, cloud seeding or contrail formation.

The use and value of weather information, weather forecasts and weather modification as it applies to various economic activities is discussed in Chapter VI. With respect to weather modification and agricultural production, however, it is pertinent to note, as pointed out by Kirkbride and Trelogan (1966), that those concerned in modifying weather need to exercise caution in using existing statistical data in the economic analysis of the effects of weather modification, for the available statistical data are generally aggregative and were not collected to relate to logically defined areas with crop production orientation. Moreover, the data is unlikely to conform precisely to areas either subject to weather modification or those likely to receive the economic benefits associated with a particular weather modification effort. It seems desirable, therefore, to utilize controlled experimentation that requires data collection adapted to the specific experimental areas in order to produce meaningful results of the economic efforts of weather modification on agricultural production. Such data will of course only relate to that specific area. It is also most important to take into account regional interdependence. For example, frost in Florida usually increases the price of citrus fruits for growers in California and Arizona. Accordingly, as Curry (1966) indicates, it is most unlikely that any sizeable increase in production in the west of the United States as a result of rain augmentation (by irrigation, or weather modification) will leave prices paid to eastern farmers unaffected. Similar arguments may also be offered for other crops in other areas. It is clear, therefore, that weather does have a value in terms of agricultural production, but it is a value which must be evaluated more closely if we are to gain the full benefit of any future weather modification activities.

2. Forestry

Several aspects of forests and forestry are of importance in a study of the value of the weather. Forest canopies, for example, influence the hydrology of different areas by reducing the amount of precipitation that reaches the ground, and by decreasing the amount of evaporation from the ground. A forested area also acts as a natural environment for wildlife, and in many areas provides income through tourism and timber. In addition, the influence of weather and climate on forests, the influence of forests on the local climatic conditions, and the importance of both man and nature with respect to forest fires are significant facets of weather and forestry.

a. Climate and forests

In most areas the 'natural' forest reflects the normal climatic conditions. However, as can be seen in New Zealand and in some other areas, the 'natural' forests do not necessarily reflect present-day climatic conditions (see for example: Holloway, 1954; Cumberland, 1962; and Holloway, 1964). This is usually because the present-day climatic conditions are different from those 500 to 1,000 years ago, but also because in some areas, such as Vancouver Island, forest fires 100 to 400 years ago destroyed all but the most hardy trees, the new forests established being in many ways different from the natural forests. Nevertheless, on a world scale the association between present-day climatic conditions and present-day forests is generally well established.

The relationships between the local climatic conditions and the local vegetation, whether natural or man-planted, are of considerable interest to biogeographers. However, the effect of vegetation, including forests, on the local climatic character is probably of greater significance in any assessment of the value of the weather. One aspect of this is the influence of a forest cover on rainfall–runoff relations, and Lawson (1967) reports that in order to explain fully this phenomenon, one must know what proportion of precipitation reaches the soil, both directly as throughfall and indirectly as stemflow. In studies described by Lawson, throughfall and stemflow were determined for a pine–hardwood stand in the Ouachita Mountains of Arkansas, and over a five-year period throughfall averaged 85% of gross precipitation, the remainder being intercepted by the forest canopy. Thus with an annual precipitation average of over 42 inches, experimental data indicated that the total interception amounted to approximately six inches per year. Interception rates of 15 to 20% were also reported by Rogerson and Byrnes (1968) in Pennsylvania, Semago and Nash (1962) in Missouri, and Boggess (1956) in Illinois.

The economic implications of these and related studies are illustrated in a report by Fahey (1964), who considered interception and throughfall under a man-planted pine plantation in the domestic water supply catchment area of the city of Dunedin, New Zealand. Fahey demonstrated that the pine plantation significantly reduced evaporation, a factor which would have obvious beneficial effects on some urban water supplies especially during a hot and dry summer.

In addition to the effect that forests and shelter belts have on the local climate, and the relationships that exist between climatic conditions and 'natural' vegetation, there is a relationship between tree growth and climate. One evidence of this is tree rings, which have been studied in considerable detail by several investigators (see, for example, Fritts, 1962; Fritts, 1965; and Glock, 1963). An associated feature of tree rings, notably total tree growth, is of specific importance in any evaluation of the value of weather. For example, if one compares tree growth in New Zealand to that in British Columbia, one essential difference is that plantations of man-planted trees in New Zealand mature in about 25 to 35 years, whereas the corresponding period in Canada is at least 60 years, and in many cases over 100 years. The value of New Zealand's mild climate to the forest industry of New Zealand is therefore considerable, since in a period of 100 years three or four tree crops can be harvested, compared with two or only one crop in many other forest producing areas.

b. Weather and forest fires

In many areas of the world the traveller is confronted with signs such as 'Use Your Ashtray' and 'Prevent Forest Fires'. The suggestion put forward by the various state forest services is that 'Only You Can Prevent Forest Fires'. But, in some areas, this is far from the truth, for weather and not man is often the major cause of forest fires. For example, in British Columbia, the Forest Service reported that in 1958, 1,150 fires were caused by lightning (27% of all fires), these fires causing damage of $6.4 million or 77% of the total damage by all fires. In some years, lightning-caused fires are less important, but in the north-west of the United States and in south-western British Columbia, lightning is very often associated with more than 50% of all forest fires.

A further aspect of weather and forest fires are those atmospheric conditions which create fire hazard conditions, notably low humidity, hot drying winds, low rainfall conditions and sunshine. Taylor and Williams (1967) also note that weather is a major *variable* in the behaviour of all outdoor fires, and that in the forest, unanticipated changes in the weather can turn a moderate wildfire into a holocaust. Such behaviour, they say, can occur under meteorological conditions similar to those which produce severe thunderstorms and tornadoes. The authors give as an example the Hellgate Fire which occurred on April 18, 1965, in a mountainous area of the George Washington National Forest in western Virginia. The fire was characterized by running (rapid horizontal spread), crowning (burning from treetop to treetop), convection columns (vertical development of the fire with towering columns of smoke), spotting (setting of new fires by flying sparks and embers), and fire whirlwinds (small, tornado-like whirls of fire). They further note that spotting occurred more than half a mile in advance of the fire front, and that some fire whirlwinds had diameters of 15 feet and heights of 150 feet.

In economic terms forest fires are often very destructive, and this is perhaps best demonstrated by some of the forest fires in Australia. Whittingham (1964) reports in detail on one of these fires, the disastrous Dandenong bushfires of January 1962, which destroyed more than £2 million of property and forest timber. Three hundred homes were completely destroyed in the fire, and Whittingham describes how weather conditions contributed in one way or another to the start and spread of the fire. In addition, the weather in the form of rain was in part responsible for bringing the fire under control on the morning of January 17.

An essential aspect of the effect of weather and forest fires is the provision of special weather forecasts for specific forest services or companies. Cleary (1965) indicates some of the requirements of the Forest Service in the State of Victoria (Australia), and he suggests that advice of the development at any time of conditions conducive to the occurrence of serious fires is essential, such advice being required so that prohibition of lighting of fires can be imposed and forest operations can be closed down. Operational advice is also required for the planning of operations for the control of both bush-fires and prescribed burning.

An associated feature of forest fires is the effect of smoke from the forest fires on visibility and the radiation balance, which can have significant effects. For example, smoke from forest fires, if moved by winds over an urban area, may reduce the normal solar radiation by as much as 30 to 60%. This reduced sunlight may in turn cause lowering visibilities, increasing use of power for artificial illumination, and probable delay in land and air transportation. A specific case of the reduction of solar radiation through dense smoke is reported by Peterson and Drury (1967). In this instance, on a reconnaissance flight on July 10, 1964, north and east of Yellow-knife, N.W.T., unusually dense smoke was observed over a wide expanse of the tundra, the smoke causing a decrease of approximately 25% in the amount of solar radiation reaching the ground.

The value of the weather associated with the forest industry is perhaps best demonstrated by the activities of the United States Government in its Fire-Weather Services. White (1968) notes that the Environmental Science Services Administration maintains 43 fire-weather offices across the United States, staffed by personnel who are especially trained to support forestry interests. The Fire-Weather Service is equipped with 20 unique truck-mounted mobile weather stations, each with two-way radio and radio facsimile communications and a sizeable number of what are called 'suitcase' units of specialized meteorological observing equipment which can be deployed to major fires for immediate operational support. In addition, during actual forest fires, Fire-Weather Service meteorologists provide many kinds of special forecasts required for the immediate fire area. The range of information and services demanded of fire-weather meteorologists, however, goes far beyond these activities. In particular, notes White, the decisions faced daily by forest and range management interests such as closing forest areas to public use, suspending logging operations, issuing burning permits, manning lookout towers, activating aircraft fire patrols, deciding on disposition of manpower and equipment, arranging for food and supplies, the parachuting of men and equipment and the release of fire fighters are all vitally dependent upon access to specialized weather information.

c. Weather modification and forest fires

Our forestry resources are so dependent upon weather conditions and so subject to certain extremes of weather, such as drought and lighting, that it is not beyond the realm of possibility that we will one day face the need to use operational weather modification techniques to protect our forests. (White, 1968.)

So stated R. M. White, Administrator of the Environmental Science Services Administration, in an address delivered at the Joint Conference of the American Meteorological Society and the Society of American Foresters, in Salt Lake City, on March 13, 1968, and they serve to illustrate the importance of weather modification in any programme concerning weather management and forests.

Forest fires, whether started by man or lightning, create significant economic and social problems. For example, such fires cause damage to watersheds, wildlife, grasslands and outdoor recreation facilities. In addition, fires can cause loss of life, and the destruction of homes and other property. Moreover, the supply of forest products and the economic well-being of many forest-resource-based industries are often affected by fires. Specific examples of the damage caused by lightning are given by Barrows (1966). He notes, for example, that during the period 1955–1965 there were instances of a single lightning fire burning more than one million acres, including more than five million acres destroyed by fire in Alaska in 1957. The high cost of controlling forest fires is an additional cost that has to be taken into account, and if one considers the province of British Columbia a comparison may be made between the 1967 cost of fire suppression of over $6,000,000, and the less than $1,000,000 spent in 1968. The summer of 1967 was a particularly dry one, whereas in 1968 unusual rainfalls from the middle of August considerably reduced the fire hazard. Consequently, the difference of $5,000,000 between the 1967 and 1968 seasons may be considered as a very approximate index of the value of this rain to the forest industry of British Columbia. It is therefore not surprising that man should look for ways of modifying those atmospheric variables which are associated with forest fires, notably lightning and the lack of moisture.

Lightning storms and the resulting fires have several distinctive characteristics which complicate fire control efforts. Barrows (1951), Fuquay (1962) and Taylor (1964) give details of these characteristics, and Barrows (1966) notes that they include the tendency for the thunderstorms to be modest in size, short-lived, but widely spread over mountainous country. In addition, the cloud bases are high (averaging 12,000 feet above sea level), and the storms are dry with about 80% producing less than 0·2 inches of precipitation. The most critical feature, however, is the tendency for very large numbers of fires to occur in a short period of time, Barrows giving an example of the 335 lightning-caused fires in the National Forests of western Montana and northern Idaho on July 12, 1940.

Three important lightning problems associated with lightning fires which might be eased or prevented by a successful programme of weather modification are suggested by Barrows. First, the reducing of the very large number of lightning fires which often occur in a short period of time; second, the prevention of lightning fires at critical places; and third, the prevention of lightning fires during periods of critical fire weather and high fire danger.

A solution to these problems requires considerable research, and Project Skyfire was a natural outcome of the lightning-caused forest fires. In brief, Project Skyfire originated in 1952 at the Northern Forest Fire Laboratory, Missoula, Montana. Its aims were twofold first, to gain basic knowledge of the way lightning starts forest fires; and second, the development of means for preventing lightning fires through weather modification. By 1956, the first experiments on lightning suppression were completed, and Fuquay and Boughman (1962) indicate that analysis of four years of field experimental data of cloud seeding showed 33% less cloud-to-ground discharges from treated storms than from untreated storms.

It is clear that more research is necessary before any substantial lightning suppression programme should be undertaken on a regional basis. However, results to date are promising, and lightning suppression together with rainfall augmentation will no doubt become powerful suppression agents of forest fires in the 1970's, or

1980's. Not all improvements in the atmospheric aspects of forest fires will, however, result from weather modification. Williams (1967), in discussing the future role of meteorology in fire control, suggests, for example, that at least as far as Canadian foresters are concerned, several avenues are open. For example, an improved danger rating scale, the use of automatic reporting stations, special fire-weather warnings, greater accuracy in long-range weather forecasts, and the use of radar to locate and track atmospheric disturbances, are all developing frontiers in research dealing with the weather and forest fires.

3. Fishing

A third sector of the primary industry that is weather-sensitive and at the same time economically important in many countries is fishing. The direct effect of adverse weather conditions on the fishing industry is obvious, and it is common knowledge that during adverse weather conditions, many fishing boats are either unable to leave port, or are forced to return to port earlier than anticipated. The weather-related factors in turn are associated with the retail price of fish, and if adverse weather conditions continue for any length of time, the decreased catch of fish would have significant economic effects on the fishing community.

Another weather-related factor is the effect of colder than normal, or warmer than normal, ocean temperatures on the total fish population, and hence the probable fish catch. One aspect of this is examined by Bell and Pruter (1958), who state that a review of the literature suggests the need for a re-examination of the bases for some of the reported temperature–fish productivity relationships, since in some instances adequate provision does not appear to have been made for changes in the amount of fishing, the economic conditions, or in the efficiency of the fishing fleets. For example, they suggest that it is doubtful that extremely cold or unusually warm air temperatures during one month or over a limited period of the year would be immediately reflected in mid-water or bottom temperatures in the oceanic regions where halibut spawn and where the eggs and larvae appear to spend their early existence. On the other hand, the authors comment on a paper by Ketchen (1956), who developed certain temperature and 'catch' data for some Pacific Coast marine fisheries. These showed that long-term changes in air and water temperatures had been occurring in the Pacific area but that possible important effects of such climatic changes on the stocks of fish may have been masked by the influences of economics and the fisheries themselves. Nevertheless, Ketchen indicated that suggestive correlations existed between temperature and 'catch' of some species, and that the indicated relationships could not be ignored. One important difficulty in establishing such correlations is the measurement of the 'catch', for unlike a wheat field with its unique farmer, the ocean, at least outside the '12-mile' limit, is open to all fishermen. Thus the catch data on which some climate–catch associations are based may not necessarily reflect the total catch of the area.

It is evident that before satisfactory relationships can be established between climatological changes and the productivity of fish stocks, a great deal more must be known regarding both the environment and the fish themselves. Moreover, although the value of weather and climate to the fishing industry may not be quantified in any meaningful manner, it seems reasonable to suggest that the atmospheric and oceanographic environments must be considered in any evaluation of fishery resources such as that suggested by Crutchfield (1962). In addition, it should be noted that the effect of weather on fishing does not necessarily apply only to the commercial fishing industry, for as any angler will testify, under the right kind of conditions the fish really do bite better!

C. SECONDARY ACTIVITIES
1. Manufacturing

In comparison with agriculture the number of studies completed on the effect of weather and climate on manufacturing is minimal. Indeed, apart from general references to manufacturing in textbooks on the applied aspects of meteorology and climatology, the interest of both atmospheric scientists and the manufacturing industry in the effect of weather and climate on manufacturing is noticeable only by its absence. Nevertheless, a number of general associations are known.

a. Effects of weather and climate on industry

The diverse impact of weather and climate on industry may be broadly categorized into two broad divisions: those influencing the location of industry, and those affecting the operations of the industrial plant once an industry has been established. Associated with these factors are the design of the plant, and the planning of its operations.

Initially, the importance of climatic factors rather than weather problems faces industry. For example, such factors as the availability and flow of water, the seasonal conditions affecting heating, cooling and the dehumidifying of the air, and the atmospheric stability or instability of the site in relation to air pollution, are directly related to capital investments. Labour, fuel supply, raw materials, markets, land cost, transportation facilities and local or state taxes, must also be taken into account, and several of these components do not appear to have a direct association with climatic conditions. Climatological components which influence storage, warehousing and the operation of transportation facilities are also important. In addition, as pointed out by Landsberg (1960), there are the less noticeable effects of climate such as those affecting the efficiency and health of workers, the weathering of stockpiles of raw materials and finished products, and the losses caused by evaporation of volatile fluids. Furthermore, almost all industries with outdoor activities are to a greater or lesser degree subjected to climatic variations, the transportation, construction, aircraft manufacturing, ship building and strip mining industries often being affected by freezes, fog, snow, ice, lightning and gales.

Weather conditions are also reflected in almost every phase of operation of manufacturing, and Critchfield (1966) notes the impact of various weather conditions such as storms, which cause workers to arrive late for work, hamper essential outdoor activities, cause damage to goods and equipment or interrupt power. Wind and low relative humidity are also sometimes important in that these conditions may increase the fire hazard at manufacturing plants, especially if combustible or explosive materials are being processed or stored.

The optimal conditions for a few selected manufacturing industries have been suggested by Grundke (1955/56). These are shown in Table III.2, and they can be considered from both a climatic viewpoint in respect to the siting of an industry and the planning of its operations, and also from a weather viewpoint, in that the day-by-day weather conditions will affect the amount of 'air conditioning' required. Landsberg (1960) also quotes some typical operating temperatures for outside industries such as a range of −20 to 100°F for the iron and steel industry, 20 to 90°F for refineries and the chemical industry, and 40 to 90°F for outdoor assembling and painting. The motion-picture industry, with its more stringent and specific weather requirements, also serves to illustrate the value of weather conditions, notably in this case sunshine, light and visibility. Both weather forecasts and climatic studies are therefore essential in determining the best time and place to photograph outdoors, if the costs of motion-picture production are to be minimized.

TABLE III.2: *Optimal indoor operating climatic range for selected industries (after Landsberg, 1960; after Grundke, 1955/56)*

Industry	Temp. (°F)	R.H. (%)
Textile industry:		
Cotton	68–77	60
Wool	68–77	70
Silk	71–77	75
Nylon	85	60
Orlon	70	55–60
Food industry:		
Milling	65–68	60–80
Flour storage	60	50–60
Bakery	77–81	60–75
Candy	65–68	40–50
Process cheese production	60	90
Miscellaneous industries:		
Paper manufacturing	68–75	65
Paper storage	60–70	40–50
Printing	68	50
Drug manufacturing	68–75	60–70
Rubber production	71–76	50–70
Cosmetics manufacturing	68	55–60
Cosmetics storage	50–60	50
Photographic film manufacturing	68	60
Electric equipment manufacturing	70	60–65

Studies on the specific relationships between variations in the weather and variations in industrial output, or efficiency, are few. An expanding field of industrial meteorology has however developed, particularly in the United States, and Boyer (1966) notes that the growth of industrial meteorology offers opportunities for meteorologists and for those businessmen who are alert to the advantages of being 'weatherwise'. He states:

Industry does have weather problems and weathermen are the ones most likely to figure out the correct answers. I have seen an entire chemical plant frozen solid. I have worked in a brewery where summertime production was limited by the cooling capacity of the plant which in turn was determined by the weather. How many meteorologists were employed in the design of these plants? NONE. Many design engineers know little about the difference between surface temperature and air temperature. Air conditioning engineers are often shocked when confronted with observational data on thermal lapse rates in the lower 100 feet of the atmosphere. These things are well known to meteorologists and they are important in the design, construction, and operation of industrial plants.

Thus, the problems of industrial weather *are* real. Furthermore, Boyer notes that a prominent weather service in the north-eastern United States is paid on a bonus–penalty arrangement, wherein each correct forecast is worth $1,200 and each error results in a $1,500 penalty to the meteorological consulting firm.

Climatic determinism, as suggested by Huntington (see Huntington, 1945), has, as is pointed out in Chapter IV, received little support during the past two decades. However, although many oppose any suggestion of climatic control, it is nevertheless evident that the location of industry, and the migration of industry within a nation, is at least partly related to climatic conditions. Wilson (1966) notes, for example, that the distribution of economic activity in the United States has been traditionally explained in terms of the interaction of four principal factors, access to capital, raw materials, labour supply and markets. However, the final decisions on the actual site are now becoming increasingly based upon cultural and amenity factors. Among these is the decision of many people to move for other than economic reasons, and Wilson has suggested that at least within the United States, some families migrate in order to find a more pleasant place to live. The most important amenity sought is a milder and usually snow-free winter, and this kind of migration is generally destined for the Pacific Coast, the South-west, Florida or the Gulf Coast. One result of this migration is, of course, the accumulation in these areas of a growing labour supply, much of it skilled, or professional, and it is this potential labour supply (there in part because of the more favourable climate) that is attractive to new industry.

It is, of course, extremely difficult to judge what role climate plays in the decisions of management concerning the location of their industrial plant or other activity, and Wilson indicates that the part climate has played in the development of Tucson over the years can be interpreted in different ways by different people. Nevertheless, there is some evidence available, says Wilson, that suggests that climate is not a completely unimportant factor, especially in economic activities of the 'footloose' type. Casaday (1952) notes, for example, that Tucson's salutary climate ranks next to its favourable labour conditions as an attraction to industry. In a special study completed by Casady, a total of 25 firms, or nearly 75% of those surveyed, cited some aspect of climate as an important reason for locating in Tucson, seven of the 25 firms reported having located in Tucson in part because of the beneficial effects of the climate upon the health and efficiency of their employees.

b. Perception of the effects of weather on manufacturing

In the United States approximately 30% of the national income is derived from manufacturing. However, in spite of this contribution, neither the use by industry of weather data and weather forecasts, nor the manufacturers' perception of their own weather sensitivity, has been studied to any appreciable degree. Indeed the results of a recent report by the U.S. Weather Bureau (U.S. Department of Commerce, Weather Bureau, 1964) suggests that manufacturing has the least economic benefit potential of any weather user classification studied. This lowest rank is bestowed on the manufacturing industry on the basis of two assumptions: first, in terms of the meaningful costs or health effects, manufacturing was only moderately weather-sensitive; and second, even with advance information of weather conditions, manufacturers are apparently relatively inflexible in their ability to take preventive action.

In view of these findings, the study by Bickert and Browne (1966) on the perception of the effects of weather on five eastern Colorado manufacturing firms is significant in assessing the value of the weather to the manufacturing industry. Two aspects of the problem are discussed: first, the perceived and real effects of weather on various functional areas of manufacturing firms; second, the present and potential utilization of weather information by such firms. The five manufacturing firms were selected for the study on the basis of their hypothesized sensitivity to weather in various aspects of their operations. The companies studied, as well as their major product lines and market territories, are shown in Table III.3.

Production and quality control were the two operations most weather-affected in Firm A. In particular, variations in temperature cause changes in dimensions and tolerances, high humidity increases the labour and cost in product maintenance and storage, and in the spring and autumn high humidity also causes many metals, particularly magnesium, to corrode and rust. Dips and oils are used by this

TABLE III.3 : *Colorado companies reviewed in the weather-perception study*
(after Bickert and Browne, 1966)

Firm	Major product lines	Market territory	Employment
A	Precision mechanical components for aero-space industry	United States and export	Over 1,000
B	Beer	Western United States	Over 300
C	Brick and ceramic products	Mountain states	Over 100
D	Ski apparel	United States	Over 200
E	Consumer and industrial non-durables	United States and export	Over 1,000

firm to prevent rust but these have to be supplemented by a large amount of additional hand labour to remove the rust that occurs. The authors also report that in one of Firm A's other plants, located in an area of higher humidity than Colorado, an additional two men are employed to protect finished parts at an annual cost of $15,000.

The manner in which weather affects Firm B (a brewery) is reported to be principally at the source of its raw materials: barley, rice and water. In addition, severe drought severely restricts the plant's water supply, a vital ingredient in any brewery process. The weather effects on internal plant operations are on the other hand relatively inconsequential. Severe winter weather causes the most concern, freezing temperatures necessitating the use of insulated railroad cars for shipping the beer.

The third firm studied, a brick and ceramic manufacturing plant, appeared to be particularly weather-sensitive, weather factors influencing nearly all aspects of the firm's operation. Bickert and Browne report that from a production standpoint, snow and cold present a number of problems. For example, grinding efficiency is lowered with wet clay, thus decreasing both output and quality. In addition, a drop in temperature below 9°F necessitates a change from natural gas to propane, which results in an increase of $300 per day in the operation of the kilns.

The manufacturer of ski apparel, Firm D, was also sensitive to weather conditions, particularly as they affected the problem of marketing. The consumer products of Firm E (consumer and industrial durables) were also reported to be particularly weather-sensitive. For example, a substantial proportion of its sales occur in the winter, and hence a lack of snow, ice and freezing conditions during November when a considerable amount of advertising is used, can seriously inhibit sales.

In their conclusions, Bickert and Browne indicate that although the initial awareness among the five manufacturers of the effects of weather on their operations was found to be minimal, the actual effects of weather variables, upon examination, were found to be considerable, particularly in terms of costs to the companies concerned. Thus it is reasonably clear that if the study by Bickert and Browne is taken to be representative of manufacturing firms in general, then the managements' perception of the effect of weather is regrettably minimal. A programme of education to inform management that better decisions could be made *if* better use was made of weather information and weather forecasts, would appear therefore to be justified.

2. Construction

Three broad aspects of the association between weather and climate on construction include their influence on structural and architectural design, the effect in economic terms of the impact of weather on the day-to-day operations of the construction industry, and the influence that weather and climate has, or more accurately should have, on urban and industrial planning.

Many aspects of the effect of the atmospheric environment on the construction industry are well documented by Critchfield (1966). For example, a given structure must be so designed as to withstand stresses caused by climatic factors such as temperature fluctuations, wind force, humidity changes, or loads of snow and ice. Of these, temperature is one of the most important climatic elements in any engineering or construction design because of the expansion and contraction of materials with changing temperatures. The maximum expected wind speeds are also among the stress factors considered in designing bridges, towers and buildings, suspension bridges being particularly vulnerable to wind, for they tend to sway and to develop destructive undulations. A notable example of this was the destruction by wind of the first Tacoma Narrows bridge at Tacoma, Washington, on November 7, 1940. Two other important atmospheric factors are humidity and excessive rainfall. For example, chemical and physical deterioration of paints increases under humid conditions, while masonry construction usually disintegrates more rapidly under moist conditions. Extremes of precipitation must also be taken into account; dams, canals, pipes and bridges all requiring that their construction withstand the maximum flood stages likely within a certain 'return period' of the event.

There are also those day-to-day and hour-to-hour weather factors which influence the actual speed of construction. Of these, rainfall is perhaps most serious, along with temperature extremes in the more continental type of climates. Precipitation, for example, as would be expected, generally has an adverse effect on construction, heavy rain often ruining newly-laid concrete, as well as slowing down or stopping the use of heavy machinery. In these and similar circumstances the 'cost' of weather is therefore high.

a. Climate, weather and architectural design

Amongst a number of studies on the relationship between climate and architectural design the contribution by Aronin (1953) is particularly noteworthy for specific climate–design applications. In addition, a *Bibliography of Weather and Architecture* (Griffiths and Griffiths, 1969) should be noted. In general, however, buildings and structures are usually designed to withstand the probable combinations of climatic extremes, and to make indoor conditions comfortable and healthful regardless of the weather conditions outside.

The five relevant weather factors in architectural design are temperature, sunshine, humidity, wind and precipitation, the selection of the design data usually being made in accordance with the probability of the occurrence of the most severe weather conditions that will justify a specific design. The probability value to be used depends upon a number of things, including the degree of climatic control required, how much the structure itself is expected to control, and the use that is to be made of mechanical aids such as heating or cooling equipment. In addition, account must be taken of both the daily and the seasonal variations in sunlight, and any drainage or roof stress problems that could arise under extreme rainfall or snowfall conditions.

In today's society, considerable interest is shown in the use of glass in buildings, and this has generated a considerable difference of opinion between architects. For example, Stephenson (1964) suggests that according to the Canadian Building Digest, the ideal arrangement for solar control is to design the building with the glass facing north; the second best arrangement is to have the glass facing south with the provision of both a canopy shade and a venetian blind.

In contrast, the purpose in the design of the 'solar house' is to attain more efficient *use* of the sun for light and heat. The effect of such a house on lighting is self-evident. The heat can, however, be utilized in two different ways: first, by trapping the sun inside the house, using large-paned windows or two or more glass layers; and second, by using the sun's heat to warm a liquid or solid solution, which is then employed as a fuel to heat the house. This latter effect is particularly significant from an economic viewpoint, and reports from the owners of 'solar houses' suggest substantial fuel savings over conventional methods. Thus, although solar radiation, particularly as a means of heating, has been largely ignored in building design, there can be no doubt of its increasing significance in the future. It is appropriate to point out, however, that the solar radiation 'problem' can be eliminated altogether by building a structure without any windows or exposed glass areas at all, as has already been done, but whether this is working in harmony with nature or against nature is another matter.

Although climate does appear to have a significant influence on the architectural design of buildings and houses, it is perhaps pertinent to comment that in some areas buildings obviously designed for a particular kind of climate are used with little modification in other areas. For example, house design in Canada does not vary to any great extent from one area to another, in spite of the fact that the climate of coastal British Columbia is significantly different from that of the Prairie Provinces. In this case, it appears that it is more economical to build houses in Vancouver and Victoria, British Columbia, with the same general design as those which are built in the colder parts of Canada. However, some believe that this is done at the expense of not living in harmony with the climate, for it is reasonably clear that considering the local climatic conditions, house designs in coastal British Columbia should be more California style and less Prairie style. Moreover, houses in many parts of the world appear to be built so that they look attractive from the street, little consideration being given to such things as windows or open areas with a sunny outlook, or patios and sun porches which are not exposed to sea breezes (at least in those areas where the sea breeze is cold). Thus it may be argued that a much greater value for weather and climatic conditions can be obtained if architects, builders and home-owners had a greater appreciation of the advantages and disadvantages of their local atmospheric environment.

b. Economic impact of weather on the construction industry

The impact of weather on the construction industry in the United States has been reviewed in a number of recent studies, the interest in such studies being justified since weather produces a severe operational and economic impact on the construction industry. This, according to Cullen (1966), is because of three factors, namely: a large percentage of the construction operations are sensitive to adverse weather conditions; building construction is often vulnerable to ordinary weather occurrences such as winds, snows, thunderstorms, etc., as well as to some extraordinary ones such as hurricanes, tornadoes, etc.; and building materials and systems are susceptible to a greater or less extent to the degradative effects of the natural weathering processes.

The amount of money involved in the construction industry is impressive, and Russo (1966) indicates that in 1964 the total value of the industry in the United States was about $88 billion, representing more than 10% of the gross national product. Of this total volume of construction, an estimated 45% ($39·7 billion) was spent in weather-sensitive areas of construction (Table III.4). The dollar loss due to weather was quantitatively evaluated at a minimum of $3 billion annually, with the maximum possible dollar loss being estimated to be as high as $10 billion. This wide range, it should be noted, results from a highly speculative estimate of the decreased construction volume due to seasonal weather effects.

One of the conclusions reached by the study was that if specific weather information is made available to the construction industry and is appropriately used, then, with the present forecast accuracy, a potential annual savings of $0·5 to $1·0 billion is possible. This, it should be noted, represents approximately 10 to 17% of the estimated weather-caused loss. In addition, the maximum savings achievable if the weather forecasts for the shorter (0–24 hour) time periods were 100% accurate are estimated to be $0·8 to $1·3 billion, or $300 million more than is obtainable with present forecast accuracy.

An associated study of the effects of weather on specific construction operations was made by the U.S. Department of Commerce (1966). Table III.5 shows that weather sensitivity is an important factor through almost the entire sequence of construction operations from planning operations like surveying, through assembly operations like concrete paving, to finishing operations like landscaping

TABLE III.4: *Distribution of total annual construction volume and the proportion considered potentially weather-sensitive ($ billions) (after Russo, 1966)*

Construction category	Annual volume	Potentially weather-sensitive				Total sensitive	% of annual volume
		Perish-able material	On-site wages	Equip-ment	Overhead and profits		
Residential	17·2	0·96	1·62	0·07	2·14	4·8	27·9
General building	29·7	1·93	4·08	0·22	2·67	8·9	30·0
Highways	6·6	1·67	1·63	0·77	0·73	4·8	72·7
Heavy and specialized	12·5	1·88	3·12	2·50	2·50	10·0	80·0
Repair and maintenance	22·0	2·67	4·00	1·39	3·14	11·2	50·9
Total (rounded)	88·0	9·1	14·4	5·0	11·2	39·7	45·1

and painting. In addition to the critical weather limits noted in Table III.5 are those of freezing rain which was critical to all operations surveyed, a low temperature/high wind ratio equivalent to a chill factor of 1,000–1,200 (see Fig. IV.1 on p. 199), temperatures above 90°F, and a temperature–humidity index of 77 or more. Temperature inversions were also critical in demolition, clearing, excavation and quarrying, and flooding and abnormal tides were critical in pile driving, dredging, erection of coffer dams and the installation of culverts and incidental drainage.

Many of these critical weather conditions, when they do occur, prevent any useful work being done, and large losses of income may result. The ability to forecast more accurately the critical weather conditions is therefore important, especially in view of the fact as reported by Cullen (1966) that an estimated $1,000 million in lost wages each year can be attributed to seasonal weather fluctuations, and that equipment valued at some $700 million is idle during the winter months due to adverse weather conditions. Cullen also notes that in the category of property damage due to intermittent adverse weather occurrences, the monetary losses range between $500 million and $1,000 million yearly.

The effect of weather on the construction industry of England

TABLE III.5: *Critical limits of weather elements having significant influence on construction operations (abridged from U.S Department of Commerce (1966) publication* Weather and the Construction Industry)

Operation	Rain	Snow and sleet	Low temperatures (°F)	High wind (mph)	Dense fog	Ground freeze	Drying conditions
Surveying	L	L	0 to −10	25	×	—	—
Demolition and clearing	M	M	0 to −10	15–35	×	×	—
Temporary site work	M	M	0 to −10	20	×	×	—
Delivery of materials	M	M	0 to −10	25	×	—	—
Material stockpiling	L	L	0 to −10	15	×	—	—
Site grading	M	M	20 to 32	15–25	×	×	×
Excavation	M	M	20 to 32	35	×	×	×
Pile driving	M	M	0 to −10	20	×	×	—
Dredging	M	M	0 to −10	20	×	×*	—
Erection of coffer dams	M	L	32	25	×	×	×
Forming	M	M	0 to −10	25	—	×	—
Emplacing reinforcing steel	M	M	0 to −10	20	—	×	—
Quarrying	M	M	32	25–35	×	×	×
Delivery of pre-mixed concrete	M	L	32	35	×	×	—
Pouring concrete	M	L	32	35	—	×	×
Stripping and curing concrete	M	M	32	25	—	×	×
Installing underground plumbing	M	M	32	25	—	×	×
Waterproofing	M	M	32	25	—	×	—
Backfilling	M	M	20 to 32	35	×	×	×
Erecting structural steel	L	L	10	10–15	×	—	—
Exterior carpentry	L	L	0 to −10	15	—	—	—
Exterior masonry	L	L	32	20	—	×	×
External cladding	L	L	0 to −10	15	—	—	—
Installing metal siding	L	L	0 to −10	15	—	—	—
Fireproofing	L	L	0 to −10	35	—	—	—
Roofing	L	L	45	10–20	—	—	×
Cutting concrete pavement	M	M	0 to −10	35	—	×	—
Trenching, installing pipe	M	M	20 to 32	25	—	×	×
Bituminous concrete pouring	L	L	45	35	×	×	×
Installing windows and doors, glazing	L	L	0 to −10	10–20	—	—	—
Exterior painting	L	L	45 to 50	15	×	—	×
Installation of culverts and incidental drainage	M	L	32	25	—	×	×
Landscaping	M	L	20 to 32	15	×	×	×
Traffic protections	M	M	0 to −10	15–20	×	×	—
Paving	L	L	32 to 45	35	×	×	×
Fencing, installing lights, signs, etc.	M	M	0 to −10	20	×	×	—

Note: L indicates light; M indicates moderate.
* Indicates water freeze.

and Wales has also been studied, and Broome (1966) considers in particular the problem of weather forecasting and the contractor. The ideal forecast, he says, would give information at two levels, each for a particular purpose. First, at the tender stage when the cost of a project is being estimated, it would be most useful to know the number of days each month when precipitation would exceed, say, 0·10 inch. This information should cover about twelve months ahead, as many contracts do not start for two to six months after the tender is received. Other useful information would be the number of days during which the air temperature would fall below freezing for more than three or four hours. The second level of forecasts required would be the daily forecasts for the next seven days, which indicated the amount of rainfall, the period during which the air temperature was below freezing, and the number of hours with a wind speed over specific limits.

It is clear from the studies cited that weather conditions, both actual and forecast, are of considerable value to the construction industry. Prior to 1960 little applied research had been done in this area, but the studies completed in the 1960's have indicated that this is a rewarding field for applied meteorology. For example, it is appropriate to note that on January 12, 1967, a meeting was held at the Chamber of Commerce of the United States in Washington, D.C., which brought together for the first time important representatives of the construction industry, architects and engineers, together with those representing weather forecasting, climatology and hydrology. Beebe (1967) commenting on the meeting said that the seminar showed that there is, and has been, a very definite lack of communications between the construction industry, which is very weather-sensitive, and the meteorologist. However, the seminar did establish that the construction industry has considerable interest in weather information and in the contributions that can be made by consulting meteorologists, factors which auger well for the future application of weather forecasting and possibly weather control to the construction industry.

c. Climate and town planning

Consideration of the effect of climate on town planning in the main involves two factors, first in the initial siting of new towns or new industries, and second in the location of industries or new settlements as part of already established areas. In the first case, the initial siting of a new town or industrial complex should be planned

E

with the knowledge of the climatological conditions, and some of the 'new' towns that have been developed in various parts of the world in the twentieth century have in some cases utilized climatological information, above all other information, in their planning. However, most towns and cities are located where they are today because of historical, political, geographical or economic reasons, and the climatologist has had little say, if any, in choosing the site.

Planning, however, does not end once the site has been chosen, and it is in the area of urban growth and development that weather and climatic information have their most immediate value. For example, the siting of the residential sections of an urban area on the windward side, and industrial establishments on the usual lee side of a development, and the benefits of parks for recreation have long been recognized. In addition the planning of street widths and street directions to take account of the prevailing cold and warm wind directions is highly desirable, as anyone who has stood at the corner of Portage and Main Streets in Winnipeg will testify. Grassed and treed areas are also important in decreasing the dust content of cities, and in decreasing the maximum temperatures in their vicinity. Planning based on applied climatological data can also aid in the size and spacing of buildings. In cold climates, for example, it is highly desirable from a fuel consumption viewpoint that houses be built closely together, and Landsberg (1960) cites some data relating to the fuel consumption for one-family houses in different settings which clearly demonstrate the correlation between increased exposure and increased fuel consumption.

Irrespective of the planning that is so necessary for the successful development of urban and industrial areas, there remains the important factor that in any urban area new microclimates are constantly being created, with consequent alterations in the water and heat balance. One facet of the microclimates so formed is atmospheric pollution, a factor which must be taken into account in both short- and long-term planning if man is to live in harmony with his urban environment. Indeed, it is clear that in many areas the 'value' of the weather as it is associated with industry and air pollution is becoming a vital factor in the planning of our cities. Moreover, it is also clear that successful planning will only be accomplished if it is done with proper consideration for the atmosphere, as well as the economic and social environment.

D. TERTIARY ACTIVITIES

1. Transportation

Almost all segments of the transportation industry (air, water, land and pipeline) are affected by the atmospheric environment. The day-to-day weather conditions, for example, affect ship, bus, railway and airline schedules, and adverse weather conditions increase the number of accidents in most kinds of transportation. The climate is also important, and climatological data can assist planners associated with design problems in transportation facilities and transportation equipment, from automobiles and highways to spacecraft and their associated ground facilities.

There are generally three kinds of weather associated with transportation. First is the influence of the actual weather conditions encountered, such as the weather on a flight path from place A to place B. Second is the use, particularly in planning, of weather forecasts; and third is the importance of weather modification, the clearing by man of fog at airports being a notable example. As will be appreciated, the airline industry is perhaps most affected and influenced by all three aspects of the weather, but other segments of the transportation industry, including long-distance road transport, city taxi services, city bus services, transportation by pipeline and the routing of ships, are all weather-sensitive.

a. Aircraft, airlines and air transportation

As air transport enters the supersonic era of the 1970's the economic effects of weather phenomena will probably increase, not because of lack of skill or technology but because 150 to 500 passengers will be transported in aircraft worth up to $30 million. From the weather forecasting viewpoint, problems may also arise in the ability to forecast with sufficient accuracy temperatures in the supersonic transport climb region. Kreitzberg (1967) discusses this problem and notes from a study by Nelms (1964) that a temperature excess of 11·1°C in the region between 25,000 and 52,000 feet would necessitate consumption of 4,350 lb of extra fuel or 2·3% of the total fuel consumed on a flight. Thus, if temperatures in the climb region were incorrectly forecast, the extra fuel required could involve the unnecessary displacement of perhaps 20 passengers and baggage if gross weight is a critical factor.

In earlier decades of the airline industry when aircraft instruments were few and aircraft relatively fragile, the weather used to delay a large percentage of flights. Since then the number of flights has increased considerably, along with an increase in the number of delays and cancellations. However, the *percentage* of flights delayed today is very much lower than it used to be, primarily due to technical and organizational (control) improvements especially in combating reduced visibility (see Table III.6). The critical factor, however, is that a delay now involves a $5 million to $20 million aircraft (compared with a $50,000 aircraft before World War II).

An analysis of airline expenditures (Table III.7) by Barry (1965) shows that fixed costs range from 68 to 88% of operating expenses and these continue regardless of whether or not a trip is operated. In addition, aircraft utilization is between six and eight hours per day with a maximum of twelve, with the rest of the 24-hour day being set aside for maintenance, servicing and flight preparation. Moreover, when a flight is cancelled virtually all revenue is lost but costs are only reduced by an estimated 12 to 40%. Further, since the average passenger aircraft today breaks even at 28 to 35% of its maximum capacity, and major airlines appear to have load factors

TABLE III.6: *Movements at selected United States airports**

Airport	Arrivals†			Delayed flights per year**	
	/week	/day	/hour	1%	2%
Atlanta, Ga.	8,633	1,233	51	4,488	8,976
Boston, Mass.	10,841	1,548	65	5,637	11,274
Buffalo, N.Y.	3,968	567	24	2,064	4,128
Chicago, Ill.	19,215	2,745	114	9,989	19,978
Denver, Colo.	4,866	691	29	2,532	5,064
New York, N.Y.	37,505	5,358	223	19,500	39,000
Omaha, Neb.	2,858	408	25	1,487	2,974
Reno, Nev.	8,687	1,241	52	4,519	9,038
Salt Lake City	2,320	331	14	1,206	2,412
Seattle, Wash.	7,853	1,122	47	4,082	8,164
Washington, D.C.	19,663	2,809	117	10,223	20,446

* All airlines are included. Data derived from International Edition, *Official Airline Guide*, January 1967, and North American Edition, *Official Airline Guide*, October 1967.
† Double all figures for total movements.
** Number of delayed flights per year assuming 1% or 2% delays due to weather.

(percentage of available seats filled) of 60 to 65%, it is evident that a small percentage of trips cancelled could turn profit into loss.

Most of the major costs in relation to air transportation have been mentioned. However, there are others that in some cases are important or even mandatory to the company. Many airlines, as a matter of public relations, will pay hotel and meal expenses of delayed passengers as well as re-transportation to the airport, and such costs in the case of a large jet aircraft could total from $1,000

TABLE III.7: *Analysis of airline expenditure* (*after Barry, 1965*)

Management Organization	Salaries, pensions, etc.	5%
Operatives	Wages, pensions, etc.	31%
Equipment	Capital equipment, depreciation, buildings, aircraft, offices, maintenance equipment, etc., consumable, fuel, oil, tyres, food	38%
Raw materials	Passengers, freight, publicity, reservations, storage, etc.	2%
Contracting	Marketing, advertising, etc.	15%
Financial	Interest on equipment	9%
		100%

to $2,000 per day depending on the number of passengers involved. The costs of weather-delayed airlines is therefore considerable, and it must also be remembered that there are many costs which an airline cannot 'reimburse', such as a contract lost or a timely meeting missed by a businessman.

The weather phenomena that most frequently influence airline operations are those involving visibility and runway conditions. In addition, there is the effect that tornadoes, thunderstorms and other severe storms have on aircraft, but this effect can usually be considered small due to the short durations and small areal coverage involved. Moreover, present-day communications, weather radar and improved forecast ability usually allow airlines to cope with these conditions, and the costs of the necessary equipment are not excessive when the savings are considered. The frequency of fog, blowing snow, slush, water and loose snow and ice on runways has, however, a far greater effect on operations, this being partly due to their more common occurrence and their general areal extent. In addition, a more severe and increasingly frequent weather event with high-altitude aircraft is clear-air turbulence; a considerable amount of damage in some aircraft being attributed to such turbulence. Another effect of weather on airline operations is the surface temperature on the runway, this affecting take-offs and at times restricting loads and hence airline revenue.

The weather factors of greatest importance are, however, those that are forecast, since the forecast weather – right or wrong – must be taken into account in most flight planning, and flight planning basically determines such things as fuel load, route taken, passenger and freight load, and alternative airports, all of which affect the costs of operation. However, if the forecasts and especially the terminal forecasts are sufficiently accurate, they may to some extent reduce costs of operation, since unnecessary fuel or unnecessary restricted loads would not be required.

The accurate forecasting of visibility, fog clearance and fog formation at most airports is in many cases difficult. Nevertheless, accurate visibility and fog forecasts are especially important for jet aircraft because of their high fuel consumption, and the many other significant costs which occur when holding or diverting occurs at low altitudes, and of course accuracy in visibility forecasts will become more critical for the supersonic airliners.

The 'cost' of weather to airline operations is of particular interest, one pertinent study being made by United Research Incorporated (1961) on a forecast of losses incurred by commercial air carriers in the United States due to their inability to deliver passengers to destination airports. The conclusions reached by the report indicated that United States airlines incurred losses (in the early 1960's) of approximately $55 million a year, due to flight disruptions in periods when weather conditions were below authorized limits. These losses were of three forms: first, cash losses when airlines were forced to cancel, delay or divert their schedules: second, short- and long-term costs resulting from landing accidents; and third, losses from the reduction in the demand for air travel as a result of its unreliability in bad-weather periods. It should be noted that the costs considered were of an incremental and direct nature only, and the authors point out that if flight depreciation expenses and all indirect and overhead expenses were also allocated to the aircraft time consumed in delays, diversions and cancellations, the costs would have been considerably higher. Much more information on the costs of weather to the airlines is given in the original report, but Table III.8 summarizes some of the details relevant to this study of the value of weather.

Most aspects of weather are fortunately only an economic hazard to airline companies, and the direct physical destruction by weather of an aircraft *en route* or on landing is rare. However, there have been many airline disasters, some of which are believed to be directly associated with adverse weather conditions. For example, on March 5, 1966, just 27 minutes after taking off from Tokyo Airport, a jet aircraft broke apart 3,000 feet above the summit of Mount Fuji, due it is believed to violent air turbulence, killing 124 persons; and another smaller airliner encountered a blinding snowstorm on March 1, 1964, near Tahoe Valley, California, and smashed into an 8,700-foot ridge of the Sierra Nevada range killing all 85 persons on board. Regrettably many other weather-related airline tragedies could be cited. It should nevertheless be appreciated that these crashes are only a few among the millions of flights safely completed each year all over the world. Moreover, as better methods are found to give an even more accurate picture of the weather, pilots will be able to lessen the chances of weather associated air crashes.

TABLE III.8: *Estimated cost of flight cancellations in the United States due to weather: 1959 and 1962 (after United Research Incorporated, 1961)*

Costs	1959	1962
	$	$
Passenger revenue loss due to flight cancellations*	14,989,000	22,730,000
Expense of operating non-revenue ferry mileage†	605,000	628,000
Passenger service – interrupted trip expense‡	2,443,000	3,116,000
Duplicate reservations, ticketing and accounting expense§	652,000	978,000
Gross costs	18,689,000	27,452,000
Less: Savings in aircraft operating expense	12,165,000	18,941,000
Net costs	6,524,000	8,511,000
Number of flight cancellations	54,309	61,101
Cost per flight cancellation	120	139

* Represents that portion (47%) of total passenger revenue booked on cancelled flights which is lost to other modes of transportation and to discontinuance of travel.
† Estimated as 5% of non-revenue mileage; non-revenue mileage was estimated as 1·6% of revenue mileage projected for U.S. domestic carriers in *National Requirements for Aviation Facilities, 1956–1975.*
‡ Based upon 1956–1959 cost per flight cancellation.
§ Estimated as 2% of passenger revenue lost to airlines as a result of flight cancellations.

In the future improvements in operational planning and greater accuracy in weather forecasting can be expected. In addition, automatic landing and flight systems (see, for example, Brady, 1964) will become almost independent of cloud and visibility, and are likely to assume greater importance. However, the proper use of good weather forecasting will continue to be an essential part of the airline industry, and to know the weather well will continue to save many lives.

b. Fog dispersal at airports

Of the various weather elements that affect air transportation, none has such a marked effect as fog. Moreover, despite the fact that in the United States (and many other countries) generally 97 to 98% of all scheduled flights are completed each year, about 1% or more than half of the flights remaining are not completed because of fog· This is of course a very small percentage, but as Beckwith (1966) points out, its importance is highlighted when it is related to the more than 800 million miles flown by airlines in the United States during 1964. Moreover, in the same 12-month period nearly 82 million passengers were the prime generating source of airline revenue grossing $4·25 billion.

The losses and inconvenience caused by fog at airports can usually be reduced by three methods: first, provision of electronic landing aids which lead to safe 'zero–zero' landings; second, accurate terminal weather forecasting; and third, clearance of fog through weather modification. All these methods have improved considerably during the 1960's, but there are limitations with each method. For example, instruments and equipment for 'zero–zero' landings will, at least in the foreseeable future, be available only at the larger airports, and even if perfect forecasts were available, a perfect forecast of fog would not eliminate the inconveniences to the airlines and their passengers. It is natural, therefore, that weather modification in the form of fog clearing has been developed.

Two associated factors with regard to fog clearing by man should, however, be emphasized. These are, first that a successful fog modification programme must depend on the ability of weather forecasters to forecast accurately both the formation and the dispersal of fog, and second, the 'right' of airlines or airport authorities to clear fog from airports could possibly be questioned on the basis that fog clearing may not necessarily be beneficial to all concerned. For example, a fog-closed airport usually results in vast disruptions at considerable cost and inconvenience to the travelling passenger. Yet these same disruptions give rise to considerable 'benefits' to restaurants, news-stands, taxi companies, bus companies and airport hotels. The question therefore arises as to what extent these associated airport facilities should be 'allowed' to make economic 'gain' out of bad weather conditions. In other words, should a business which caters for the travelling passenger be allowed to make economic gain out of the naturally occurring adverse weather conditions, at the expense of the travelling passenger and the airlines both of whom would suffer losses, or should it be the other way around?

The history of fog dispersal at airports is discussed by Beckwith (1966), and he mentions the relatively successful programmes that have been started at Medford, Oregon, and Salt Lake City, Utah, using dry ice and silver iodide. Fog dispersal, however, is usually feasible only under supercooled conditions (see, however, the paper by Osmun (1969) on warm fog dispersal) and, except for a few airport sites, this represents a relatively small percentage of the total fog conditions. The work done in France by the Paris Airport Authority using liquefied propane is reviewed by Halacy (1968), who says that so successful and reliable was the system that it was classed semi-operational in November 1964, and on Christmas Eve of that year it permitted take-offs and landings from runways that would have otherwise been closed. An operational system has now been designed. The use of liquefied propane to clear fog is also commented upon by Hicks (1967), who describes the results of tests in the Hanover–Lebanon, New Hampshire, area during the winter of 1964/65, and in Greenland during the summer of 1965. Hicks noted that the entire system is inexpensive in initial costs and operating expenses, a cost of $20 per hour being estimated to be sufficient to keep an airport approach zone clear of fog.

The costs and benefits of any weather modification are of importance. Beckwith (1966) has examined these with respect to fog dispersal and Table III.9 shows that the benefit/cost ratios are of the order of five to one. The total benefits are of course very difficult to estimate, for there are considerable hidden costs and benefits especially if the airport is a feeder airport for other airports. A study by Deemer (1963) for United Air Lines showed, for example, that in the four-month period from December 1, 1962, to March 31, 1963, irregular operations due to *all* weather factors over the United Air Lines system caused a revenue loss and added expenditures of more than $9 million, this total being based on the effects of cancellations and of delays of more than 30 minutes, specified

TABLE III.9: *Summary of supercooled fog seeding – winter 1963–1964 (after Beckwith, 1966)*

	Salt Lake City	Medford	Total
No. of seeding flights	8	8	16
Successful	8	7	15
Marginal	0	1	1
Failure	0	0	0
No. of seeded days	5	8	13
Total seeding time	$6^h\ 26^{min}$	$6^h\ 29^{min}$	$12^h\ 55^{min}$
Longest flight	55^{min}	75^{min}	
Shortest flight	26^{min}	32^{min}	
Average flight	49^{min}	49^{min}	
Weight of dry ice dropped	655 lb	2,775 lb	3,430 lb
No. of operations permitted	24	15	39
Landings	14	6	20
Take-offs	10	9	19
No. of passengers accommodated	516	124	640
Total cost of seeding	$3,300*	$525	$3,825
Total benefits (est.)	$16,000	$3,000	$19,000

* Including stand-by charges.

under either minimums, winds preventing take-offs and landings, runway conditions or downline weather. The 'downline weather' factor is at times the most important and is the effect of delays, or of not being able to take off or land at scheduled stops upline because of weather-closed airports downline. An important connecting terminal such as Chicago O'Hare Airport, for example, when closed by weather, produces extensive delays and cancellations backing up to both coasts, the total cost of which is very difficult to assess. However, Deemer estimated that of the $9 million loss reported above, the portion attributable to below-minimum weather and cross winds was $3,400,000, with the downline effects costing an additional $4,000,000.

This report was concerned with only one airline, but Beckwith (1966) comments that since United's share of U.S. airline revenue passenger miles (the usual airline activity index) currently runs about 20%, these figures may be multiplied by a factor of five to obtain equivalent total United States airline costs. The resulting fog- and wind-caused losses and added expenses on a national basis for these four winter months would therefore be approximately $37 million.

Dispersal of cold fog appears to be successful, safe and relatively inexpensive, and, as has been pointed out, is now being carried out operationally at several airports. Many questions remain unanswered, however, and it is suggested that at least four kinds of studies should be completed before any large-scale fog-dispersal programmes are initiated. These include: first, the determination of the air traffic actually affected by adverse weather due to cold fog; second, the determination of the value of air traffic affected, in terms of cost to commercial air carriers, to general aviation and to the public; third, the determination of the cost and effectiveness of cold fog dispersion at the airports, and the comparative cost and effectiveness of instrument landing systems and/or weather forecasts to overcome adverse weather; and fourth, the determination of the gross and net benefits to commercial air carriers, to the public and to the economy as a whole from cold fog dispersion, and the comparative cost and effectiveness of the use of instrument landing systems and/or the availability of more accurate weather forecasts.

c. Water transport

For centuries man has been aware of the effect of the weather upon water transport, and although modern navigational aids and improved safety procedures have considerably reduced marine disasters, it is still not uncommon for ships to be lost at sea with little if any trace. Moreover, the power of the sea is such that in certain circumstances man unfortunately can only stand and watch.

In earlier times, sailing ships were very dependent on the prevailing wind conditions. Today, the importance to shipping of the prevailing winds is minimal, and although strong winds associated with tropical revolving storms and other severe cyclonic storms can at times be destructive, the main weather elements affecting water transport are low temperatures and fog.

Low temperatures influence several operations including loading and unloading activities, extra protection for both cargo and passengers often being required. In addition, if ice forms, navigable waterways may become impassable. Forecasts of both the opening and closing of waterways are therefore most important, for it allows shippers and others to increase their revenues without any unnecessary increase in their costs. Problems relating to ice of course occur on many rivers, canals, lakes and harbours, and the net effect is to produce a seasonal rhythm in waterborne traffic with, in most cases, unwelcome variations in the length of the shipping period from year to year and place to place. Of possibly greatest importance in any study of ice and water transport is the effect of ice on the Great Lakes, for not only do the Great Lakes constitute the largest reservoir of fresh water in the world, but near their shores are located some of the most heavily populated sections of both the United States and Canada. Snider (1967) comments on the importance of ice forecasting on the Great Lakes, and it may be suggested that a knowledge of ice climatology is of considerable economic importance.

Fog is another important factor affecting water transport, and although lights, fog-horns, radio, radar and electronic devices for sounding ocean depths have considerably lessened the danger of collision, or running aground, dense fogs still bring harbour traffic to a standstill and contribute to many marine accidents.

Precipitation is not a major factor affecting shipping, but loading and unloading operations are often disrupted if perishable cargo is involved. At some harbours special loading equipment is used to counteract wet weather conditions. In New Zealand where 'all-weather' loading facilities are available at some ports, it is estimated that of the total number of days spent on the coasts by ships, the cargo is not worked on about 20 to 25% of the days. Explicit information regarding the cause of this lost time is not always available, but adverse weather conditions at ports which do not have special loading equipment no doubt account for a significant proportion.

Most modern forms of water transport are less likely to be seriously inconvenienced by weather conditions than the water transport modes of three or four decades ago. However, it is clear, at least from the writings of one naval officer, that weather is still very important. For example, Admiral Moorer, Commander-in-Chief U.S. Atlantic Fleet, at a speech to the American Meteorological Society's National Conference on Marine Meteorology (Moorer, 1966), stated:

> Today, in 1966, the greatest peacetime threat to the safety of the five hundred ships of the Atlantic Fleet, their commanders, and the men who man them is weather. Primarily horizontal and vertical wind speed and shear, and wave and swell conditions....
>
> Even today, to come close to the center of a mature hurricane or typhoon is synonymous with disaster. No matter how splendid or how powerful our ships.

Moorer also noted the importance of surface, sub-surface and above-the-surface weather for naval operations and clearly pointed out that while weather has always been a factor in the prosecution of wars, as military technology has advanced and become more complex military operations have become even more sensitive to the environment. In recent years, in fact, this has been particularly true of natural catastrophes attributable to weather elements, the impact of such events invariably resulting in the counting of the dead and missing, dollars wasted and the irreplaceable loss of resources.

d. Weather routing of ships

A natural outcome of man's response to weather and sea transportation has been the provision of specialized weather forecasts which enable a ship to gain all the advantages of the prevailing weather conditions, but to avoid most of the disadvantages. The technique is known as weather routing, and several papers on the subject have appeared, including Evans (1968), Verploegh (1967), Frankcom (1966), Cummins (1967), Spicer (1967) and Canham (1966).

The technique of weather routing, according to Evans (1968), is to project the wave pattern ahead as far as possible along the normal route of the vessel from the point of departure to that of destination, utilizing the three- to five-day forecast charts of surface weather and wave height, modified perhaps by local prognosis. Estimates are then made of the vessel's speed as a consequence of encountering predicted wave heights based upon previously determined performance data of the ship under similar conditions. By repeating this process and projecting ahead, the ship's least-time track is computed, which may be adjusted from time to time.

The principal justification for weather routing is the increase in margins of safety and added comfort to cargo and crew that will arise from the avoidance of heavy weather. However, it is obvious that the economic advantages of weather routing will play an increasingly important part, and will presumably eventually determine what specific kinds of information a weather routing service should provide. For example, it is clear that the successful avoidance of high seas will eliminate much heavy weather damage, and also damage to cargo from this cause. In addition, a considerable saving of fuel is possible, and Frankcom notes that the annual saving in fuel costs for all British ships operating over the North Atlantic is likely to be not less than £2 million if weather routing was adopted on a widespread scale.

The experience of the Mobil Shipping Company in commercial weather routing is related by Spicer (1967) who compared the hourly cost of operating an oil tanker (£21 per hour of vessel time for a tanker of 19,000 tons deadweight, to £102 per hour for a tanker of 95,000 tons deadweight), with the cost of commercial weather routing. Excluding radio-message charges, the weather routing charges were £54 for each trans-Pacific crossing, £41 for each trans-Atlantic crossing and £18 for each crossing South American to United States, except £41 in the hurricane season. Thus when the cost of weather routing is compared with the hourly value of a 95,000-ton deadweight tanker, it is obvious that weather routing only has to assist the master in saving a half-hour in vessel time to pay for the service for the trans-Pacific voyage. Spicer notes, therefore, that the master of such a vessel who reported a saving of at least 12 hours using weather routing in a trans-Pacific voyage is saying in effect that the vessel saved £1,170.

United States Military Sea Transport ships also use weather routing. A review of their operations is given by Cummins (1967) and he indicates that an analysis of the ship weather routings issued by the Fleet Weather Ship Facility, Norfolk, Virginia, showed that of the 800 routes which were valid for evaluation, 63% of the ships completed the voyage without slowing for adverse weather.

The use, by ships, of weather information provided by private weather consultants, rather than using the weather information provided by the various government weather services, also shows the value that shipping companies place on accurate weather information. One of the companies involved is the Allen Weather Corporation, and Allen (1960) gives some details of this company's operations. In the Atlantic, for example, a saving of 15 days for a 10-knot Liberty ship has been effected, and during each winter savings of five to seven days are not unusual. The average saving for all passages eastbound and westbound in the Atlantic and the Pacific, winter and summer, is in excess of 24 hours. Allen also indicates that special studies have shown that the threat from heavy weather damage has been reduced by 46% for vessels utilizing recommended routes. Other benefits also accrue such as fuel savings, and ultimately insurance costs.

Although optimal weather routing is still in its first stages of development, and despite the fact that some routed ships have actually taken longer to arrive at their place of destination than un-routed vessels, statistical evidence has proved, according to Evans (1968), that routed ships generally experience a reduction in transit time and also a reduction in cargo and hull damage. This applies especially to ships of the same company when employed on a particular service, and Evans suggests that it must be of increasing importance as larger and faster ships are introduced, fleets reduce in number of units, and reliance for profitability upon tight time schedules and a minimum of heavy weather delay and damage become more universal. Weather routing will probably, therefore, eventually become a standard operating procedure for ships as it already is for aircraft, and it may be noted, according to Verploegh (1967), that in a number of countries, including the Federal Republic of Germany, the Netherlands, Norway, U.K., U.S.A. and U.S.S.R., a regular ship routing service has been established.

e. Road transport

A major consideration in the relationship of weather to road transport is safety, since poor visibility, slippery surfaces and gusty winds greatly increase the hazards of driving, a factor which is clearly borne out by accident statistics. Increased traffic in the public sector also occurs during times of adverse weather conditions, particularly bus and taxi services in cities. However, at least as far as the London taximan is concerned, the weather, as Whitten (1947) points out, is far from being the cabman's harvest the general public imagines, for although a heavy shower in the mid-afternoon creates a heavy demand for cabs and taxis, fog produces so many headaches and possibilities of wrong turns for the cabby that profit is negligible.

It is generally recognized that the factors tending to reduce traffic include rain, low temperatures, snow and fog, while sunshine and high temperatures usually tend to increase it. Few studies on the specific effect of weather on road transport have been published. Tanner (1952), using data at five fixed points over England for a year, showed, however, that temperature is more important in the winter than in the summer, and that sunshine and rainfall are more important in the summer than in the winter.

A more recent approach to the effect of weather on traffic flow was developed by Maunder and Halkett (1969a), the purpose of the study being to determine to what extent the volume of urban and inter-urban traffic flow is influenced by the weather, and which weather factors are significant at various times of the year, and on various days of the week. The two areas considered were Victoria and Vancouver, British Columbia, and among the conclusions reached were: (1) inter-urban traffic in Victoria appears to be more sensitive to weather than does urban traffic in Vancouver; (2) warmer and sunnier than normal conditions appear to be associated with increased traffic flow, whereas colder and wetter conditions than normal appear to be related to decreased traffic flow; (3) traffic flow appears to be most sensitive to the weather at the weekends and on Tuesdays. The sensitivity at the weekends is not surprising since many people use vehicles for recreational purposes at this time, for which warm and sunny weather is preferred. However, the sensitivity of traffic flow to the weather on Tuesdays more than any other weekday cannot be accounted for without further study. These conclusions, it should be noted, are subject to the limitations of the study, the most important of these being that the study was concerned only with two counting stations in two cities, that the period of the study was limited to two years for the urban counting device (in Vancouver) and one year for the inter-urban device (in Victoria), and that two important weather factors, snow and fog, were not taken into account.

Another relationship between weather and highways is that of the many maintenance problems and flood damage caused by various elements of the weather. Information of this type is usually readily available in most countries, and a brief account of weather as it affected maintenance and repair work during a season in New Zealand gives an indication of the weather's importance. The New Zealand National Roads Board in reporting on the 1959/60 season said, for example, that the weather in Northland was 'typical and generally interfered with the highway programme'. On the other hand, good weather in Taranaki resulted in resealing being more advanced than at any time in the past, while in Auckland unsettled weather in February and March hampered the earth works programme. Similarly, in the Gisborne area, flood damage and heavy rain in February interrupted progress on sealing though better conditions earlier in the season gave a good sealing year on the whole. Similar examples could be given for other areas, and for other seasons, the notable contrasts between one area and another, and between seasons, emphasizing the importance of the weather hazard faced by those responsible for the maintenance, repair and construction of highways.

One aspect of weather and road traffic that is of particular concern to highway safety officials is the effect of weather conditions on traffic accidents. Table III.10 gives, as an example, the weather conditions at the actual time of the 44,187 accidents in British Columbia in 1966, and it shows clearly that 'rain' conditions were particularly important. For example, 26% of the 44,187 accidents causing property damage, injury or death occurred under rain conditions, and over 4% under snow conditions. Thus, although the majority of accidents occurred when conditions were clear, it is obvious that adverse weather conditions play a much larger part in traffic accidents in British Columbia than the normal expectancy of such conditions would suggest.

A similar study completed in Australia (Robinson, 1965) compared accidents and weather conditions in Melbourne during 1960. The analysis showed that 4,062 casualty road accidents occurred on the 161 rainy days of 1960, compared with 3,945 similar accidents on the 205 non-rainy days. The annual average daily number of casualty accidents was therefore 25 during rainy days and 19 during rainless day, thus giving an increase of 30% to the 'credit' of the rainy days.

TABLE III.10: *Traffic accidents and weather conditions in British Columbia, 1966 (after Statistical Summary, Motor Vehicle Accidents, Motor Vehicle Branch, Victoria, B.C.)*

Weather conditions	Total	Fatal	Personal injury	Property damage only	Percentage of total
1. Clear	23,504	257	7,000	16,247	53·2
2. Rain	11,359	71	3,259	8,029	25·7
3. Cloudy	6,208	94	1,738	4,376	14·0
4. Fog or mist	687	7	189	491	1·6
5. Snow	2,020	9	386	1,625	4·6
6. Smoke or dust	107	0	26	81	0·2
7. Not stated	302	7	92	203	0·7
Totals	44,187	445	12,690	31,052	100·0

Column header above data: *Number of Accidents*

A major contributing factor to impeding traffic is snow, and although engineers have found it practical and economical to melt snow as it falls, using special heaters under footpaths, driveways, runways and ramps, snow still presents a formidable problem to many municipalities and traffic control authorities. Snow tyres, for example, cost the citizens of Rapid City, Casper and Cheyenne combined more than $600,000 annually (Rooney, 1967). Added to that are private and public expenditures for tyre chains, shovels, snow brooms, scrapers, sand, salt and numerous ice-melting compounds. The costs are therefore high, but Rooney points out that, as a technically advanced urban society, the United States has at its disposal the organizational and inventive abilities to deal successfully with snow. The snow menace, however, still severely hampers activity in many U.S. major centres, and the question posed by Rooney is not how these inventive abilities can be used, but 'why do we not use them?' Perhaps the organizational and inventive ability does exist, but snow still brings to a standstill many U.S. and other world cities. For, in a case such as the snowstorm which hit New York in February 1969, closing most businesses including the New York Stock Exchange, it is not difficult to see that man can do very little except to forecast well in advance the occurrence of such conditions.

The effect of weather on the pedestrian is also important, two major weather factors being precipitation and sunshine. For example, the glare from dazzling white pavements can be most pronounced, and Griffiths (1966) suggests that the use of coloured pavements, such as green or pink, would reduce this annoyance, and at the same time lead to an increased absorption of sunshine which would be especially useful in winter for melting snow. An associated factor is the protection of pedestrians in shopping areas from rainfall, and it is interesting to note the contrast between the protection offered (through permanent verandahs and awnings) to pedestrian shoppers in Australian and New Zealand cities, with the open sidewalks of most North American cities, which offer little if any protection from sun, snow or rain.

The automobile, in its use and operation, is subjected to many weather conditions. Such conditions have a direct effect upon the automobile's design and construction, and an engineer cannot therefore ignore the climatic factor when designing an automobile because nearly all its functional parts are influenced by the ever-changing climatic elements. Among the major weather elements is wind, which constitutes a most significant effect upon the body design and steering geometry of a vehicle, and engineers strive towards a design that, when encountered by a wind, cannot make the car deviate perceptibly within the reaction time of the driver. The production cost of automobiles is also affected by weather conditions, many of which are now taken very much for granted. For example, precipitation necessitates windscreen wipers, which cost about $40, and sunshine causes a glare in the driver's eyes which can be eliminated by a tinted windscreen at $25, or by sun visors at about

$10. Hot and humid weather can also be partially eliminated, but at some cost, by the installation of a $150 to $300 air conditioner. Many accessories are also needed to adjust to winter conditions, such as a $100 heavy-duty heater and defroster. It is evident, therefore, that weather can cause the cost of production of automobiles to be substantially increased, particularly if very hot and humid conditions as well as extreme cold conditions are to be successfully combated.

f. Transportation by rail

Flooding streams, swollen by rain, melting mountain snow and a slide, have closed the Canadian Pacific Railway line and several traffic routes. A westbound CPR freight crashed into a slide that covered the mainline 15 miles west of the eastern British Columbia community Friday night. A small river, apparently blocked by the slide, flooded both the railway and the trans-Canada Highway.

So stated a Canadian press report on June 3, 1967, and as any newspaper editor will testify, such a story is not unusual. Indeed, in some areas of the world snowdrifts have caused passenger trains to be stranded for days, and few railway lines in the world have not at some time in the past suffered from the vagaries of the weather. Thus, although railways would seem to be little affected by weather because of their permanent all-weather track systems, railways, as Critchfield (1966) points out, are far from immune to weather hazards, and their operations must be constantly geared to weather changes. Severe storms, for example, may damage tracks, bridges, signals and communication lines; floods, heavy snowfall and earth slides are related menaces, and heavy snow and avalanches are particularly troublesome in mountains. In some areas, for example, it has been expedient to build expensive tunnels to avoid the severe weather of high altitudes. Low visibility due to fog or precipitation are added weather factors which call for extra caution and decreased speeds, and some passenger and freight schedules can be disrupted by the effects of such weather on tracks and equipment. Weather may of course cause dislocation at any time of the year, and in extreme cases bridges can be washed away and major subsidence and landslides occur, the resultant diversions and rebuilding adding greatly to the running costs of railways.

Weather conditions which deviate from the optimum point can in fact affect railway systems in two major ways. First are those conditions which hinder the safe and rapid movement of the trains, and probably the most obvious and most costly weather factors detrimental to railway operations are those caused by low temperatures, which cause problems of track expansion and contraction. For example, long lengths of rail, not properly fastened down, can easily loosen or buckle as their length decreases or increases, and in the case of manually operated switches and semaphore signals, such action can cause settings to change and as a result may lead to accidents.

A set of more serious problems can occur with ice, for any water left on the railbed as a result of poor drainage will freeze where it lies and this may in turn cause operational problems with the switches and heave the tracks up with its expansion. To counteract the freezing in switches, anti-freezing compounds can be spread on the slides along which the switching rails move, or in some cases expensive heaters can be placed in operation. Freezing may also directly affect the train itself, and high-powered heat sheds are now being used in some areas for loaded mineral carriers, to counteract ice-bonding.

Apart from the effects of weather which may determine the actual safe movement of the train, there are various economic effects which are associated with freight and passengers. For example, to maintain passenger cars at a comfortable temperature, expensive heating and air conditioning apparatus must often be installed. Similarly, freight, like passengers, has to be protected from the weather, but to differing levels. Perishable goods, for example, must be kept at specific temperatures; accordingly, railways have invested heavily in insulated heated or refrigerated freight cars. Humidity, too, must be kept at a specified level for both food products and for metals which could easily be corroded by condensation of water vapour, while the possibility of expansion of liquid freight must also be taken into account with some apparent payload space being reserved for possible freight expansion.

Few specific studies have been made on the 'value' of weather to transportation by rail. However, two studies have been made (Cameron, 1947; Henley, 1951) on railway operations in Britain, and it is clear that weather does have a considerable economic value to railway operations. Hopefully, research into the economics of weather/railway operations will show economically useful benefit/cost ratios.

2. Utilities

The demand for services provided by public utilities such as fuel oil, electric power, radio, television, telephone, water and gas are all affected by variations in the weather and the climate. The normal seasonal climatic variations, for example, create associated seasonal variations in the supply of water for hydro-electric generation plants, and also seasonal variations in the demand for fuel oil, gas and electricity. Possibly of even greater economic significance are the day-by-day variations in the weather. Air conditioners, for example, can create a critical demand for the supply of electric power in some hot and humid urban areas, particularly in the south and east of the United States. The maintenance of services in adverse weather conditions is also a factor of some concern to utility companies, particularly since the demand for some services increases considerably in critical weather conditions.

a. Weather and fuel consumption

The heating required to maintain a comfortable indoor temperature depends on controllable factors such as the design and size of the building and also the 'uncontrollable' factors of the external climate, the most important of which are temperature, wind and humidity. For example, if the outside air is cold the wall temperatures of a building will be low, and thus it is necessary to keep the indoor air temperature a few degrees higher than 'normal' in order to compensate for the radiative loss of heat. Additional fuel is consumed because of this loss of heat. In fact, the difference between the temperatures indoors and those of the outside air is clearly associated with the amount of heating required, the concept of the heating degree day therefore being particularly useful in heating requirement calculations. Conversely, in the summer, if the outside temperatures are high, it is desirable to cool buildings because they acquire part of the outside heat.

Other weather features influencing the heating requirements of buildings, and hence fuel consumption, are evaporation and wind, the latter being particularly important in areas subject to severe winters. Moreover, the actual siting of a building has an appreciable effect on heating requirements, and it is estimated that the heating requirements for buildings outside a built-up area are 15 to 25% more than similar structures within the built-up area.

The amount of fuel consumed per month in houses heated by

oil is closely associated with local climatic conditions. For example, data supplied to students by fuel dealers in Victoria, British Columbia, revealed that the average Victoria house (1,400 sq. ft) consumes on the average 800 gallons of fuel each year, the monthly totals varying from six in July and seven in August, to over 100 gallons in each of the months December to March. These fuel consumptions may be compared with over 610 heating degree days (65°F base) in each of the months December to March, and less than 160 in each of the months July and August. A more detailed study of fuel consumption in Victoria, British Columbia, was also made based on a random selection of five fuel oil accounts from 30 randomly selected oil accounts. The five accounts were compared against the degree days below 65° for the same period. Two calculations were made: degree days per gallon of fuel, and gallons per degree day, the number of degree days being recorded between the dates of oil tank refills and divided by the number of gallons of fuel oil used over the same period. The period studied extended from December 1965 to April 1967, and with fuel oil retailing for 20·6 cents per gallon, it was shown that each degree day costs the average home-owner in Victoria 2·8 cents, equivalent to an annual cost of $156 based on the 5,579 average (1931–1960) number of degree days. Based on this data, it is therefore evident that the average home-owner in Victoria has an annual fuel oil cost of only half that of his counterpart on the Canadian Prairies.

The importance of climatic fluctuations on fuel requirement in Britain has been discussed by Manley (1957), who assessed the total 'monthly degree days' below 57°F, and suggested that it provides a valid index which enables a summary of past vicissitudes to be presented covering upwards of 250 years. Manley found that for each 'heating season' the sum of the month–degrees presented a pattern of wide and erratic variations, from 74 in the mild winter and spring of 1821/22, to 137 in 1739/40 and 130 in 1878/79. Writing in 1957, Manley sounded a note of warning about the possible effects on fuel demands of a repeat of the 1878/79 season, a warning which no doubt was well taken in the severe winter of 1962/63.

Although meteorologists have published little relating to the value of the weather and how it is associated with the heating industry, it is pertinent to note the remarks of Richard (1968), who suggested that the engineering profession has been accomplishing its 'mission' in spite of the calibre of the environmental support given. He also indicated that the heating industry is working on a total energy computer programme which will incorporate 'typical year weather data' into a total building concept which includes room size, wall coefficients and internal heat sources.

b. Weather and gas consumption

In many parts of the world heating requirements of houses, factories and office buildings are furnished by gas, rather than electricity or oil. The natural gas industry of the United States, for example, is not only an important factor of the utilities division of the economy, but it is also particularly weather-sensitive.

As with electric power companies, the value of weather lies principally in connection with the consumer demand, but in the case of the natural gas industries it is particularly important to be able to anticipate in advance what the consumer demand will be. Critchfield (1966), for example, noted that the delivery of natural gas over long distances cannot ordinarily be increased as rapidly as can electrical service. It is therefore very important to be able to anticipate future needs in the light of weather forecasts, since the demand for gas resulting from colder temperatures must be met by dispatching gas from a remote source before the low temperatures occur. Consequently, gas supply companies constantly refer to the meteorological conditions and forecasts for the area being served, and several gas companies have found it profitable to employ meteorologists who can interpret weather conditions specifically in terms of the problems of gas distribution.

Although it is clear that weather has a measurable effect on the natural gas industry, few studies on the value of weather to this industry have appeared in the literature. One study that is relevant to the discussion, however, is that of McQuigg and Thompson (1966), who considered the flow of natural gas through the distribution system of a company serving Columbia, Missouri. This particular study is considered in detail in Chapter VI, and Fig. VI.2 shows that, compared with the average daily gas demand in Columbia of 14·5 million cubic feet, 'average annual loss functions' can vary from 10 to 70 million cubic feet, depending upon the standard deviation of the forecast temperature, and the standard deviation of a 'disturbance term'.

Temperature is of course the most important weather element affecting gas consumption, and the work of McQuigg and Thompson shows the 'value' of 'good' and 'bad' temperature forecasts. There is also a well-defined relationship between actual temperature and consumer demand; Critchfield (1966), for example, shows a graph based on U.S. Weather Bureau investigations which indicates that for a typical large city, the amount of gas piped to consumers increases from 160 million cubic feet with an expected temperature of 43°F, to 200 million cubic feet at 22°F, and 240 million cubic feet if the temperature is expected to be 1°F. Thus demand would appear to increase 50% for a drop in temperature from 43°F to 1°F. Such a 50% increase would in a typical large city be close to 100 million cubic feet of gas, which at 4 cents per 100 cubic feet, would 'cost' something in the neighbourhood of $4,000,000.

c. Weather and electric power

Weather conditions have an important influence on the generation, transmission and distribution of electric power. The weather affects the load on the plant, and in addition, it may cause damage to equipment and hence interrupt the supply of electricity to the consumer. The heating and lighting components of the total load are also closely associated with weather conditions, since the lighting load is a function of the daylight illumination and, apart from 'abnormal' weather conditions, it varies throughout the year in a predictable manner. The heating load on the other hand varies with the air temperature, and to a lesser extent with the wind, sunshine and humidity, and although it too varies in a predictable manner throughout the year, it can be irregular as a result of sudden cold spells.

One comprehensive study on weather variations and the electric power industry is that of Nye (1965), who considers the value of services provided by the Australian Bureau of Meteorology in the planning operations of the Electricity Commission of the State of Victoria. In the first instance he notes that careful study of the climatic conditions affecting a catchment is vital to the choice of location of a hydro-electric undertaking. In particular, attention should be given to precipitation, including the incidence of snowfall and the duration of snow cover, the frequency of droughts and the behaviour of streams within the catchment. Second, the frequency of strong winds and ice conditions is considered in the design of high-voltage line structures, and maximum temperatures are taken into account when the degree of clearance of line-sag during summer months is forecast. Nye also points out that an electricity supply system is required to respond instantly and unfailingly to the ever-changing demands of its consumers, and weather conditions have a very large bearing on this demand. In this specific electric supply situation, the author reported that the data received from the Bureau of Meteorology enables the Electricity Commission to predict power requirements more accurately, particularly at the peak-hour periods around 8 a.m. and 6 p.m., and to assist in this, the loading engineer at the system control centre is in touch with the Bureau every three hours for information on anticipated temperatures, wind speeds, cloud cover and rainfall.

In many areas, notably Japan, New Zealand, Canada, Norway and Sweden, the weather and the climatic conditions also have a significant effect on the production of electric power, since a considerable proportion of the electric power in those countries is generated by water. Arakawa (1959), for example, shows that in Japan extra stand-by steam-generated plants are necessary to counteract the seasonal variations in rainfall that occur, this being in spite of the effort which is made to hold sufficient water reserve from wet seasons to provide water-generated power during the dry seasons. Arakawa also notes that since water-generated power is usually cheaper to produce than steam-generated power, it is important, at times of localized drought, to be able to transfer energy from a place of sufficient water supply to a place of less water. Moreover, the value of accurate weather forecasts one season ahead can be readily appreciated.

The importance of short-range forecasts should not be overlooked, however, and Davies (1960), in commenting on grid system operation in Britain, notes that the difficulty of making accurate load estimates arises from the fact that in Britain the demand is highly sensitive to changes in the weather. For example, at around freezing point the increase in demand per 1°F sustained fall of temperature can be as much as 160 megawatts, and at still lower temperatures this can rise to 200 megawatts with no signs of demand saturation. Other meteorological elements such as wind, cloud, fog and precipitation also cause considerable variations in electricity demand. At freezing point, for example, a 25-knot wind is found to increase the demand in Britain by about 700 megawatts as compared with a calm day, which is equivalent to a drop in temperature of 4–5°F. Cloud also has an appreciable effect on the lighting load and can cause a variation of more than 1,200 megawatts. For example, dark clouds over London often increase the demand by 350 megawatts.

Several additional studies have been published on the effect of weather on various aspects of electric power production, transmission and consumption, including Dryar (1949), Stephens (1951) and Fleishman (1954). A study of load dispatching and weather in Philadelphia was made by Dryar (1949), who found that in general there is a 2% increase in load for every 5°F drop in temperature from 65°F, that sky conditions can cause a spread of approximately 20% in the systems load, and that the effect of wind on the system load is an increase of 2% for each 5 mph in wind speed. Stephens (1951) discusses a method for analysing weather effects on the power consumption in a part of Pennsylvania. He notes the importance of studying the network in order to appreciate the different purposes for which electricity is used within the network, and the manner in which these uses of electricity vary from time to time. Associated with this is the importance of distinguishing the effect of weather and non-weather on power consumption at different times during the day. Fleishman (1954), for example, gives data appropriate to two sample hours of an electric power system, and shows that the increase in demand due to a 1°F decrease in temperature varied from 1,114 kW at 6 a.m., to 2,254 kW at 9 a.m., but that the non-temperature-affected load for the same hours were 51,000 kW and 104,000 kW respectively.

The studies cited above clearly indicate the significant effect that weather and climatic conditions have on electric power production and consumption. Most studies emphasize the effect of cold winds and cloud on power consumption, and it is clear that these weather factors are very important. However, an increasing demand for power is occurring in those parts of the world where air-conditioning equipment is commonplace, one interesting aspect of this being explored in a study by Johnson, McQuigg and Rothrock (1969), who considered the modification of the outside temperatures, and its possible effects on electric power consumption and production costs. In the analysis, electric power loads are related to temperature by standard regression techniques, and then historical and experimentally generated series of average daily temperatures are subjected to a modification scheme based upon the formation of contrail clouds. Electric power loads from both modified and non-modified historical and experimentally generated series are then included by the authors in a linear programming model which includes costs and generating capacities for 14 large utility companies in the United States mid-west. The costs of generating electric power under modified and unmodified temperature conditions are further used as a basis for providing some insights into the implications of temperature modification for the electric power industry.

The study initially focused on the relationship between hourly load data and the average daily temperatures in the supply area. The coefficients of determination of the regressions of the daily temperature and load data during July and August for the years 1962, 1963 and 1966 are shown in Table III.11, all the regression coefficients on temperature being positive and statistically significant at the 1% rejection level.

A series of 'modified' temperatures was then produced using a simulation model developed by Nicodemus and McQuigg (1969), a model based on the reduction of solar radiation arriving at the surface of the earth through the 'creation' of contrail cirrus clouds. The model is described in Chapter V, and an important feature is its ability to create a time series of daily temperatures that are

TABLE III.11: *Coefficients of determination obtained from regressions of peak hourly electric loads on average daily temperatures at 14 electric companies for July and August 1962, 1963, 1966 (after Johnson et al. 1969)*

Company name	Company number	Coefficient of determination		
		1962	1963	1966
Union Electric	1	·87	·90	·90
Illinois Power	2	·81	·77	·92
Central Illinois Public Service	3	·72	·69	·88
Iowa & Illinois Gas and Electric	4	·74	·94	·94
Missouri Public Service	5	·77	·81	·90
St Joseph Light and Power	6	·87	·79	·81
Iowa Electric Light & Power	7	·59	·52	·24
Commonwealth Edison	8	·88	·94	·94
Kansas Gas & Electric	9	·87	·94	·79
Empire District Electric	10	·71	·79	·90
Kansas City Light & Power	11	·51	·79	·94
Central Illinois Light	12	·77	·74	·88
Iowa South Utilities	13	·61	·71	·85
Iowa Public Service	14	·76	·81	·85

reasonably similar to what might be produced by a deliberate attempt to modify summertime temperatures in the central United States. For example, decreases in afternoon maximum temperatures are generated in the simulation model on about 20 days during the months of July and August, the 'resulting' change in surface temperature being from 3° to 5°F on most of the 'modified days'.

A production cost model was also formulated, the objective function of the model being to minimize the cost of supplying a given set of electric power demands by the 14 electric power companies. Next, the generation of the modified and non-modified historical and experimental load series was accomplished by combining the results of the temperature–load regressions with a series of generated temperatures. Johnson *et al.* then calculated the production costs of electric power in 1962 and 1963 by substituting daily electric loads for observed and modified temperatures into a linear programming model (Table III.12). This shows that of the 51 days (Sundays and

TABLE III. I 2 : *Costs of pooled and non-pooled power production for observed and modified hourly electric loads (July, August 1962) (abridged from Johnson et al. 1969)*

Date	With pooling power*			Without pooling power†		
	Actual cost ($)	Modified cost ($)	Difference in cost ($)	Actual cost ($)	Modified cost ($)	Difference in cost ($)
1962						
July 6	37,891	37,867	24	56,024	56,001	23
,, 11	38,168	36,401	1,767	56,582	54,849	1,733
,, 19	36,651	36,511	140	54,995	54,861	134
1962						
Aug. 6	39,744	39,528	216	59,604	59,429	175
,, 7	41,668	41,322	346	61,968	60,753	1,215
,, 9	36,473	36,430	43	55,368	54,880	488
,, 10	34,782	34,652	130	54,502	54,381	121
,, 23	42,982	41,620	1,362	62,972	61,651	1,321
,, 27	35,055	34,618	437	52,756	52,301	725
,, 29	40,359	39,413	946	59,422	58,489	933

* Transmission costs between all 14 electric power production and distribution areas equal zero.
† Transmission costs between all 14 electric power production and distribution areas set equal to the highest per unit cost of power production.

holidays excluded) in July and August of 1962, ten had conditions necessary for successful modification of surface temperatures. By contrast, in 1963 (not shown in Table III.12) modification occurred on 20 of 53 possible days in July and August. Power production costs for all modified days are reported for two transmission cost assumptions. First, it is assumed that transmission costs are zero, the actual modified electric power production costs and differences for this situation being shown in Table III.12 under the heading 'pooling power'; second, it is assumed that transmission costs are prohibitively high, that is that the cost of transporting power is just equal to the highest generation cost at each of the 14 locations. Results for this experiment are reported under the heading 'without pooling power'. The average differences in the modified and non-modified hourly production (for 1962 and 1963) costs are $405 if it is assumed that power is pooled, and $529 if the power is not pooled. It should be noted that these are hourly costs, which should be multiplied by a factor of about 12 to obtain daily production cost differences, since daily load curves for hot summer months are at or near the peak of periods from about 10.00 a.m. to 10.00 p.m.

Similar results are given in the original paper for the costs of producing power on very hot days using an experimentally generated and modified 80° and 85°F series, some of the results using the 85° series being shown in Table III.13. This shows that the hourly cost differences varied from $2,027 to $50,425, with an average of $22,540.

TABLE III.13: *Costs* of non-pooled power production for hourly loads based on an experimentally generated and modified 85° series† (abridged from Johnson et al. 1969)*

	Cost ($)			Temperature	
Date‡	Actual	Modified	Difference	Average§ actual	Average§ modified
530608	106,876	104,849	2,027	76·7	75·3
610730	96,940	91,865	5,075	82·1	81·8
530629	123,640	114,043	9,597	81·3	79·5
550820	141,910	127,563	14,347	85·1	82·2
620820	136,583	121,161	15,422	79·8	77·1
530830	132,214	111,569	20,645	83·9	80·6
540828	131,074	109,248	21,826	82·2	79·4
540730	137,601	114,747	22,854	82·8	80·3
540815	108,875	85,095	23,780	82·7	77·4
550725	118,272	91,301	26,971	80·5	74·9
540711	163,198	131,527	31,671	85·9	81·3
630701	143,191	107,044	36,147	84·2	77·8
540827	143,910	103,562	40,348	79·7	75·2
540706	106,682	56,257	50,425	83·1	75·1

* In order of difference in cost.
† Transmission costs between all 14 electric power production and distribution areas set equal to the highest per unit cost of power production.
‡ Year, month, day.
§ Averages of average modified and observed daily temperature at the 14 distribution areas.

In their summary, the authors indicate that the implications of the results for the electric power production industry are clear, in that the sensitivity of power demands to temperature can be quantified. Furthermore, the results of the programming model suggest that benefits from modification on high average temperature days can be substantial, a factor which may well prove highly beneficial to power-producing companies in the next decade if weather modification in the form of cloud formation is proceeded with on a regional scale.

d. Weather and communications

Most kinds of communication services are affected in a small measure by weather conditions. For example, telephone services have many of the weather-related problems that electric power distribution companies are faced with. In some cases the transmission lines can be put underground, usually at considerable cost, but a considerable amount of telephone and telegraph traffic is still subject to delay in times of adverse weather conditions. Further, as with electric power companies, the demand for communication services is often much higher in bad weather. Consequently, as noted by Critchfield (1966), communications media must be geared to meet maximum traffic and poorest operating conditions at the same time. Microwave transmissions eliminate some of these problems but often introduce new difficulties because transmissions are affected by temperature and moisture, and other conditions in the atmosphere.

Other forms of communication such as radio and television are usually not affected by normal weather conditions, but as is well known lightning and other electrical disturbances in the atmosphere can interfere with transmissions and reception, and short-wave radio communication is often seriously affected by conditions in the upper atmosphere. The forecasting of adverse conditions in the upper atmosphere is therefore of considerable value to radio communication in several parts of the world.

A third form of communication is the newspaper industry, which is normally not very weather-sensitive. However, even here, distribution of newspapers especially in bad weather can provide many problems. Perhaps, also, the value of the weather can at least be partly determined by the space given in newspapers to weather forecasts, weather data and weather maps, and of course, as every newspaper editor knows, weather stories especially about a reader's home town are always newsworthy.

e. The value of wind for power

Wind as a source of energy is not a new idea. Yet it has been almost neglected for some decades as more 'compact' and dependable methods of producing power have come into extensive use. During the past twenty years, however, considerable research efforts have been made with the aim of developing wind-driven electrical generators called 'aerogenerators' of different sizes and designs according to the purposes envisaged.

In spite of some disadvantages as a source of power, energy produced by the wind has certain advantages which make it attractive. For example, it is free, and over a period of time could be considered 'inexhaustible'. It also has the merit of conveying itself, without any transportation problems, to places where its exploitation can be useful. Moreover, since the annual average wind speed at many places is more constant from year to year than rainfall, upon which water power depends, its use as a source of power is very attractive. Wind, however, cannot usually be relied upon to occur with adequate strength at any particular time; it must therefore be considered as a random source of power whose energy cannot be stored directly, like that of water, in a reservoir.

Although the use of wind for power has not been widely recognized, except for the use of windmills in various parts of the world, it is clear that where wind power is backed up by other power sources, its applicability should be seriously considered. Moreover, although nuclear power and the other traditional forms of power supply are probably much easier to 'administer', perhaps the 1970's, with the added probability of large-scale weather modification, will focus man's attention on the energy of motion in the atmosphere, in the same way that radiant energy has been successfully used in some areas in the 1960's.

f. Weather and water consumption

Almost every year there is a serious shortage of water in many parts of the world, such shortages occurring in places having annual rainfalls ranging from less than 10 inches to more than 100 inches. The reason for the shortages is partly climatic, partly political, partly economic and partly engineering, but the overriding problem is that in these instances there is too much demand for the available water (see, for example, Landsberg, 1967). In some cases this demand can be controlled by local regulations, and in many instances such regulations are more of a nuisance than a hardship. On the other hand, as pointed out in Chapter II, droughts and similar long periods without rain can cause considerable economic and social repercussions, particularly in agricultural and industrial areas.

One aspect of water consumption that is partly related to weather conditions is the water used by households, an interesting facet of this being the method used for charging for the water. In some areas, for example, there is little or no direct charge, or there is a flat rate which is charged irrespective of the quantity of water used. On the other hand, many communities charge for the actual quantity of water used, and in such a case natural or even man-made rainfall can have considerable 'benefits' to the householder. Moreover, a number of authors (e.g. Hudson, 1964; Simmons, 1966) have indicated that metering is an effective device for reducing demand for water, and Howe and Linaweaver (1967) cite some data (Table III.14) for 10 metered areas and eight flat-rate areas in the western United States. For example, the average householder's use of water shows little variation between the metered and the flat-rate areas. A considerable difference does exist, however, in the use of water in sprinklers, the average annual demand in the flat-rate area being more than double the usage in the metered areas. Accordingly, it is reasonably clear that water, like many other products, is only really valued when it is charged for, and in analysing the data in Table III.14, Howe and Linaweaver note that the greater efficiency of sprinkling application in the metered areas can be seen in the ratios of summer sprinkling to summer potential evapotranspiration.

If the weather is considered to have any value with regard to water consumption, then it would appear that perhaps the most relevant measure of its value is the decrease in consumption of water for 'non-essential' uses, such as the water used in a sprinkler system. By way of an example, the case of Victoria, British Columbia, the 'garden city of Canada', may be noted. In most years, particularly in the summer, some control is placed on sprinklers in this city, and two extracts from the *Victoria Daily Times* in August 1967 illustrate the importance of lawn and garden watering. On August 8, 1967, the newspaper reported that a total ban on sprinkling faces Greater Victoria within the next two weeks unless citizens impose voluntary restrictions or there is a major rainfall. Mayor Stephen, chairman of the Greater Victoria Water District, described the situation as extremely critical. He stated: 'Unless citizens cut back drastically on water consumption, or unless there is a major rainfall, we shall have

TABLE III. I 4: *Water use in metered and flat-rate areas**
(after Linaweaver et al., 1966)

	Metered areas	Flat-rate areas
Annual average†		
Leakage and waste	25	36
Household	247	236
Sprinkling	186	420
Total	458	692
Maximum day†	979	2,354
Peak hour†	2,481	5,170
Annual‡		
Sprinkling	12·2	38·7
Potential evapotranspiration	29·7	25·7
Summer‡		
Sprinkling	7·4	27·3
Potential evapotranspiration	11·7	15·1
Precipitation	0·15	4·18

* October 1963 to September 1965.
† Gallons per day per dwelling unit.
‡ Inches of water.

to impose a ban on sprinkling.' Mayor Stephen noted that the usual water reserve held in major storage reservoirs totalled five billion gallons, which under normal conditions is enough to serve the Victoria area from April to October. However, in 1967 a billion gallons a month was being used with the result that on August 11 an almost total ban was imposed on the use of water for gardens and lawns in Victoria, watering only being allowed on Thursdays. In view of these restrictions a study was subsequently made by Maunder and Westaway (1968) of the effect of rain and sunshine on water consumption in the Victoria area, the principal finding being that the total water consumption decreases from around 30 million gallons per day on a sunny rainless day to a low of about 15 million gallons on a cloudy day with a few showers. A particularly interesting feature was the fact that it appeared that consumption decreased because of cloudy conditions and not necessarily heavy rain, the effect of only 0·10 inch of rain together with cloudy conditions being sufficient to effectively reduce garden and lawn watering to a minimum. Thus in the case of the garden city of Victoria, the study suggested that the 'normal' householders' use of water was about 15 million gallons per day which increased to 30 million gallons per day on some sunny

F

and rainless days. Thus, the 'value' of cloud and showers to both the householders and the water supply authorities in Victoria can be readily appreciated, and it is reasonable to suggest that some kind of weather modification to produce cloudy conditions in the Victoria area would be of considerable value, not only to the consumer of the water, but also to the supply authorities who would not have to enforce water restrictions. Whether the tourists would like it is of course another question.

g. Weather modification and hydro-electric power

As pointed out in the previous section, weather modification would probably have considerable value in reducing the demand for water in garden- and lawn-loving cities. Weather modification could also be of value to other utilities, and the influence of jet-aircraft-produced contrails on power requirements for air conditioners has also been mentioned. However, perhaps the most practical use of weather modification in the utility field lies in hydro-electric power generation, and Eberly (1966) discusses in some detail weather modification and the operations of an electric power utility. He notes, for example, that weather modification in the form of increased precipitation offers one means of improving the efficiency of the hydro-electric portion of a power system, since in years of less than normal river flows, the hydro-electric power capacity may be idle, with the result that the power utility would have to turn to other sources of power such as thermal generation, or import power from other regions.

A test programme of the Pacific Gas and Electric Company, the largest investor-owned power utility in the western United States operating 67 hydro-electric plants with an installed capacity of 2,226,000 kW, is described by Eberly, who also outlined some of the factors that would need to be taken into account in appraising the economic value of weather modification against the alternatives. If an increase from cloud seeding could be depended upon, for example, it might be possible to introduce some efficiencies in the operations of a hydro-system, since water levels in reservoirs at the end of the dry season are usually maintained at a level which will provide a reasonable assurance that the lakes will be full by the end of the following wet season. Thus, if the company did not have to guard against dry years being so dry, the water levels in the reservoirs could be reduced, which would have the added effect of having available more storage space when the wet years came.

Second, an increase in runoff from cloud seeding should provide additional energy from a hydro generating system, the increased generation possibly satisfying some of the growth requirements of the company, which could delay capital expenditures for new thermal plants. For example, Eberly states that averaged over two 'recent' years, this particular company's hydro system produced $1 \cdot 097 \times 10^{10}$ kWh from $2 \cdot 1 \times 10^6$ kW of installed capacity, whereas for the same years the thermal portion of the system produced $1 \cdot 868 \times 10^{10}$ kWh from $4 \cdot 7 \times 10^6$ kW of installed capacity. The ratio of kWh produced from the installed plant capacity for the system is $3 \cdot 96 \times 10^3$ kWh per kW. Now a 10% increase in hydro generation would be equivalent to $1 \cdot 097 \times 10^9$ kWh, which Eberly suggests would involve an investment in thermal generating plants of $25,000,000. This is of course a gross simplification of reality, for, as Eberly points out, generation capacity must be built to meet both energy and capacity peak demands. It is nevertheless at least an approximate value which may be placed on increasing hydro-electric power generation by 10% through weather modification.

The investigation of the potential benefits from cloud seeding for several different watersheds is reviewed by Eberly, who assumes that the 'increased precipitation' from seeding results in a 10% increase in runoff, that the increased runoff results in a 10% incremental increase in daily runoff, and that the value of the increased runoff is in the fuel saved by not using thermal generation. Type I watershed is relatively large (500 square miles), mostly forest-covered, with a reservoir storage capacity of over one million acre feet. Annual runoff averages 625,000 acre feet which is substantially less than the storage capacity. The computed values of a hypothetical 10% increase in runoff are shown in Fig. III.4, values being plotted for the approximate range of natural flows in the watershed. Thus, in a dry year when the natural flow is about 300,000 acre feet, an increase of 10% would be valued at about $210,000, whereas in years when the natural flow becomes greater, the *rate* of increased value decreases because all of the water cannot be fully utilized. In the smaller watershed (Type II), the annual runoff averages 270,000 acre feet; the effect of a 10% increase for different runoffs is illustrated in Fig. III.4. An important feature of this curve is its asymptotic shape, maximum return being possible when the runoff is a little over 300,000 acre feet, since above this the possible return decreases because a larger portion of the runoff is coming with spill, for which the return is zero.

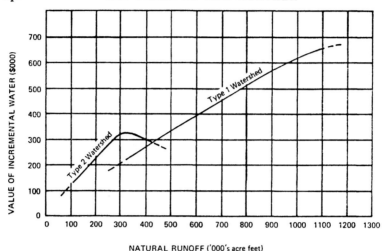

NATURAL RUNOFF ('000's acre feet)

FIG. III.4. *Value of a 10% increase in runoff as a function of the natural runoff for two different watersheds (after Eberly, 1966)*

The benefit/cost ratios of any weather modification programmes are of course an essential part of such programmes, and using various assumptions the benefit/cost ratios for weather modification in the two watershed types discussed is shown in Table III.15. Thus, if a

TABLE III.15: *Benefit/cost ratios for a hypothetical 10% increase in runoff in two watersheds (after Eberly, 1966)*

Watersheds	Dry year	Normal year	Wet year
Type I	4·6 : 1	9·0 : 1	14·1 : 1
Type II	3·0 : 1	8·3 : 1	8·0 : 1

10% increase from cloud seeding could be achieved, the benefit/cost ratios on the Type I watershed would vary from 4·6 : 1 to 14·1 : 1 depending on the climatic conditions.

In summary, it can be noted, following Eberly, that on a large watershed less than a 2% increase in average runoff might cover the cost of a cloud seeding project, whereas a 10% increase or more would be required on some small watersheds. Further, the economic return that might result from cloud seeding will reach a maximum on many watersheds when the runoff is near normal, since during wet years the increase in the runoff that comes when the reservoirs are spilling cannot add to power generation.

3. Commerce

Although a glance at most newspapers will indicate the importance of weather conditions to the world of commerce, few if any substantial studies have been completed on the specific effects of weather on the business community. Indeed it seems almost paradoxical that the value of the weather to business and commercial activities seems to be ignored not only by the meteorologist and climatologist, but also rather surprisingly by the businessmen themselves. There are exceptions, of course, and the relatively flourishing consulting activities performed by meteorologists in the United States is at least partial proof that some business enterprises are willing to pay good money to know more about a particular aspect of the economy that influences their business, namely the weather and its variations.

a. Weather and business

Many economic, social, psychological and environmental factors influence the state of economic activities at any given place in time, and of the environmental factors to be considered, those concerned with the atmospheric environment are of paramount importance. For example, Critchfield (1966) states:*

> ... climate and weather are partially responsible for determining fluctuation in sales, and the type of goods or services in demand. ... In spring the sale of gardening supplies and outdoor sports equipment increases sharply during the first few warm days. ... The housewife buys more salad vegetables and 'lighter' items for family meals. ... On a sunny day in summer, the sales of soft drinks, ice cream, cold meats, and other picnic items are likely to be great. Amusement parks will do a rush business, as will service stations, especially on a weekend. Motion picture theatres may play to empty houses as the populace treks to beaches or mountains; but, if the day is oppressively hot, their air conditioning will draw good crowds regardless of the quality of the film. With the first cool spells of autumn, sales of fuels begin to rise and the demand for furnace repairs or installation grows. The first threat of freezing weather is accompanied by a rush of motorists to garages and service stations for anti-freeze. ... The effects of cold weather and storm on business extend into the winter. Fuel consumption rises, and after a hard freeze, the plumber makes his rounds to thaw out pipes. ... The local grocery can expect a larger turnover when storms make travel to the larger shopping centers difficult and hazardous. The druggist who dispensed sunburn lotion to vacationers in the summer now sells it to skiers along with an infinite variety of cold medicines. ...

* Reproduced with permission: Howard J. Critchfield, *General Climatology*, 2nd edn, Copyright 1966, Prentice Hall, Inc., Englewood Cliffs, N.J.

In addition to the effect of weather and climate and its many variations at specific places, are the regional climatic differences which as is well known have a considerable influence on the type and volume of goods sold. For example, the sales of air conditioners are normally much greater in the south-east of the United States than in coastal British Columbia. In the same way, winter tyres and chains for automobiles would normally find a much greater potential market in the north-east of the United States than in the south-west. Such examples are of course obvious, but what is not so obvious is how businessmen make decisions relating to the seasonal buying and selling of goods. In what way, for example, is the wholesaler of air conditioners influenced by a forecast of below-normal temperatures in one part of his distributing area and above-normal temperatures in another part of his area? Similarly, on what basis do places serving food and drink, whether these be at a football stadium, outdoor recreational area, or downtown in a large city, associate weather conditions with the type of food and drink sold? Does weather information – either actual conditions or forecast conditions – influence their buying? Further, one could ask in what way the advertising business is influenced by weather conditions?

Some of the various weather items mentioned above are undoubtedly taken into account by the business community, although to what degree is not generally known. However, businessmen do realize that economic activities follow a recurring seasonal pattern during the year; for example, the retailer is aware that there will be increased business around Christmas, for which he must plan his purchasing and personnel changes. Similarly, the contractor buys materials and hires additional workers for the increased construction activity that inevitably comes during the summer months, and the farmer's expenditures rise in the spring and autumn because of planting and harvesting costs. Bankers also recognize these and other patterns of seasonal activity and they plan for an uneven deposit inflow and demand for loans during the year. In the same manner wage earners in industries with high degrees of dependence on seasonal activity realize that their incomes may not flow evenly during the year.

The seasonal patterns briefly discussed above usually follow a relatively systematic pattern from year to year, and procedures have been developed to measure these fluctuations. McLeary (1968) notes that modern techniques of seasonally adjusting economic data are highly sophisticated and refined, but the overall approach in-

volves a process for averaging the data for the same months over
several years to arrive at a typical measure of activity for that
month. The normal level of business activity measured for each
month is then considered in relation to the average monthly level
for the entire year. The re-
sulting seasonal factors are
then used to adjust economic data for normal seasonal
activity, in order to reveal
the fundamental changes
that are of concern to the
economist and business analyst. The seasonal index thus
offers a convenient indicator
of the normal level of activity during a certain month
as well as the strength or
weakness compared to other
months.

Several possibilities for
study arise in looking at
seasonally adjusted economic
data, but pertinent to this
discussion is the seasonal
variation in wages, personal
income, profits, rents and
transfer payments as discussed by McLeary. These are
major indicators of general
economic conditions, since
the level or rate of change in

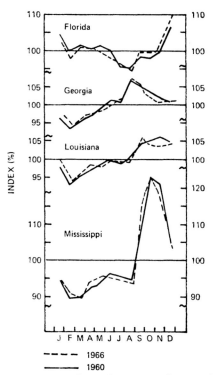

FIG. III.5. *Seasonal pattern of personal
income in Florida, Georgia, Louisiana and
Mississippi (after McLeary, 1968).*

business activity is reflected by the total income produced. Seasonal
personal income indexes for four contrasting states within the Sixth
Federal Reserve Bank District of the United States are shown graphi-
cally in Fig. III.5. Each area has of course its own distinct seasonal income pattern related to the resources from which its income is derived,
and Table III.16 shows the components of personal income in four
states – Florida, Georgia, Louisiana and Mississippi – and their range
of seasonality. Those sectors directly related to weather or climatic
conditions – construction, agriculture, trade, and property – generally experience the most volatility in income during the year,

TABLE III.16: *Shares of total personal income by selected components* and range of seasonality† in four states (after McLeary, 1968)*

Component	State			
	Florida	Georgia	Louisiana	Mississippi
Manufacturing				
% Share	11	20	13	18
Seasonal range	10	4	7	6
Construction				
% Share	5	4	7	4
Seasonal range	14	30	17	30
Agriculture				
% Share	3	4	4	11
Seasonal range	142	173	136	214
Trade				
% Share	13	12	11	9
Seasonal range	15	7	7	5
Services				
% Share	10	7	7	7
Seasonal range	11	3	5	4
Property				
% Share	18	11	15	12
Seasonal range	33	21	20	23

* Shares of total income are based on U.S. Commerce Department's annual figures for 1967.
† Seasonal ranges were calculated from monthly income estimates and their components for 1966.

according to McLeary, and it will be noticed in Table III.16 that in every state agriculture is the most changeable, ranging from 136% of the normal of 4% (of total personal income) in Louisiana, to 214% of 11% in Mississippi. In three of the states, the overall importance of agriculture is small, and hence seasonality due to the climatic conditions is not especially important. However, in Mississippi where 11% of the total personal income comes from agriculture, the 214% seasonal range is a major factor. Accordingly, in Mississippi, climate and especially 'climatic variations' in the form of week-to-week weather variations could very well have a significant effect on the total personal income for that state. Similar observations also apply to other areas of the world, such as Iowa, New Zealand and Indonesia, where a large proportion of the total personal income comes from agriculture. Construction income is also related to climatic conditions in many areas, as indicated by the 30% seasonal range in Georgia and Mississippi.

The recognition of the seasonal nature of economic activity in any area is, as pointed out by McLeary, of prime interest to the banking community. Commercial banks, for example, are often faced with seasonal demands for credit, and these demands do not always coincide with seasonal variations in new deposits. In particular, banks often find that the demands for loans vary with changes in business conditions in their communities, and as has been suggested at least a part of these variations in business activities are related to seasonable weather conditions.

The importance of weather in business decision-making is considered in a paper by Hallanger (1963). This paper is reviewed in Chapter VI, but a specific comment by Hallanger is relevant at this point. He stated:

> The meteorologist, working as a team with your people [business], must become familiar with your operation. Only then is he in a position to identify the true weather problems. Only then can he provide the appropriate weather information in the most useful form. Only then will the return exceed the cost by the greatest amount.

Thus, as some meteorologists but regrettably few businessmen appreciate, the value of weather which includes the basic weather and climatic data, the weather forecast and the modification of that weather, will be realized only when the 'consultant' and the 'patient' know the full circumstances of a particular operation. And as Hallanger (1963) points out, 'the payoff' in such circumstances 'could be substantial'.

As demonstrated above, the specific effects of weather on business are often difficult to obtain, and for much of his information the economic climatologist must rely on newspaper or business magazine reports. They, nevertheless, tell a most useful story, and the following extract from *Business Week* (February 4, 1961) indicates to some extent the value of the weather to the business community.

> . . . In Washington, D.C., a snowfall of 7 or 8 in. threw a wet blanket over Inaugural Eve Celebrations – people couldn't get to the parties. The same storm cut thousands of suburban Philadelphians off from their jobs . . . Lack of a thaw forced many communities – and companies, too – to truck snow away, a slow and costly process ($1.45 a cu. yd. in New York City) that soon exhausts snow removal budgets. . . . Getting hit hardest by the

weather are retailing first, then construction. In 10 Federal Districts, department store sales in the four weeks ended Jan. 21, ranged from 1% to 9% below the year-ago. Merchants don't blame the weather alone but the combination of weather and recession. . . .

Outdoor construction normally continues through most of December, but not this year. The premature halt of work dropped construction employment nationally to just over $2\frac{1}{2}$ million in December, the drop of 320,000 from November is about double the normal season decline. In Boston, unemployment compensation checks rose sharply just after the Dec. 12 storm . . .

Wintry weather helps some businesses of course. Hotels made money on commuters who couldn't get home. Sales of heating oil soared 10% or 20% above normal, and auto chains and snow shovels were completely sold out in some communities. . . .

Lack of snow in Omaha caught the . . . Steel Works with a lot more snow plows on hand than it could sell, yet in the East stores couldn't keep up with demand for anything of household size that would move snow. . . .

In New York, hardy theatergoers found that almost any very cold or stormy evening was the time to pick up seats for a hit show. . . .

Similar weather-related events have, of course, occurred in other places and at other times. In fact, the effect of weather variations are perhaps the most well known and yet the least understood of all the factors which influence our economy. Indeed, it would appear that dividends are almost there just for the asking provided the world of business and the world of meteorology would only get together and really begin to appreciate how, and in what specific way, the atmospheric environment influences our activities and mode of living.

b. Retail trade and the weather

Many factors influence the demand for goods, but whether these are primarily economic, sociological or psychological, it is generally recognized that the prevailing atmospheric conditions do have a considerable overriding influence. A basic factor is the 'urgency' to shop and this involves decisions related to the prevailing weather conditions, any alternative to shopping that the person might have, and the type of goods required. Some of these considerations are therefore taken into account by planners even before a store has emerged from the blueprint stage, for it is clear that extremes of temperature and excessive precipitation depress the desire to go shopping. Under certain circumstances, then, stores close to parking facilities or directly connected to parking facilities are favoured over those in which the shopper must walk many blocks from car to store. Shopping centres offering covered malls are also attractive to shoppers during inclement weather. A further step in overcoming the weather conditions is the shopping centre in which a major department store is connected to smaller shops by an enclosed 'air-conditioned' mall so that the shopper is not only protected from precipitation, but also from heat or cold. However, the comfort of the shopper is only one facet of the complicated retail trade business and it is clear, irrespective of the type of shops or shopping centres available, or the means of transportation to and from such shopping centres, that weather conditions as well as the seasonal climatic characteristics influence to a considerable extent the type of goods sold, and hence the amount of money paid by consumers for goods and the amount of profit made by the retailer.

Department store executives should, therefore, utilize knowledge of the relationship between weather and trade in planning day-to-day sales, for by correctly assessing weather forecasts, the days on which there is likely to be the largest volume of customers can be predicted. This is useful in determining those days on which maximum sales personnel should be available, and perhaps for deciding which days would be most suitable for holding unadvertised sales. Long-range forecasts are also helpful, and the improvements in long-range forecasting expected in the 1970's should enable those responsible for the purchasing of the stores merchandise to become more efficient buyers, since such forecasts are a guide to the kind and quantity of seasonable merchandise that is likely to be required.

Although little research has been done on the specific effects of weather variations on retail trade, the studies completed are of considerable interest. Steele (1951), for example, assessed the effects of weather on the daily sales of three departments of Younker Bros., Inc., Des Moines, Iowa. The period covered in the study was the seven weeks before Easter, 1940–1948 inclusive. Steel suggests that weather might affect the sales of a retail store in four ways: first, the weather could be of such a nature that it is for one reason or another 'uncomfortable' to go to the store; second, the weather could produce situations that would physically prevent people from going to the store, as in the case of snowdrifts over roads and streets; third, the weather may have psychological effects on people that may change their shopping habits; and fourth, some kinds of merchandise may be more desirable during a period in which certain types of weather prevail.

In the study, five weather elements (the amount of precipitation between 6.00 a.m. and 5.00 p.m., the depth of snow cover at 6.30 p.m., temperature at 12.30 p.m., wind velocity between 12 noon and 1.00 p.m., and the amount of sunshine during daylight hours) were associated with 'seasonal adjusted' sales data through a multiple regression analysis. The results showed that when Easter was early the amount of snow cover was the critical factor, whereas when Easter was late the precipitation factor was more significant. The overall results also indicated that 88% of the variance in the sales of the store were accounted for by the weather variables.

In another study, Zeisel (1950) examined the importance of weather to the beer industry in the state of Rhode Island. It was concluded that for every 1·0% temperature change, consumption of beer changed 1·1% above or below the expected level. Zeisel suggests that a similar analysis could also be made for carbonated beverages and iced tea sales, and from preliminary data given in his paper, it would appear that such sales, especially iced tea, would in fact show a much greater association with temperature changes than is the case with beer sales.

Another study relating sales to temperature was done by Linden (1962), in which sales of women's winter coats in New York department stores, as reported by the Federal Reserve Bank, were associated with temperature through a correlation analysis. Linden notes that although research has contributed an extensive literature on the psychology of the buying impulse, surprisingly little attention has been given to so obvious and important a determinant of consumer demand as the weather. It is clear that each category of goods has a unique set of weather complexes, a factor which renders some weather–retail studies of doubtful significance, since they are concerned, through necessity of available data, with either total retail trade or sections of the retail trade, such as department stores or hardware stores. The study of Linden concerning the sales of women's winter coats in New York department stores is therefore of considerable importance.

In the study, the proportion of fall–winter total-season sales that were made in each of the four months September to December were calculated for each year, and Fig. III.6 shows for September and October the share of total-season sales accounted for by the indicated month and the average temperature in the identical period. As will be seen, there is a noticeable correlation between the relative amount of business done in September and that month's mean tem-

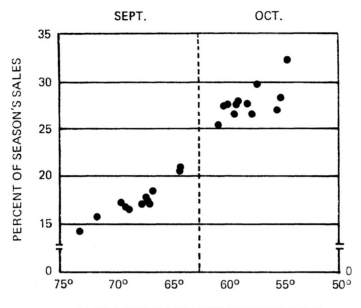

SEPT.   OCT.

AVERAGE MONTHLY TEMPERATURE (°F)

FIG. III.6. *The relationship between the average temperature in September and October, and the relative contribution of those months to the total September to December sales of women's winter coats. Based on 12 years' data in New York department stores (after Linden, 1962).*

perature. For example, the normal temperature in September in the New York area is about 67°, and department stores realize about 18% of their women's winter coat sales in this month. In 1964, however, the mean temperature was a near-record 73° and only slightly more than 14% of the season's business was done in that month. On the other hand, in the historically cool September 1956, which had an average temperature of only 64°, the proportion of women's coats sold totalled 31%. From the experience of the twelve years, Linden concluded that with each full °F of deviation from the normal September temperature, there is an inverse deviation of about 0·6% in September's contribution to the total women's winter coat sales in the season.

As pointed out by Linden, some knowledge of the weather reflex of a particular product line could be extremely useful in improving marketing tactics. For example, sales results, especially in the early period of the season, may be compared against the weather record, a comparison which could well be a critical prerequisite for accurate short-term demand forecasting. Similarly, salesmen's calls at retail outlets might be arranged to follow a period of weather which is especially favourable for a product, for it is likely that inventories would be lowest at that time, and that the merchant would be well disposed towards re-ordering.

There are of course many other ways in which weather knowledge might be used to increase sales. Much depends on the item, and in some instances promotion might specifically stress the particular buying impulses that are triggered by the climate. Or, alternatively, attributes of the item that would promote its sales in off-season periods might be emphasized. But as Linden effectively states: '. . . whatever tactics may be most appropriate for a particular product, it is a fact that weather has a powerful effect on demand. Moreover, with a little research and some imaginative application, marketers can make it a profitable ally.'

One of the concerns of the various Federal Reserve Banks in the United States is the interpretation of the month-by-month 'economic data' for their area. One key problem is 'explaining' the variations which take place from the 'seasonally adjusted' data. For example, Petty (1963) at a meeting in Chicago in 1962, quoted the case of January and February 1962, in which the interpretation of the movement in the Federal Reserve Bank of Chicago department store sales and the corresponding U.S. Department of Commerce data for total retail sales was considered. As Petty relates:

> We had to make some judgment as to how much of the poor showing to blame on 1) severe weather or 2) the consumer's persistent reluctance to buy. The relatively sharp drop-off in sales in areas hardest hit by heavy snows and extremely low temperatures suggested that abnormal weather was again exerting its influence on consumer buying.
> As someone has remarked: 'Most seasonal adjustments for last February "took out" only two inches of the ten-inch snowfall we had.'

Analyses of sales experience which consider only temperature, precipitation, sunshine, wind, humidity or various combinations of these, clearly omit the most important factors that determine the desire and ability of consumers to buy, such as the holdings of cash and other liquid assets, anticipated future income, price expectations and present ownership of goods. Nevertheless, within limits the state of the weather can explain a good part of the short-run variation in sales.

Many of the studies done in the United States in relating weather variations to retail trade have been completed by the National Industrial Conference Board, and two papers by Linden (1959a, b) are of particular importance. In the first paper, Linden examines the effects of adverse weather on customer traffic in department stores, and uses as a tool the weekly data for New York city for the 24-consecutive-month period January 1957 to December 1958. Each week's sales during the latter twelve months of the period were compared with the sales in the related week of the preceding year, the difference was then matched with the variation in weather conditions over the corresponding time. Sales information was obtained from standard Federal Reserve Bank releases, and weekly weather values were based on Weather Bureau records. The weather characteristics of each business day was appraised in terms of their likely effect on store sales, the variables considered including rain, sunshine, temperature and snow on the ground, as they were experienced during business hours. Then, depending on the actual weather for each of these conditions, the days were classified within a five-class value range, from a plus two (excellent shopping day) to a minus two (very bad shopping day), with an 'o' position denoting marginal weather.

A graphical record of the changes in department store sales and weather in New York City over fifty-two consecutive weeks ending with the final quarter of 1958, is given in Linden's paper, sales data for each week in 1957/58 being compared with the corresponding sales data for each week in 1956/57. Linden in commenting on the relationship between weekly weather and weekly sales, states that they 'demonstrate a parallel behavior with a frequency markedly greater than would be anticipated on the basis of simple chance'. Moreover, with the 1957/58 weather being worse than the year before, sales were also poorer in three-fourths of the observations. In addition, after adjusting for non-comparable weeks and for cases where shopping weather was about the same, the data indicated that sales and weather moved in the same direction four times more frequently than they moved in the opposite direction.

Daily sales records for two years from a sizeable New York department store, which, in addition to its downtown outlet operates several branches in the related suburban area, were also studied by Linden, who suggested the following weather/traffic associations. (1) On wet days, traffic in downtown stores decrease on the average approximately 8% as compared with cloudless ones; the loss in the suburbs running fractionally higher. (2) The duration of rain makes a big difference in the toll it takes. For example, with rain occurring during three hours or less, sales volume in the parent outlet decreases by a little more than 5% as compared with good shopping weather. However, with four to six hours of 'wetness' the deficit exceeds 8%, and in a complete 'washout' the loss is over 15%. (3) The intensity of the rain does not seem to be very important; an hour or so of heavy showers often being less inhibiting to business than half the precipitation spread over thrice the time. (4) The time of the rainfall makes some difference. For example, sales are most vulnerable when the morning is wet, a little less so when rain comes only in the afternoon, or in the evening on night openings. (5) Rain on Monday dampens business a little less than on other days of the week, and Saturday traffic also appears to withstand the rain with more success. (6) The second of two consecutive days of rain is generally more attractive to retail trade than the first, bad weather carried over from the day before being most likely to affect the earlier hours.

Some of these rain-related events as they apply to retail sales can be seen in Fig. III.7, which shows for 51 business days in a large New York department store, the daily sales of umbrellas and women's raincoats. Linden notes that 14 of the 51 business days examined had rain some time during store hours, and such days accounted for about half of all the umbrellas sold during the four months studied. On a particularly wet day the sales of umbrellas were three times as heavy as they were on good days. For women's raincoats, rain-weather sales make up 40% of all volume, a wet day being twice as busy as a dry one. Also shown in Fig. III.7 are the days on which there were raincoat and umbrella promotions, and Linden, in commenting on this, stated that although the facts are admittedly scanty, bad weather appears to do a considerably better promotional job in this business than man-made media.

A question of considerable interest to economists and others concerned with the overall sale of goods, rather than the day-by-day or week-by-week sales, relates to the permanency of lost sales. That is, how permanent is the loss of business experienced on bad shopping weather days, or 'is the sale that is lost today, gained tomorrow?' According to Linden, the evidence suggests that the specific item not bought is not always purchased when better shopping weather comes along. Perhaps even more important is the fact that bad shopping weather means fewer trips to the store, a factor which is usually connected with lower total buying. That is, a single exposure of the buying impulse, is usually less rewarding –

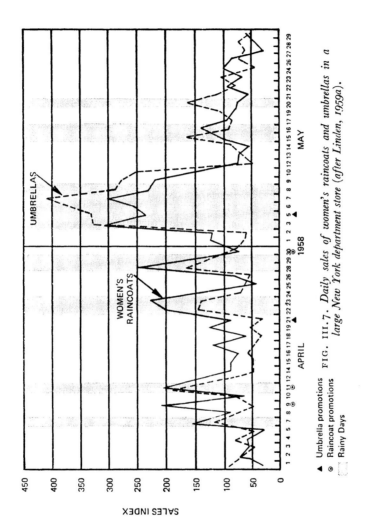

FIG. III.7. *Daily sales of women's raincoats and umbrellas in a large New York department store (after Linden, 1959a).*

▲ Umbrella promotions
◉ Raincoat promotions
▦ Rainy Days

from a retailer's point of view – than two exposures. As an example, Linden quotes the case of the near record losses sustained by New York stores in the mid-February 1958 blizzard, losses which failed to result in any detectable swell in buying when the snows melted and the sun returned. The Federal Reserve index for the crippled week, for example, was off 25% compared to expectations for the period, but in the week following the storm it merely returned to normal, although the weather was fair and Easter was early.

The application of the known weather retail sale relationships could be a profitable venture for many businessmen in the retail trade. It is of course quite conceivable that many companies are now using these relationships in their day-to-day and week-to-week activities; it is nevertheless reasonably clear that much more could be gained by a better appreciation of the atmospheric influences at work in the retail trade. For example, Linden (1959b) notes that a rough check of department store lineage in three leading New York newspapers indicates that there had been (at least in 1958/59) little co-ordination of promotional activity with the weather. In fact, there was more advertising in the record warm month of September 1957, than in the historically cool September 1956, and as already pointed out, retail sales in cool Septembers are usually well above average.

Much more research is of course required in the field of weather and retail trade before any real conclusions can be reached that will affect all aspects of retail trade. However, some work is proceeding in this area, analyses of climatic variations and retail trade in Canada (Maunder, 1968c) and the U.S. being considered in Chapter V, as an example of a national application of econoclimatic data. The U.S. Bureau of the Census has also shown some interest, and some of the early exploratory work of the Bureau is detailed in a paper by Shor (1964). In this analysis monthly estimates of the irregular component of the seasonally adjusted retail sales series were developed for the eight-year period May 1953 to April 1961 for total retail trade and for several kind-of-business categories. The data applied to the United States as a whole and they were compared with individual monthly 'averages' of selected weather factors developed for the same period.

Simple and multiple correlation analyses were carried out, and Table III.17 gives a selection of the results obtained from the multiple regression analysis. The data given applies to 'all retail stores' and, as can be seen, some of the multiple correlations, but by no means all, were significant at the 5% level. Analysis for the various categories of retail trade were also considered, and Shor in commenting on the results noted that in the lumber group a 'high' correlation between weather and retail sales appears to occur in February and March and to a lesser extent in September and December. Similarly, in several fall and winter months 'high' correlations occurred between weather and gasoline and drugstore sales.

The difficulties in using national retail trade data and national weather averages is discussed by Shor, who interestingly enough came up against the same problems as a study done in Canada (Maunder, 1968c) on retail trade in that country. A major problem is the inability of the available sales data to reflect the effect of weather with any precision.

The use of weather information by retail trade managers, and the extent that they appreciate the weather–sales relationship that undoubtedly exists, has not been subject to any large-scale study. Johnson (1965), however, discusses the results of a questionnaire designed to gain some idea of the use of the facilities of the Australian Bureau of Meteorology by the retail trade in the state of Victoria. Included in the replies were that short-range and medium-range forecasts were required for display changes, impulse selling, timing of advertisements, opening of seasonal promotions such as mowers and hoses, and the gauging of stocks of perishables. The long-range forecasts, when used, were for the purpose of buying and advertisement planning. The goods affected included rainwear, clothing, blankets, footwear, bed socks, hot-water bottles, sports wear, draught-stoppers, food, swim wear, underwear, drink chillers, working clothing, hoses, sprinklers, paint, door mats, plastic carpet covering, fans, air conditioners and hats. However, only six of the respondents stated that they received forecasts from the Bureau direct, the remainder obtaining them from newspapers, TV or radio. In commenting on the replies, Johnson says that even though the sample was small and confined mainly to larger stores, the evidence suggested that retailers do not use the Bureau's services as much as might be expected bearing in mind the important effect of weather on retail merchandising. On the other hand, the positive evidence from the questionnaire conveyed the impression that those who use the Bureau's services were reasonably well satisfied with them.

From the studies reviewed it is reasonably clear that weather variations are at least one cause of fluctuations in retail trade sales.

TABLE III.17: *Multiple correlation of selected weather data with irregular component of retail sales, by month. United States: All retail stores (abridged from Shor, 1964)*

Combinations of weather categories	Jan.	Feb.	Mar.	Apr.	May	June
(1) Average daily maximum temperature, average daily snow depth,* and, separately, each of the following 6 categories:						
Average daily total precipitation amount	·83	·94†	·87	·77	·83	·14
Percentage of days with measurable precipitation	·72	·90	·83	·82	·87†	·19
Average daily number of hours (or part of an hour) with measurable precipitation for the hours ending between:						
9.00 a.m. and 12 noon	·70	·98‡	·84	·76	·90†	·12
1.00 p.m. and 6.00 p.m.	·81	·94†	·83	·74	·93‡	·29
9.00 a.m. and 6.00 p.m.	·78	·96‡	·82	·75	·93‡	·19
9.00 a.m. and 9.00 p.m.	·77	·95†	·82	·76	·91†	·17
(2) Average daily maximum temperature, percentage of days with measurable snow depth, and, separately, each of the following 4 categories:						
Average daily total snowfall amount	·78	·97‡	·78	·74	NA	NA
Percentage of days with measurable snowfall	·73	·96†	·87	·74	NA	NA
Average daily number of hours (or part of an hour) with measurable precipitation for the hours ending between:						
9.00 a.m. and 6.00 p.m.	·84	·98‡	·78	·75	NA	NA
9.00 a.m. and 9.00 p.m.	·82	·97‡	·79	·76	NA	NA

* 'Average daily snow depth' omitted for May and June.
† Significant under assumption of normality, at 5% level.
‡ Significant under assumption of normality, at 1% level.
  NA: Not calculated.

Much more research is needed, but it appears that the cost of doing this research and the cost of using the weather information and weather forecasts would be far outweighed by the benefits, for as noted in the following extract from *The Economist* (July 2, 1960), weather does mean money to many businesses.

> Shoppers who spent more in April than ever before became harder to please in May and sales, seasonally adjusted, fell from $18·9 billion to $18·4 billion. This may mean no more than that a late Easter robbed Mother's Day of some of its usual tribute; certainly

the unseasonably cold weather made many people postpone their buying of the summer clothing, the air conditioners, the fans, outdoor furniture and other hot-weather goods which in recent years has kept the cash registers jingling through the dog days. . . .

c. Advertising and the weather

Apart from their value as amiable chit-chat, comments about the weather represent a marketing tool. At present, nearly 30 private consultant companies are keeping sharp weather eyes out for advertisers whose products and services are weather-sensitive.

So stated a report by Danzig (1965) published in the July 5, 1965, issue of *Advertising Age*, and it pinpoints the fact that advertising is, or at least should be, geared to 'favourable' weather forecasts. One interesting aspect of this is the use that some companies make of long-range weather forecasts as a mailing piece to their customers. For example, Danzig quotes the case of Weather Trends, Inc., a leading company in long-range forecasting and weather marketing research, which sends out more than 200,000 pieces of weather-orientated mail each month, on behalf of specific customers. The Industrial Division of Corn Products, for example, has for a number of years sent weather forecasts, prepared by Weather Trends, to its customers, notably canners, food nutritionists, candy manufacturers and soft-drink bottlers.

Direct advertising is also very weather-sensitive, and Wulp (1957), writing in *Printers Ink*, quotes the case of a leading manufacturer of suntan lotion in the United States. Apparently, the J. B. Williams company, manufacturer of Skol, committed itself to a heavy and rigid schedule for magazines and TV, in the summer of 1956, $15,000 being spent for a spread in *Life* that appeared just before Memorial Day (May 30) weekend, and $63,000 went for a TV spectacular on June 17. Unfortunately, however, the weather throughout the summer was so bad that there was no great need for people to buy suntan oil. Thus, in this case, if the company had known that the summer was going to be poor, or more importantly had known in what places the weather was expected to be especially poor, it could have focused its advertising at a regional or local level rather than at a national level. Several problems do of course arise in gauging the impact of advertising, and in some cases the advertising space or time must be booked several weeks, or even months, in advance. Nevertheless, since the right thing at the right time is so important in advertising, it is perhaps not inappropriate to suggest that a greater attention be given by advertisers to the value of the long-range weather forecasts, which will undoubtedly become more accurate and possibly more detailed in the decade of the 1970's.

A final comment could be made with regard to the use of climatic data rather than weather forecasts to aid in an advertising campaign. An interesting case, discussed by Landsberg (1960), is the timing of advertisements for anti-freeze in the area of Worcester, Massachusetts. In this case, climatological data were available which showed the average first date of occurrence of specific temperatures (32°, 28°, 24°, etc.) and the standard deviation of the dates (Table III.18). Accordingly, the mean first freeze date is October 7, with 67% (± one standard deviation) of all cases during the past record being between September 25 and October 19. Similarly, 67% of all cases of the first 28° temperature occurred between October 5 and October 27. Thus, if the first occurrence of a 28° temperature is a reliable guide for the buying of anti-freeze, it follows that the first week of October might be the most advisable period to use for an advertising campaign for this product in Worcester.

TABLE III.18: *Mean first dates of various low autumn temperatures at Worcester, Mass. (after Landsberg, 1960)*

Temperatures, °F	32	28	24	20	16
Date	Oct. 7	Oct. 16	Nov. 1	Nov. 15	Nov. 26
Standard deviation, days	12	11	16	15	10
Range of dates*	Sept. 25/ Oct. 19	Oct. 5/ Oct. 27	Oct. 16/ Nov. 17	Nov. 1/ Nov. 30	Nov. 6/ Dec. 8

* ± One standard deviation of mean date.

d. Weather and the commodities market

As is well known, prices of stocks on the New York and other stock exchanges are a reflection of many things, including the 'state of the economy', what buyers think of the market, the bank rate, unemployment and, in addition, what prominent investors, politicians and bankers say or do not say. Accordingly, just like the weather, stock prices vary considerably, and also just like the weather, they vary because of a complex set of interrelated factors. Few stock market experts would of course consider that there was any real link between weather variations and the prices on the stock exchange, and it is not the intention of this author to suggest that they are wrong in their thinking. However, in another kind of stock market – the commodities market – prices do vary with the weather, and in this case it is not only the actual weather that is important but also the weather that is reported to have occurred (which may not necessarily be strictly accurate), and the weather that is expected to occur.

The commodities market is the exchange where commodities such as wheat, corn, soybeans, cattle, eggs, potatoes, silver, sugar, cocoa and frozen orange juice are bought and sold; not directly, it should be added, but rather the rights to buy (and hence sell) a certain quantity of these goods for delivery at a stated time in the future.

The prices for the commodity 'futures' are related to a number of things, but possibly of most importance is the anticipated demand for, and the anticipated supply of, the particular commodity at some future date. Consequently, it can be seen that there is at times a very close relationship between the weather conditions and the price at which 'futures' in a commodity will be sold. Moreover, as already pointed out, it is not only the actual weather or report of the actual weather that is important but also forecasts of the expected weather conditions in the various producing areas – wheat, or cattle, or soybeans as the case may be – that influences the price which buyers of the commodities are willing to pay. Information on the commodities market is published daily in many newspapers, and Table III.19 shows a selection of the prices for 'futures' in wheat, corn, broilers and cocoa on the U.S. Commodity Market for July 7, 1969. As can be seen, the prices on the day quoted all rose, and the following extracts from *The Wall Street Journal* for July 8, 1969, indicates some of the reasons for these changes.

Actual weather, reports of weather and forecasts of weather, therefore, have a real 'value' as far as the commodity markets are concerned, and consequently in many cases the prices that the consumer will pay for the commodities. Unfortunately, however, little research, outside the commodity dealers, has been done on the specific relationships between weather, weather information, weather forecasts, and 'futures' prices. It nevertheless appears to be a most worthwhile field for study, not, it should be added, for speculation purposes, but more importantly for the insight it would give into how weather information and weather forecasts are valued. Moreover, an historical study of weather forecasts, and 'futures' prices, may well provide some key as to how and in what way weather forecasts – particularly as they apply to agricultural commodities – have increased in value through time.

TABLE III.19: *Futures prices: Monday, July 7, 1969 (abridged from The Wall Street Journal)*

	High	Low	Close	Change	Season's High	Season's Low
Chicago – Wheat*						
July	$128\frac{1}{4}$	127	128-$128\frac{1}{8}$	$+1\frac{7}{8}$ to 2	$142\frac{1}{8}$	$124\frac{1}{2}$
Dec.	$136\frac{7}{8}$	$135\frac{7}{8}$	$136\frac{5}{8}$-$\frac{3}{4}$	$+1\frac{3}{8}$ to $1\frac{1}{2}$	$143\frac{7}{8}$	$133\frac{3}{8}$
May '70	$141\frac{1}{4}$	$140\frac{1}{4}$	$141\frac{1}{4}$	$+1\frac{3}{8}$	$146\frac{1}{4}$	$138\frac{1}{4}$
Corn*						
July	$128\frac{1}{2}$	$126\frac{1}{2}$	$128\frac{3}{8}$-$\frac{1}{2}$	$+2$ to $2\frac{1}{8}$	$131\frac{1}{8}$	$112\frac{1}{4}$
Dec.	$127\frac{1}{2}$	125	$127\frac{3}{8}$-$\frac{1}{2}$	$+2\frac{3}{4}$ to $2\frac{7}{8}$	$128\frac{3}{8}$	$112\frac{3}{4}$
May '70	136	$133\frac{1}{2}$	136-$135\frac{7}{8}$	$+3\frac{1}{8}$ to 3	136	129
Broilers* – Iced (Chicago Board of Trade)						
July	34·65	33·75	34·27	+ ·52	34·65	25·90
Nov.	26·15	25·62	25·90	+ ·20	26·15	23·75
May '70	28·05	27·80	28·02-·05	+ ·12 to ·15	28·15	26·75
Cocoa*						
July	43·10	42·75	42·85	+1·00	47·99	27·20
Dec.	40·65	40·50	40·55	+ ·90	45·30	28·10
May '70	39·47	39·43	39·47	+1·00	43·80	34·65

* Cents per bushel for wheat, corn; cents per pound for broilers; cents per pound for cocoa.

Cocoa, grain, broiler, egg and silver futures markets rose sharply yesterday.

The Dow-Jones commodity futures index advanced 1·08 to 140·71, the highest since Feb. 3. It was the sharpest gain since Aug. 8, 1968. . . .

On the Chicago Board of Trade, corn prices pushed ahead $3\frac{1}{4}$ cents a bushel. The wheat market had a gain of two cents and price improvement for oats, rye, and soybeans was more than one cent. . . .

Declining world stocks and uncertain new crop prospects stimulated demand for cocoa contracts. Traders received reports that recent heavy rains made flowers fall off cocoa trees in Brazil, and this is expected to affect the cocoa crop harvest starting next October. . . .

Wet weather was also a consideration for buying grain contracts in U.S. and Canada.

Rains halted harvest of wheat in the U.S. Southwest and Midwest and in the Ohio Valley. The moisture and cool weather have slowed progress of new corn and soybean crops.

Reserve stocks of feed grains, especially corn, are low and traders believe that if new U.S. crop corn production should be low, larger than normal stocks of wheat would be fed to livestock. Others thought that if the wet weather should reduce quality of new wheat, the grain would be diverted to livestock feed.

e. Weather and insurance

The weather is a very important factor in many areas of the insurance industry. Hendrick and Friedman (1966) state that in the United States property valued at more than $500 billion is insured against storm damages (that is about $2,500 per person), and that the annual paid claims range from $0·5 to $1·0 billion. With this much money involved it is evident that weather and insurance are closely associated; it follows then that meteorologists can assist insurance companies in establishing the degree of risk, and hence the premiums to be levied, together with expert advice in claim settling. For example, the probability of individual loss as well as multiple loss must be considered as well as expert evaluation of the severity of weather. To the underwriters of an insurance company, the meteorologist can therefore be of great help. He is influential in the acceptance, rejection or re-insurance of risks, he influences the type and adequacy of the spread of risks, he makes possible extended coverage, and in addition the meteorologist's research helps to explain the reasons for and causes of losses.

Many kinds of weather-related insurance are written each year, but possibly of most direct concern to the policyholder is crop insurance, since crop failure through adverse weather conditions is an ever-present risk in most kinds of farming in many areas of the world. Fortunately, because of improvements in technology, the risks of crop failure attributable to natural hazards such as drought, disease or insects are smaller now than a decade ago, but the fact remains that when failure does occur today it is usually much greater because of the increase in costs per acre of producing crops. Many natural risks are of course insurable, and two general types of crop protection are often available to farmers. The first is crop–hail insurance, which is offered in the United States by both stock and mutual companies, and the second is all-risk crop insurance, which is offered in the United States by the Federal Crop Insurance Corporation (F.C.I.C.), an agency of the U.S. Department of Agriculture.

Crop–hail insurance is the predominant type of insurance purchased by farmers in the United States, and in 1968 farmers purchased about $2·9 billion of such coverage (Table III.20), a substantial increase over the $1·0 billion in force in 1950. Corn and soybean crops account for about 40% of the coverage in the

United States. Premiums on these policies amounted to nearly $110 million in 1964, while losses paid to farmers totalled about $67 million.

TABLE III.20: *Crop insurance in the United States (after Federal Reserve Bank of Chicago, Business Conditions, May 1966)*

	1950	1955	1960 (million dollars)	1962	1964
Private					
Coverage	1,057	2,067	2,495	2,651	2,918
Premium	40	77	103	108	110
Indemnity	17	45	59	81	67
Federal					
Coverage	240	309	266	357	543
Premium	14	22	18	22	34
Indemnity	13	26	16	24	30

In contrast, the premium/indemnity ratio is much less with the F.C.I.C.-insured crops, but it should be noted that the U.S. Congress appropriates a sufficient amount of funds each year to cover the administrative and operating expenses of the F.C.I.C.

Weather insurance is of course not only confined to the United States, as the following comments by *The Economist* (March 2, 1963) indicate:

> The insurance companies have their fingers crossed; they are hoping the outrageous winter weather is over. Already the companies are looking back on the most expensive winter on record. The British Insurance Association estimates that total claims resulting from the snow storms and bitter cold will exceed £15 million.
>
> . . . in Liverpool claims are up by 50 percent – half on account of storm damage and half resulting from water damage following burst pipes. Claims in Manchester are up by 25 to 40 percent while in Yorkshire an increase of 50 to 100 percent is reported and companies are taking on extra staff to cope with the backlog of paper work.

A further report on losses in weather insurance was reported in the July 13, 1963, issue of *The Economist*, which states that British insurers expect to pay over £20 million for weather damage claims as a result of the 1963 winter, this coming on top of the heavy claims of 1962 when windstorm damage in the north country cost £10 million. They comment that some damage could have been avoided if policyholders had taken reasonable precautions. Nevertheless, although the British Insurance Association issues a guide to house-holders on commonsense precautions, pipes still remain unlagged and snow is left piled up in lofts after blizzards, so that ceilings are damaged when it thaws.

An interesting associated feature of weather insurance is posed by Hendrick and Friedman (1966), who comment that weather insurance and weather modification are contrasting methods of adjustment between man and his atmospheric environment. That is, if man cannot control the weather, he insures against it, in which case he takes his losses but spreads the cost among the many who run a similar degree of risk. However, Hendrick and Friedman suggest that should man gain control over those weather effects that damage his home and property, he would probably prefer to replace in in-surance against nature with control over nature, provided the costs of the controls proved to be less than the costs of the damages that would be prevented, and provided also that adverse side effects such as droughts or the loss of aesthetic values of storms did not negate the benefits. Thus, in theory, a potential impact of weather control on the property insurance industry could be the eventual elimination of weather insurance because the insurance would ultimately cease to exist.

As of today, however, man has still a long way to go before any really significant weather control is possible, and hence the estimated range of annual insurance property losses in the United States, as given in Table III.21, is relevant at least in the foreseeable future. The degree of weather control necessary for any substantial reduc-tion in insurance claims is, however, of interest, and Hendrick and Friedman consider the case of hurricane control, where a 10% change of intensity, frequency and path are all possibilities in the 1970's. They note, for example, that a 15% reduction in hurricane intensity would reduce property damage by 41% as a result of the effect of fewer claims and reduced amounts per claim. Also considered is the frequency and path of hurricanes, and on the assumption that all three types of hurricane and tropical storm control are realizable, and that each is independent of the others, the storm damage re-duction possibilities become quite substantial. The storm damage reduction resulting from the control of tornadoes, thunderstorms, hail and high winds is also assessed but a smaller loss reduction is suggested. For example, if it is assumed that 10% of potential tornado clouds could be seeded so as to effectively modify the cloud dynamics that relate to tornado formation, tornado destruction might be reduced by 15%. Similarly, a total reduction of 20% in thunderstorm wind/hail losses as a result of both frequency and intensity modifications is suggested as being possible.

TABLE III.21: *Estimated range of annual insured property losses by storm type (after Hendrick and Friedman, 1966)*

Storm type	Range of annual insured property losses (millions of dollars)
Hurricanes	250– 500
Tornadoes	100– 200
Hail/wind thunderstorms	125– 250
Extra-tropical windstorms	25– 50
Total	500–1000

The reduced losses resulting from implementation of all controls postulated by Hendrick and Friedman, are indicated to be 36% (Table III.22), an impressive saving considering the minor degrees of assumed control. As the authors point out, however, even these minor degrees of control may prove unrealistic as understanding of the atmosphere and required control technologies become clearer.

TABLE III.22: *Effects of assumed weather controls on storms in the United States (after Hendrick and Friedman, 1966)*

Storm type	Current range of annual losses ($ millions)	Reduction in losses from storm controls	Annual range of insured losses with storm controls ($ millions)
Hurricanes	250– 500	57%	108–216
Tornadoes	100– 200	15%	85–170
Hail/wind thunderstorms	125– 250	20%	100–200
Extra-tropical cyclones	25– 50	0%	25– 50
Totals	500–1000	36%	318–636
Potential total savings (based on current losses)			182–364

The ability to modify damage-producing storms, as well as the ability to forecast more accurately their occurrence, especially one year ahead, would of course have significant implications to the insurance industry. For example, an improved tornado warning system would undoubtedly help to lessen the chances of severe damage and presumably subsequent claims. Of possibly much greater importance, however, is the relationship between the government (who presumably would control weather modification) and the insurance industry, and as pointed out in Chapter VII the political, legal and social problems of weather modification are many and complex. In fact, it may even be suggested that it would be better to leave the atmosphere alone, at least as far as modification of it is concerned, for to quote Hendrick and Friedman:

> . . . a much more comprehensive view must be taken of the atmosphere as it contributes to the total quality of man's environment and its role in his ecosystem. It is not clear what these damaging elements – hail, wind, lightning – in their natural environmental settings contribute to the total human experience. For instance, how can proper account be taken of natural weather scenery, the stimulation, the unique experience of living with weather in its more robust forms? Does the natural wilderness of weather provide some essential environmental uncertainty? The impacts of natural weather are in need of further diagnosis in the context of man and his total environment. Man may instinctively prefer, and indeed need, the occasional uncontrollable and unpredictable violent and sublime beauty of the storm.

f. Weather and the tourist industry

It is generally believed that the tourist industry is highly sensitive to weather conditions. It is not known in any detail, however, as to what extent any particular tourist region is affected by the weather in that region, by the weather in another tourist region or by the weather in the tourists' areas of origin. Neither is it generally known as to what extent tourists are influenced by short- or long-range weather forecasts.

In discussing the Australian holiday-maker, Tucker (1965) suggested that there are several avenues for promoting meteorological services which might be explored. For example, he notes that only hours separate most Australians from the major resort areas, with the result that the travel industry is finding an upsurge in 'impulse' travel. That is, a growing number of people are able, with a week's or two weeks' notice, or even less, to go north in search of the sun, or south from the hotter climates. The exodus from the southern states of Australia to the Queensland 'Gold Coast' in winter, for example, is well known, and Tucker suggests that extended forecasts for resort areas such as the Great Barrier Reef, Central Australia and the south coast of New South Wales could do much to stimulate interest in travel to wider areas of Australia.

Similar observations could be made for many other areas in the world. For example, mention should be made of the importance of seasonal variations that highlight the value of the 'right' kind of weather to the tourist industry of Florida, of Switzerland, of Rio de Janeiro or of Central Australia. Moreover, few hotel owners in Miami Beach need to be reminded of what an unseasonably cool winter can do to their trade, and a study of the travel and hotel section of the Sunday edition of the *New York Times* in January, when conditions in Florida have been unseasonably cool, is at least an index of how closely the tourist trade is tied to the weather conditions. On the other hand, that same situation would no doubt offer many bargains for the tourist who is not concerned with the fact that Miami Beach was having its coldest January on record. In fact, as Tucker points out, meteorology is not alone with its highs and lows, and one of the hard facts of the tourist industry is that it has pronounced peaks during a year.

Regrettably, little specific information of exactly how weather affects the tourist industry is readily available, and little, if any, research on the topic has been published in the professional journals. A preliminary study was therefore made by Maunder and Halkett (1969b) in order to answer some weather-related questions as they apply to the tourist industry in Canada. The study included a series of interviews with federal, provincial and municipal government officials involved in the tourist industry, a search of the literature on the tourist industry, an examination of camp-site occupance data for Banff National Park in relation to weather data, and a series of interviews with tourists in Victoria, British Columbia.

The general consensus of the government officials interviewed was that the weather has a very definite influence on the tourist industry. However, although some of those interviewed could state specific examples of the effect of weather on the industry, none of these examples were substantiated by statistical evidence. For example, most of those interviewed said that poor weather discouraged tourists but none was able to give any indication of how long the 'bad' weather had to last before tourists left an area, or to what extent tourists sought indoor alternatives to outdoor activities during periods of 'bad' weather.

A rapidly growing literature on tourism and tourist industries was consulted, but little if any of the literature reflected any serious consideration of meteorological or climatic factors, most of the literature being concerned with economics and planning, or with market and advertising analysis. Nevertheless, it is reasonably clear that the effects of weather and climate should be primary considerations in economics and planning for the tourist industry, and it is possible that the tourist weather literature does not reflect this because much of the basic research has not yet been done. The third approach to the question of weather and the tourist industry in Canada involved a study of the occupancy of the Lake Louise camp-site in Banff National Park. Many difficulties were encountered, however, one factor being that it is not known what non-meteorological factors may have prompted visitors to enter or leave the camp-site. Another unknown factor, and probably the most important, is that it is not known where the people who left the camp-site (possibly as a result of poor weather) went. For example, did they abandon their holiday and go home; did they move to a large urban centre; or, did they move to a hotel or motel to await better weather? The answers to these questions are important, for it is quite possible that a day of rain at Lake Louise provides economic benefits for the urban centre of Banff when campers move into the town to shop, eat in restaurants and seek indoor entertainment.

Finally, as part of a larger study of tourism in Victoria, British Columbia, a series of 32 interviews was conducted with a random sample of tourists. Part of the questionnaire dealt with the effects of weather, and the tourists were asked what the weather had been like during their stay (that is, up to the time of the interview), whether it had caused them to change their plans, and whether it would influence their decision to return to the Victoria region. Three conclusions were drawn from the results of these questions: first, that if tourists are forced by the weather to change their plans, they often only postpone a planned outdoor activity, carrying on with it later, regardless of the weather; second, that the effect of the weather on the plans of a party is a function of the weather expected by the members of the party, and the weather to which they are accustomed; and third, that the weather has little effect on the decision to return to the region although poor weather sometimes affects the decision as to the season of the next visit.

The organization of any tourist industry is complex, a factor of importance being that most of the money spent by tourists is taken by the private sector, but that advertising, promotion and market research is undertaken (at least in Canada) largely by government agencies. Moreover, although the overall value of the tourist industry to Canada is very considerable, the degree of importance of the weather to the overall tourist dollar is not known. Such information would be of great value to the industry. For example, the meteorologist in certain areas could tailor his forecast for the tourist so that the tourist could re-direct his activity or re-plan his trip, the economic climatologist should be in a position to assist with the planning of tourist areas and facilities, and probably most important, those in the tourist industry, on the basis of detailed knowledge of tourist reaction to various weather phenomena could plan, advertise and promote alternatives for the tourist.

g. Weather forecasting, weather consulting and business

As has already been indicated, the use of weather forecasting services in business generally depends on two factors: first, the interest that the businessman has in weather conditions, which is usually related to the weather-sensitivity of his business; and second, the interest that the weather forecasters show in business activities. These ideas are explored further in Chapter VI, but a few comments are pertinent at this time.

The U.S. Weather Bureau, for example, according to a report by Shoenfeld (1958) entitled 'Can Weather Bureau Forecasts Help Solve Your Sales Problems?' has enough information to excite almost any sales manager or analyst, but few ask for it. But some companies do ask for it, and Shoenfeld quotes the example of a fairly large air-conditioning company which uses (or at least did use in 1958) many of the services of the Weather Bureau including the 30-day, five-day and daily weather forecasts. For example, with air conditioners people apparently do not buy units on the first, second or even third torrid day, but only after they have endured a long period of weather with temperatures higher than those to which they are accustomed. Thus, where others try to spot rising population, money or new outlets, the sales manager of this particular air-conditioning company searches the country for the 'right' stretches of heat.

The government is, of course, not the only dispenser of weather forecasts, and perhaps the most successful applications of weather forecasts to business have come about, particularly in the United States, through direct contacts between businessmen and consulting meteorologists. Effective use of weather information in business requires a thorough understanding of the business and also of how weather affects the operation of that business. The specialized and individual tailored weather services necessary for many businesses are therefore not readily available through the public weather forecast. However, in the United States, and to a limited extent in some other countries, private meteorological consulting firms have been established to fulfil the many specialized business needs. A considerable amount of information on weather consultants is available from the American Meteorological Society, which has a certification programme for consulting meteorologists. In addition, many of the certified consulting meteorologists indicate their availability and services in the monthly issues of the *Bulletin* of the American Meteorological Society. In the April 1969 issue of the *Bulletin*, for instance, 49 consulting meteorological companies or individuals are listed, and the following brief extracts indicate the wide interests and services offered:

*Specializing in Investigations for Law Firms & Insurance Companies.
*Aviation . . . Legal & Insurance Investigations. Specializing in Weather Investigations for Aviation Insurance Underwriters.
*Specializing in Programming Jet Minimum Cost Cruise Control Flight Planning For Air Carriers and Business Jets.
*Effective Long and Short Range Specialized Forecasts, and Consulting for Business and Industry.
*Research in Atmospheric Physics, Research Equipment, Instrumentation & Systems Turbulence, Diffusion, Air Pollution Surveys, Field Investigations.
*Weather & Forestry Instruments & Supplies; Factory Representative – Automatic Recording, Telemetering & Data Processing Instruments.
*Municipal, Industrial and Aviation Forecasts; Atmospheric and Climatological Research Utilizing Computer Facilities.
*Weather Modification, Forecasting, Research, Air Pollution Surveys, Field Study Programs and Instrumentation.
*Specializing in Long-Range Forecasts, Weather/Marketing Research & Promotional Weather Services in North America and Europe.

Weather forecasting in the world of business is not necessarily restricted to the work of weather consultants; indeed over the world it is highly probable that the greatest number of weather-related

G

decisions are made in response to the official government weather forecast issued via radio, television or the newspaper. But whatever the source and whatever the cost of weather information and forecasts, it is clear that economically valuable weather-related decisions will only be made through the closest co-operation of the weatherman and the businessman. As one consulting firm of meteorologists* puts it:

> You too can now increase your profits through weather planning. Join the experts. . . .

* 'Weather Consultants of Canada' in a brochure outlining their services.

BIBLIOGRAPHY

ALLEN, L., 1960: Inside ship weather routing. *Marine News*. (March.)

ANONYMOUS, 1969: An economic survey of drought affected properties – New South Wales and Queensland 1964–65 to 1965–66. *Wool Economic Research Report*, No. 15, Bureau of Agricultural Economics, Canberra, 51 pp.

ARAKAWA, H., 1959: Hydroelectric power generation and the climate of Japan – a case of engineering meteorology. *Bull. Amer. Met. Soc.*, 40: 416–22.

ARKLEY, R. J. and ULRICH, R., 1962: The use of calculated actual and potential evapotranspiration for estimating potential plant growth. *Hilgardia*, 32: 443–62.

ARMY, T. J., BOND, J. J. and VAN DOREN, C. E., 1959: Precipitation yield relationship in dryland wheat production on medium to fine textured soils on the Southern High Plains. *Agron. Jour.*, 51: 721–4.

ARMY, T. J. and HANSON, W. D., 1960: Moisture and temperature influences on spring wheat production in the plains area of Montana. *U.S. Dept. Agric. Prod. Res. Rep.*, 35.

ARONIN, J. E., 1953: *Climate and Architecture.* Reinhold, New York, 304 pp.

ASANA, R. D. and WILLIAMS, R. F., 1965: The effect of temperature stress on grain development in wheat. *Aust. Jour. Agric. Res.*, 16: 1–13.

BARROWS, J. S., 1951: *Forest Fires in the Northern Rocky Mountains. Northern Rocky Mountain Forest and Range Expt. Sta.*, Station Paper 29.

BARROWS, J. S., 1966: Weather modification and the prevention of lightning-caused forest fires. In: SEWELL, W. R. D., ed., 1966: *Human Dimensions of Weather Modification.* University of Chicago, Dept. of Geography, Research Paper No. 105, pp. 169–82.

BARRY, W. S., 1965: *Airline Management.* George Allen & Unwin Ltd, London, 352 pp.

BASILE, R. M., 1954: Drought in relation to corn yield in the northwestern corner of the corn belt. *Agron. Jour.*, 46: 4–7.

BEAN, L. H., 1967: Crops weather and the agricultural revolution. *Amer. Stat.*, 21(3): 10–14.

BECKWITH, W. B., 1966: Impacts of weather on the airline industry: the value of fog dispersal programs. In: SEWELL, W. R. D., ed., 1966:

Human Dimensions of Weather Modification. University of Chicago, Dept. of Geography, Research Paper No. 105, pp. 195–207.

BEEBE, R. G., 1967: The construction industry as related to weather. *Bull. Amer. Met. Soc.*, 48: 409.

BELL, F. H. and PRUTER, A. T., 1958: Climatic temperature changes and commercial yields of some marine fisheries. *Jour. Fish. Res. Board Canad.*, 15: 625–83.

BENOIT, G. R., HATFIELD, A. L. and RAGLAND, J. L., 1965: The growth and yield of Corn. III. Soil moisture and temperature effects. *Agron. Jour.*, 57: 223–6.

BICKERT, C. VON E. and BROWNE, T. D., 1966: Perception of the effect of weather on manufacturing: A study of five firms. In: SEWELL, W. R. D., ed., 1966: *Human Dimensions of Weather Modification.* University of Chicago, Dept. of Geography, Research Paper No. 105, pp. 307–22.

BLAIR, I. D., 1959: Loss and damage in potatoes. *Lincoln College (N.Z.) Tech. Pub.*, 18.

BOGGESS, W. R., 1956: Amount of throughfall and stemflow in a short leaf pine plantation as related to rainfall in the open. *Trans. Illinois Acad. Sci.*, 48: 55–61.

BOURKE, P. A. M., 1963: Agricultural biometeorology. *Inter. Jour. Biomet.*, 7: 121–5.

BOYER, A., 1966: Expanding industrial meteorology. *Bull. Amer. Met. Soc.*, 47: 528.

BRADY, F. B., 1964: All-weather aircraft landing. *Scientific American*, 210: 25–35.

BROOME, M. R., 1966: Weather forecasting and the contractor. *Weather*, 21: 406–10.

BUSHNELL, J., 1925: The relation of temperature to growth and respiration in the potato plant. *Minnesota Agric. Expt. Sta. Res. Bull.*, 34.

CAMERON, D., 1947: Weather and railway operation in Britain. *Weather*, 2: 373–80.

CANHAM, H. J. S., 1966: Economic aspects of weather routeing. *Marine Observer* 36: 195–9.

CARNEY, J. P., 1950: Drought feeding of sheep. *Quart. Rev. Agric. Econ.*, 4: 11–14.

CASADAY, L. W., 1952: *Tucson as a Location for Small Industry.* Bureau of Business Research, University of Arizona, Tucson, Special Studies No. 4, pp. 22–3.

CASTLE, E. N. and STOEVENER, H. F., 1966: The economic evaluation of weather modification with particular reference to agriculture. In: SEWELL, W. R. D., ed., 1966: *Human Dimensions of Weather Modification.* University of Chicago, Dept. of Geography, Research Paper No. 105, pp. 141–58.

CHANDLER, W. H., 1957: *Deciduous Orchards.* Lee & Febiger, Philadelphia, Pa., 492 pp. (third edition).

CHANG, JEN-HU, 1968: Progress in agricultural climatology. *Prof. Geog.*, 20: 317–20.

CHANGNON, S. A., 1960: Relations in Illinois between annual hail loss cost insurance data and climatological hail data. *Crop–Hail Insurance Actuarial Association Research Report* (Chicago), 3, 18 pp.

CHANGNON, S. A., 1966: Summary of 1965 Hail Research in Illinois. *Crop–Hail Insurance Actuarial Association Research Report* (Chicago), 30, 38 pp.

CHANGNON, S. A. and NEILL, J. C., 1968: A mesoscale study of corn–weather response on cash-grain farms. *Jour. App. Met.*, 7: 94–104.

CHANGNON, S. A. and STOUT, G. E., 1967: Crop–hail intensities in central and northwest United States. *Jour. App. Met.*, 6: 542–8.

CHILDERS, N. F., 1961: *Modern Fruit Science, Orchard and Small Fruit Culture.* Rutgers – The State University, Horticultural Publications, New Brunswick, New Jersey, 893 pp.

CLEARY, V. P., 1965: Requirements of the forest service. In: *What is Weather Worth?* Bureau of Meteorology, Melbourne, pp. 37–8.

CLOOS, G. W., 1967: The trend of business. The Federal Reserve Bank of Chicago; *Business Conditions*, April, pp. 2–6.

COFFMAN, F. A. and FREY, K. J., 1961: Influence of climate and physiological factors in growth in oats. In: *Oats and Oat Improvement*. Agron. Monog., Vol. 8, Amer. Soc. Agron., pp. 420–64.

COOP, I. E., 1955: Some animal production problems of the South Island. *Proc. N.Z. Soc. Anim. Prod.*, 15: 5–16.

COOP, I. E. and HART, D. S., 1953: Environmental factors affecting wool growth. *Proc. N.Z. Soc. Anim. Prod.*, 13: 113–19.

CORNISH, E. A., 1950a: Yield trends in the wheat belt of South Australia during 1896–1941. *Aust. Jour. Sci. Res.*, B2: 83–137.

CORNISH, E. A., 1950b: The influence of rainfall on the yield of wheat in South Australia. *Aust. Jour. Sci. Res.*, B3: 178–218.

CRAWFORD, T. V., 1964: Computing the heating requirements for frost protection. *Jour. App. Met.*, 3: 750–60.

CRITCHFIELD, H. J., 1966: *General Climatology*. Prentice-Hall, Englewood Cliffs, New Jersey, 420 pp. (second edition).

CROWDER, L. V., SELL, L. V. and PARKER, E. M., 1955: The effect of clipping, nitrogen application and weather on the productivity of fall sown oats, rye grass and crimson clover. *Agron. Jour.*, 47: 51–4.

CRUTCHFIELD, J. A., 1962: Valuation of fishery resources. *Land Econ.*, 38: 145–54.

CULLEN, W. C., 1966: Weather data worth millions to the construction industry. *American Roofer and Building Improvement Contractor*, October.

CUMBERLAND, K. B., 1962: Climate change or cultural interference. In: McCASKILL, M., ed., 1962: *Land and Livelihood: – Geographical Essays in Honour of George Jobberns*. N.Z. Geog. Soc., Christchurch, pp. 88–142.

CUMMINS, W. E., 1967: Practical results of weather routeing. *Marine Observer*, 37: 23–6.

CURRY, L., 1958: *Climate and Livestock in New Zealand – A Functional Geography*. Ph.D. Thesis, University of N.Z. (unpublished).

CURRY, L., 1962: The climatic resources of intensive grassland farming: the Waikato, New Zealand. *Geog. Rev.*, 52: 174–94.

CURRY, L., 1963: Regional variation in the seasonal programming of livestock farms in New Zealand. *Econ. Geog.*, 39: 95–117.

CURRY, L., 1966: Seasonal programming and Baysian assessment of atmospheric resources. In: SEWELL, W. R. D., ed., 1966: *Human Dimensions of Weather Modification.* University of Chicago, Dept. of Geography, Research Paper No. 105, pp. 127–40.

DANZIG, F., 1965: Weather Trends find advertisers need its highly volatile commodity. *Advertising Age,* July 5.

DAVEY, A. W. F., 1962: Lessons from the drought: experience on the Massey College intensive production unit, 1961–1962. *Dairy Farming Annual,* pp. 85–94.

DAVIES, M., 1960: Grid system operation and the weather. *Weather,* 15: 18–24.

DAVIS, F. E. and PALLESON, J. E., 1940: Effect of the amount and distribution of rainfall and evaporation during the growing season on yields of corn and spring wheat. *Jour. Agric. Res.,* 60: 1–23.

DECKER, W. L., 1967: *Periods with Temperature Critical to Agriculture.* North Central Regional Research Publication No. 174, Agric. Expt. Sta., Univ. of Missouri, Columbia, Mo., 76 pp.

DEEMER, R. E., 1963: *Irregular Operations – Causes and Costs,* United Airlines Report, IE–232 (unpublished).

DINGWALL, A. R., 1953: Main crop production in Canterbury. *N.Z. Jour. Agric.,* 86: 531–44.

DOEPEL, G. C. and TURNER, H. N., 1959: Causes of variation in fleece weights. *Quart. Rev. Agric. Econ.,* 12: 50–7.

DOWSETT, C. P., 1946: Influence of rainfall on average wool clip per sheep in N.S.W. *The Farm Front* (Rural Bank), October.

DRYAR, H. A., 1949: Load dispatching and Philadelphia weather. *Bull. Amer. Met. Soc.,* 30: 159–67.

DUFAUR, R. T., 1962: *Dairy Farming for Profit.* Reed, Wellington.

EBERLY, D. L., 1966: Weather modification and the operations of an electric power utility: the Pacific Gas and Electric Company's test program. In: SEWELL, W. R. D., ed., 1966: *Human Dimensions of Weather Modification.* University of Chicago, Dept. of Geography, Research Paper No. 105, pp. 209–26.

EHLERS, W. F., 1960: Economic implications of drought probabilities for humid area irrigation. *Jour. Farm. Econ.,* 42: 1518–19.

EVANS, L. T., 1963: *Environmental Controls of Plant Growth.* Academic Press, New York, 449 pp.

EVANS, S. H., 1968: Weather routeing of ships. *Weather,* 23: 2–8.

FAHEY, B. D., 1964: Throughfall and interception of rainfall in a stand of pine. *Jour. Hydrol.* (N.Z.), 3: 17–26.

FERGUSON, K. A., CARTER, H. B. and HARDY, M. H., 1949: Studies of comparative fleece growth in sheep. *Aust. Jour. Sci. Res.,* B2: 42–81.

FILMER, J. C. F., 1960: Butterfat per acre. *N.Z. Jour. Agric.,* 101: 274–279.

FLEISHMAN, E., 1954: How outdoor temperature affects electrical system loads. *Electrical West,* June, pp. 92–6.

FLINN, J. C. and MUSGRAVE, W. F., 1967: Development and analysis of input–output relations for irrigation water. *Aust. Jour. Agric. Econ.,* 11(1): 1–19.

FRANKCOM, C. E. N., 1966: The general problem of weather routeing by the shipmaster himself or as advised by the meteorologist ashore. *Marine Observer*, 36: 186–91.

FRANKLIN, M. C., 1962: Drought. In: BARNARD, A., ed., 1962: *The Simple Fleece: Studies in the Australian Wool Industry*. Melbourne University Press, Melbourne, pp. 267–77.

FRISBY, E. M., 1951: Weather crop relationships: forecasting spring wheat yield in the northern Great Plains of the U.S. *Trans. Inst. Brit. Geog.*, 17: 77–96.

FRITTS, H. C., 1962. An approach to dendrochronology screening by means of multiple regression techniques. *Jour. Geophys. Res.*, 67: 1413–20.

FRITTS, H. C., 1965: Tree ring evidence for climatic changes in western North America. *Monthly Weather Rev.*, 93: 421–43.

FUQUAY, D. M., 1962: Mountain thunderstorms and forest fires. *Weatherwise*, 15: 149–52.

FUQUAY, D. M. and BOUGHMAN, R. G., 1962: *Project Skyfire Lightning Research*. Final Report to N.S.F., Grant NSF G—10309, Intermountain Forest and Range Expt. Sta., Ogden, Utah.

GEDDES, A. E. M., 1922: Weather and the crop yield in the northeast counties of Scotland. *Quart. Jour. Roy. Met. Soc.*, 48: 251–68.

GLOCK, W. S., 1963: Anomalous patterns in tree rings. *Endeavour*, 22(1): 9–13.

GRAINGER, J., SNEDDON, J. L., CHISHOLM, E. DE C. and HASTIE, A., 1954: Climate and the yield of cereal crops. *Quart. Jour. Roy. Met. Soc.*, 81: 108–11.

GRANT, P. M., 1960: The response of maize to different forms of phosphate under varying incidence of drought. *South African Jour. Agric. Sci.*, 4: 447–57.

GREGORY, L. E., 1954: *Some Factors Controlling Tuber Formation in the Potato Plant*. Ph.D. Dissertation, University of California, Los Angeles (unpublished).

GRIFFITHS, J. F., 1966: *Applied Climatology: An Introduction*. Oxford University Press, London, 118 pp.

GRIFFITHS, J. F. and GRIFFITHS, M. J., 1969: *A Bibliography of Weather and Architecture*. U.S. Dept. of Commerce, ESSA Technical Memorandum EDSTM 9, Silver Spring, Md., 72 pp.

GRUNDKE, G., 1955/56: Uber die Bedeutung Technoklimatischer Forschungen. *Wiss-Zeitscher. Hochsch. f. Binnenhandel*, 1(3): 19–38 (cited from Landsberg, 1960, p. 405).

HAGAN, R. M., HAISE, H. R. and EDMINSTER, J. W. (eds.), 1967: *Irrigation and Agricultural Lands*. Amer. Soc. Agron., Madison, Wisc., 1180 pp.

HALACY, D. S., JR, 1968: *The Weather Changers*. Harper & Row, New York, 246 pp.

HALLANGER, N. L., 1963: The business of weather: Its potentials and uses. *Bull. Amer. Met. Soc.*, 44: 63–7.

HARGREAVES, G. H., 1956: Irrigation requirements based on climatic data. *Proc. Amer. Soc. Engr.*, 82(IR)(1105): 1–10.

HEADLEY, J. C. and RUTTAN, V. W., 1964: Regional differences in the impact of irrigation on farm output. In: SMITH, S. C. and CASTLE,

E. N., eds., 1964: *Economics and Public Policy in Water Resources Development*. Iowa State University Press, Ames, Iowa, pp. 127–49.

HENDRICK, R. L. and FRIEDMAN, D. G., 1966: Potential impacts of storm modification on the insurance industry. In: SEWELL, W. R. D., ed., 1966: *Human Dimensions of Weather Modification*. University of Chicago, Dept. of Geography, Research Paper No. 105, pp. 227–46.

HENDRICKS, W. A. and SCHOLL, J. C. 1943: Techniques in measuring joint relationships. The joint effects of temperature and precipitation on corn yields. *North Carolina Agric. Expt. Sta. Tech. Bull.*, 74.

HENLEY, E. D., 1951: Weather and the carriage of freight by rail in Britain. *Weather*, 6: 233–6.

HEWES, L., 1958: Wheat failure in western Nebraska. *Annals Assn. Amer. Geog.*, 48: 375–97.

HEWES, L., 1965: Causes of wheat failure in the dry farming region, Central Great Plains, 1939–1957. *Econ. Geog.*, 41: 313–30.

HICKS, J. R., 1967: Improving visibility near airports during periods of fog. *Jour. App. Met.*, 6: 39–42.

HOLLOWAY, J. T., 1954: Forests and climates in the South Island of New Zealand. *Trans. Roy. Soc. N.Z.*, 82: 329–410.

HOLLOWAY, J. T., 1964: The forests and climates of the South Island: the status of the climatic change hypothesis. *N.Z. Geog.*, 20: 1–9.

HOLT, R. F., TIMMONS, D. R., VOORHEES, W. B. and VAN DOREN, C. A., 1964: Importance of stored soil moisture to the growth of corn in the dry to moist subhumid climatic zone. *Agron. Jour.*, 56: 82–5.

HOOKER, R. H., 1922: The weather and crops in Eastern England 1885–1921. *Quart. Jour. Roy. Met. Soc.*, 48: 115–38.

HOPKINS, J. W., 1935: Weather and wheat yield in western Canada. I. Influence of rainfall and temperature during the growing season plot yields. *Canad. Jour. Res.*, 12: 306–34.

HOWE, C. W. and LINAWEAVER, F. P., 1967: The impact of price on residential water demand and its relation to system design and price structure. *Water Resources Res.* 3: 13–32.

HOXMARK, G., 1928: The influence of the climatic conditions on the yield of wool in Argentina. *Monthly Weather Rev.*, 56: 60–1.

HUDSON, W. D., 1964: Reduction of unaccounted for water. *Jour. Amer. Water Works Assn.*, 56: 143–8.

HUNTINGTON, E., 1945: *Mainsprings of Civilization*. Wiley, New York, 660 pp.

HUTTON, J. B. and WRIGHT, M., 1952: A study in dairy farming trends in the Waipa County, 1941–50. *Proc. N.Z. Soc. Anim. Prod.*, 12: 76–82.

IVERSON, C. E. and CALDER, J. W., 1956: Light land pastures. *Proc. N.Z. Grass. Assn.*, 18: 78–88.

IVINS, J. D., and MILTHORPE, F. L., 1963: *The Growth of the Potato*. Butterworth, London, 327 pp.

JOHNSON, S. R., McQUIGG, J. D. and ROTHROCK, T. P., 1969: Temperature modification and costs of electric power generation. *Jour. App. Met.*, 8: 919–26.

JOHNSON, R. W. M., 1955: The aggregate supply of New Zealand farm products. *Econ. Record.* 31: 50–60.

JOHNSON, W. C., 1959: A mathematical procedure for evaluating relationships between yield climate and wheat yields. *Agron. Jour.*, 51: 635–9.

JOHNSON, W. C., 1964: Some observations on the contribution of an inch of seeding time soil moisture to wheat yield in the Great Plains. *Agron. Jour.*, 56: 29–35.

JOHNSON, W. D. N., 1965: The winds of change. In: *What is Weather Worth?* Bureau of Meteorology, Melbourne, pp. 53–60.

KEMP, W. S., 1956: Central Otago pip and stone fruit orchard. *N.Z. Jour. Agric.*, 93: 421–30.

KETCHEN, K. S., 1956: Climatic trends and fluctuations in yield of marine fisheries of the northeast Pacific. *Jour. Fish. Res. Board Canada*, 13: 357–74.

KINSMAN, K. I., 1953: Incidence of drought loss. *Quart. Rev. Agric. Econ.*, 5: 19–23.

KIRKBRIDE, J. W. and TRELOGAN, H. C., 1966: Weather and crop production: some implications for weather modification programs. In: SEWELL, W. R. D., ed., 1966: *Human Dimensions of Weather Modification.* University of Chicago, Dept. of Geography, Research Paper No. 105, pp. 159–68.

KOLB, L. L. and RAPP, R. R., 1962: The utility of weather forecasts to the raisin industry. *Jour. App. Met.*, 1: 8–12.

KREITZBERG, C. W., 1967: Mesoscale temperature changes in the supersonic transport climb region. *Jour. App. Met.*, 6: 905–10.

LANCASHIRE, J. A. and KEOGH, R. G., 1964: Grazing management and facial eczema. *Proc. N.Z. Grass Assn.*, 26: 153–63.

LANDSBERG, H. E., 1960: *Physical Climatology.* Gray Printing Co., Inc., DuBois, Penn., 446 pp.

LANDSBERG, H. E., 1967: Climate, Man and some world problems. *Scientia*, 102: 197–206.

LANDSBERG, H. E., 1968: A comment on land utilization with reference to weather factors. *Agric. Met.*, 5: 135–7.

LAVE, L. B., 1963: The value of better weather information to the raisin industry. *Econometrica*, 31(1–2): 151–64.

LAWSON, E. R., 1967: Throughfall and stemflow in a pine–hardwood stand in the Ouachita Mountains of Arkansas. *Water Resources Res.*, 3: 731–5.

LAYCOCK, D. H., 1958: An empirical relationship between rainfall and tea yields in Nyasaland. *Proc. Agric. Met. Conf. East African Agric. and Forestry Org.*, Kenya, 103: 1–4.

LEMONS, H., 1942: Hail in American agriculture. *Econ. Geog.*, 18: 363–78.

LEONARD, W. H. and MARTIN, J. H., 1963: *Cereal Crops.* MacMillan Company, New York, 825 pp.

LINAWEAVER, F. P., BEEBE, J. C. and SKRIVAN, F., 1966: *Data Report of the Residential Water Use Research Project.* Department of Environmental Engineering Science, Johns Hopkins University, Baltimore, Maryland.

LINDEN, F., 1959a: Weather in business. *The Conference Board Business Record*, 16: 90–4, 101.

LINDEN, F., 1959b: Merchandising with the weather. *The Conference Board Business Record*, 16: 144–9.

LINDEN, F., 1962: Merchandising weather. *The Conference Board Business Record*, 19: 15–16.

McLEARY, J. W., 1968: Seasonal income patterns in the south. *Monthly Review*: Sixth Federal Reserve District, Nov., pp. 150–3.

McMEEKAN, C. P., 1960. *Grass to Milk*. N.Z. Dairy Exporter Newspaper Co., Wellington.

McQUIGG, J. D. and THOMPSON, R. G., 1966: Economic value of improved methods of translating weather information into operational terms. *Monthly Weather Rev.*, 94: 83–7.

McWILLIAM, J. R., 1966: The role of controlled environments in plant improvement. *Aust. Jour. Sci.*, 28: 403–7.

MALCOLM, J. P., 1947: *An Ecological Study of the Barley Crop in New Zealand*. M.Agric.Sci. Thesis, University of New Zealand (unpublished).

MANLEY, G., 1957: Climatic fluctuations and fuel requirements. *Adv. of Sci.*, 13: 324–6.

MARTIN, J. H. and LEONARD, W. H., 1949: *Principles of Field Crop Production*. MacMillan Company, New York, 1176 pp.

MAUNDER, W. J., 1963: The climates of the pastoral production areas of the world. *Proc. New Zealand Inst. Agric. Sci.*, 9: 25–40.

MAUNDER, W. J., 1965a: *The effect of Climatic Variations on Some Aspects of Agricultural Production in New Zealand, and an Assessment of their Significance in the National Agricultural Income*. Ph.D. Thesis, University of Otago, New Zealand (unpublished).

MAUNDER, W. J., 1965b: The effect of climatic variations on oat production in Southland County 1933/34–1959/60. *Proc. 4th New Zealand Geog. Conf.*, pp. 105–11.

MAUNDER, W. J., 1966a: Climatic variations and dairy production in New Zealand – A review. *New Zealand Sci. Rev.*, 24: 69–73.

MAUNDER, W. J., 1966b: Climatic variations and agricultural production in New Zealand. *New Zealand Geog.*, 22: 55–69.

MAUNDER, W. J., 1967a: Climatic variations and wool production – A New Zealand review. *New Zealand Sci. Rev.*, 25: 35–9.

MAUNDER, W. J., 1967b: Climatic variations and meat production – A New Zealand review. *New Zealand Sci. Rev.*, 25(5): 9–12.

MAUNDER, W. J., 1968a: Effect of significant climatic factors on agricultural production and incomes: A New Zealand example. *Monthly Weather Rev.*, 96: 39–46.

MAUNDER, W. J., 1968b: Agroclimatological relations: A review and a New Zealand contribution. *Canad. Geog.*, 12: 73–84.

MAUNDER, W. J., 1968c: *An Econoclimatic Model for Canada: Problems and Prospects*. Paper presented at the Conference and Workshop on Applied Climatology of the American Meteorological Society, Asheville, North Carolina, Oct.

MAUNDER, W. J., 1968d: *Rainfall Patterns in Australia 1941–65 and their Relationship to Variations in Agricultural Productivity*. (Unpublished.)

MAUNDER, W. J. and HALKETT, I. P. G., 1969a: *The Effect of Weather on the Canadian Tourist Industry*. (Unpublished report.)

MAUNDER, W. J. and HALKETT, I. P. G., 1969b: *The Effects of Weather on Urban and Interurban Traffic Flow*. (Unpublished report.)

MAUNDER, W. J. and WESTAWAY, P., 1968: *Water Consumption and Weather in Victoria, British Columbia*. (Unpublished report.)

MILLINGTON, R. J., 1961: Relations between yield of wheat, soil factors, and rainfall. *Aust. Jour. Agric. Res.*, 12: 397–408.

MITCHELL, K. J., 1959: Results of climate studies and their implications for seasonal productivity of pasture. *Proc. N.Z. Grass Assn.*, 21: 108–113.

MITCHELL, K. J., 1960: Growth of pasture species in controlled environments II – Growth at low temperatures. *N.Z. Jour. Agric. Res.*, 3: 647–55.

MITCHELL, K. J., WALSHE, T. C. and ROBERTSON, N. G., 1959: Weather conditions associated with outbreaks of facial eczema. *N.Z. Jour. Agric. Res.*, 2: 584–604.

MODENHAUER, W. C. and WESTIN, F. C., 1959: Some relationships between climate and yields of corn and wheat in Spink County, South Dakota and yields of milo and cotton at Big Spring, Texas. *Agron. Jour.*, 51: 373–6.

MONTEITH, J. L., 1965: Radiation and crops. *Expt. Agric. Rev.*, 1: 241–51.

MOORER, T. H., 1966: Importance of weather to the modern seafarer. *Bull. Amer. Met. Soc.*, 47: 976–9.

MOULE, G. R., 1962: The ecology of sheep in Australia. In: BARNARD, A., ed., 1962: *The Simple Fleece: Studies in the Australian Wool Industry*. Melbourne University Press, Melbourne, pp. 82–105.

NAYA, A., 1967: Insects, insecticides and the weather. (Part II.) *Weather*, 22: 211–15.

NELMS, W. P., JR, 1964: *Some Effects of Atmospheric Temperature Variations on Performance of the Supersonic Transports*. NASA Report, TMX-51563, 32 pp.

NICODEMUS, M. L. and MCQUIGG, J. D., 1969: A simulation model for studying possible modification of surface temperature. *Jour. App. Met.*, 8: 199–204.

NUTTONSON, M. Y., 1955: *Wheat Climate Relationships and the Use of Phenology in Ascertaining the Thermal and Photothermal Requirements of Wheat*. Amer. Inst. Crop Ecol., Silver Spring, Md., 388 pp.

NYE, R. H., 1965: The value of services provided by the Bureau of Meteorology in planning within the State Electricity Commission of Victoria. In: *What is Weather Worth?* Bureau of Meteorology, Melbourne, pp. 87–91.

ODELL, R. T., 1959: Effects of weather on corn and soybean yields. *Illinois Agric. Expt. Sta., Illinois Research*, 1(4): 3–4.

OSMUN, W. G., 1969: Airline warm fog dispersal program. *Weatherwise*, 22: 48–53, 87.

PATTERSON, R., 1957: The influence of rainfall on beef cattle numbers. *Quart. Rev. Agric. Econ.*, 10: 16–21.

PENGRA, R. F., 1946: Correlation analysis of precipitation and crop yield data for the sub-humid areas of the Northern Great Plains. *Jour. Amer. Soc. Agron.*, 38: 848–50.

PENMAN, H. L., 1962: Weather and crops. *Quart. Jour. Roy. Met. Soc.*, 88: 209–19.

PETERSON, J. T. and DRURY, L. D., 1967: Reduced values of solar radiation with occurrence of dense smoke over the Canadian tundra. *Geog. Bull.*, 9: 269–71.

PETTY, M. T., 1963: Weather and consumer sales. *Bull. Amer. Met. Soc.*, 44: 68–71.

PHILPOTT, B. P., 1956: Wool production and fleece weights in New Zealand 1919–1955. *N.Z. Meat & Wool Boards' Econ. Serv. Paper*, 139.

PORTSMOUTH, G. B., 1957: Factors affecting shoot production in tea (Camelliasinensis) when grown as a plantation crop. II. The influence of climatic conditions and age from pruning on flush shoot production. *The Tea Quart., J.T.R.I. Ceylon*, 28: 8–20.

RICHARD, O. E., 1968: *History of Environmental Support to the Heating and Air Conditioning Industry*. Paper presented at the Conference on Applied Climatology of the American Meteorological Society, Asheville, North Carolina, October.

RIGG, T., 1959: The orchard and its production. In: *Fruit Investigations at Annesbrook Orchard, Nelson, N.Z.*, Cawthron Institute, Nelson, N.Z., pp. 5–21.

ROBINSON, A. H. O., 1965: Road weather alerts. In: *What is Weather Worth?* Australian Bureau of Meteorology, Melbourne, pp. 41–3.

ROGERSON, T. L. and BYRNES, W. R., 1968: Net rainfall under hardwoods and Red Pine in Central Pennsylvania. *Water Resources Res.*, 4(1): 55–7.

ROONEY, J. F., JR, 1967: The urban snow hazard in the United States: An appraisal of disruption. *Geog. Rev.*, 57: 538–59.

ROSE, J. K., 1936: Corn yield and climate in the corn belt. *Geog. Rev.*, 26: 88–102.

RUNGE, E. C. A. and ODELL, R. T., 1960: The relation between precipitation temperature and the yield of soybeans on the Agronomy South Farm, Urbana, Illinois. *Agron. Jour.*, 52: 245–7.

RUSSELL, E. J. and BISHOP, L. R., 1933: Investigations on barley. *Jour. Inst. Brewing*, 39: 287–421.

RUSSO, J. A., 1966: The economic impact of weather on the construction industry of the United States. *Bull. Amer. Met. Soc.*, 47: 967–72.

SANDERSON, F. H., 1954: *Methods of Crop Forecasting*. Harvard University Press, Cambridge, Mass., 259 pp.

SEARS, P. D., 1961: The potential of New Zealand grassland farming. *Proc. 3rd N.Z. Geog. Conf.*, N.Z. Geog. Soc., pp. 65–70.

SEMAGO, W. T. and NASH, A. H., 1962: Interception of precipitation by a hardwood forest floor in the Missouri Ozarks. *University of Missouri Agric. Expt. Sta. Res. Bull.*, 796.

SEN, A. R., BISWAS, A. K. and SANYAL, D. K., 1966: The influence of climatic factors on the yield of tea in the Assam Valley. *Jour. App. Met.*, 5: 789–800.

SHAW, L. A. and DUROST, D. A., 1965: The effect of weather and technology on corn yields in the corn belt, 1929–1962. *Agric. Econ. Rep.*, 80, U.S. Dept. of Agriculture, Washington, D.C., 23 pp.

SHAW, R. H. and THOMPSON, L. M., 1964: Grain yields and weather fluctuations. In: THOMPSON, L. M., et al., 1964: *Weather and Our Food Supply*. Center for Agricultural and Economic Development, Report 20, Iowa State University, Ames, Iowa, pp. 9-22.

SHELFORD, K. O., 1953: An estate investigation into tea and rainfall. *Nyasaland Farmer and Forester*, 1.

SHOEMAKER, J. S. and TESKEY, B. J. E., 1959: *Tree Fruit Production*. Wiley & Sons, New York, 456 pp.

SHOENFELD, J., 1958: Can Weather Bureau forecasts help solve your sales problems? *Sales Management*, August, pp. 38, 40.

SHOR, M., 1964: Exploratory work in measurement of the effect of weather factors on retail sales. *Proc. Amer. Stat. Assn.*, 1963, Business and Economic Section, pp. 54-8.

SIMMONS, J. G., 1966: Economic significance of unaccounted for water. *Jour. Amer. Water Works Assn.*, 58: 639-50.

SMITH, L. P., 1961: Measuring the effects of climatic changes. *New Scientist*, 12 (1961): 608-11.

SMITH, L. P., 1967a: Meteorology and the pattern of British grassland farming. *Agric. Met.*, 4: 321-38.

SMITH, L. P., 1967b: Meteorology applied to agriculture. *World Met. Org. Bull.*, 16: 190-4.

SMITH, L. P., 1968: Forecasting annual milk yields. *Agric Met.*, 5: 209-14.

SNIDER, C. R., 1967: Great Lakes ice session of 1967. *Monthly Weather Rev.*, 95: 685-96.

SPICER, H., 1967: Mobil's experience with commercial weather routeing. *Marine Observer*, 37(215): 30-3.

STACY, S. V., STEAVSON, O., JONES, L. S. and FOREMAN, W. J., 1957: Joint effects of maximum temperatures and rainfall on corn yields, Experiment, Georgia. *Agron. Jour.*, 49: 26-8.

STEELE, A. T., 1951: Weather's effect on the sales of a department store. *Jour. Marketing*, 15: 436-43.

STEPHENS, F. B., 1951: A method of analyzing weather effects on electrical power consumption. *Bull. Amer. Met. Soc.*, 32, 16-20.

STEPHENSON, D. G., 1964: Principles of solar shading. *Architecture Canada*, 41(11): 59-64.

STEVENS, P. G., 1958: *Sheep – Part I: Sheep Husbandry*. Whitcombe & Tombs, Christchurch, 200 pp.

STOREY, L. F. and ROSS, D. A., 1959: Effect of shearing time on wool – IV: Effect on the fleeces. *N.Z. Jour. Agric. Res.*, 2: 1096-1103.

STOREY, L. F. and ROSS, D. A., 1960: Effect of shearing time on wool– VI: The rate of growth of wool and its relation to time of shearing. *N.Z. Jour. Agric. Res.*, 3: 113-24.

TANNER, J. C., 1952: Weather and road traffic flow. *Weather*, 7: 270-5.

TAUHEED, M., 1948: *The Influence of Weather on the Yield of Wheat in New Zealand with Special Reference to the Meyer Ratio*. M.Agric.Sci. Thesis, University of New Zealand (unpublished).

TAYLOR, A. R., 1964: Lightning damage to forest trees in Montana. *Weatherwise*, 17: 61-5.

TAYLOR, D. F. and WILLIAMS, D. T., 1967: *Severe Storm Features of a*

Wildfire. Paper presented at the Amer. Met. Society. Conference on Weather Forecasting, Fort Worth, Texas, Nov. 6–8.

TAYLOR, M. A., 1954: *The Climate of New Zealand's Pastures.* M.A. Thesis, University of New Zealand (unpublished).

THOMPSON, L. M., 1962. Evaluation of weather factors in the production of wheat. *Jour. Soil Water Conserv.* 17: 149–56.

THOMPSON, L. M., 1964: Foreward. In: THOMPSON, L. M., *et al.,* 1964: *Weather and Our Food Supply.* Center for Agricultural and Economic Development, Report 20, Iowa State University, Ames, Iowa, p. i.

THOMPSON, L. M., 1966: *Weather Variability and the Need for a Food Reserve.* Center for Agricultural and Economic Development, Report 26, Iowa State University, Ames, Iowa, 101 pp.

THORNTHWAITE, C. W. and MATHER, J. R., 1954: Climate in relation to crops. *Met. Monographs,* 2(8): 1–10.

TIPPETT, L. H. C., 1926: On the effect of sunshine on wheat yield at Rothamsted. *Jour. Agric. Sci.,* 16: 159–65.

TUCKER, G. W. L., 1965: The weather and the holiday-maker. In: *What is Weather Worth?* Bureau of Meteorology, Melbourne, pp. 49–51.

TWEEDIE, A. D., 1967: *Water and the World.* The Griffith Press, Adelaide, South Australia, 236 pp.

UNITED RESEARCH INC., 1961: *Forecast of Losses Incurred by U.S. Commercial Air Carriers Due to Inability to Deliver Passengers to Destination Airports in All-Weather Conditions: 1959–1963.* Prepared for: The Bureau of Research and Development, Federal Aviation Agency, Contract No. FAA/BRD-309. By: United Research Inc., Cambridge, Mass.

U.S. DEPARTMENT OF COMMERCE, 1966: *Weather and the Construction Industry.* U.S. Department of Commerce, ESSA, Washington, D.C., 12 pp.

U.S. DEPARTMENT OF COMMERCE, WEATHER BUREAU, 1964: *The National Research Effort on Improved Weather Description and Prediction for Social and Economic Purposes.* Federal Council for Science and Technology, Interdepartmental Committee on Atmospheric Sciences, 84 pp.

VERPLOEGH, G., 1967: Weather routeing of ships. *World Met. Org. Bull.,* 16(3): 139.

WARD, A. H., 1951: Production per cow and production per acre. *Dairy Farming Annual,* pp. 101–11.

WATSON, D. J., 1963: Weather and plant yield. In: EVANS, L. T., ed., 1963: *Environmental Control of Plant Growth.* Academic Press, New York, pp. 337–49.

WEAVER, J. C., 1943: Climatic relations of American barley production. *Geog. Rev.,* 33: 569–88.

WHITE, R. M., 1968: Fire, water and weather. *Bull. Amer. Met. Soc.,* 49: 715–19.

WHITTEN, A. J., 1947: Weather and the cabby. *Weather,* 2: 308–9.

WHITTINGHAM, H. E., 1964: Meteorological conditions associated with the Dandenong bush fires of 14–16 January, 1962. *Aust. Met. Mag.,* 11: 10–37.

WILLIAMS, D. E., 1967: Future application of meteorology in fire control. *Forestry Chronicle,* 43: 89–92.

WILLIS, W. O., LARSEN, W. E. and KIRKHAM, D., 1957: Corn growth as affected by soil temperature and mulch. *Agron. Jour.*, 49: 323–8.

WILSON, A., 1966: The impact of climate on industrial growth: Tucson, Arizona: A case study. In: SEWELL, W. R. D., ed., 1966: *Human Dimensions of Weather Modification.* University of Chicago, Dept. of Geography, Research Paper No. 105, pp. 249–60.

WISHART, J. and MACKENZIE, W. A., 1930: The influence of rainfall on the yield of barley at Rothamsted. *Jour. Agric. Sci.*, 20: 417–39.

WORLD METEOROLOGICAL ORGANIZATION, 1963: *Protection Against Frost Damage.* Tech. Note 51, WMO No. 133, TP60, Geneva, Switzerland, 62 pp.

WORLD METEOROLOGICAL ORGANIZATION, 1967: Assessing the economic value of a National Meteorological Service. *W.W.W. Planning Report*, No. 17, WMO, Geneva, Switzerland, 14 pp.

WULP, J. E., 1957: You *can* do something about the weather. *Printers Ink*, March 15, pp. 21–2, 60–2.

ZACKS, M. R., 1945: Oats and climate in Southern Ontario. *Canad. Jour. Res.*, 23C: 45–75.

ZEISEL, H., 1950: How temperature affects sales. *Printers Ink*, 233: 40–2.

——, 1960: Chilly May for shops. *The Economist*, 196(6097), July 2, pp. 36, 39.

——, 1961: Where the winter hurts. *Business Week*, Feb. 4, pp. 21–2.

——, 1963: Cold weather costs. *The Economist*, 206(6236), March 2, p. 824.

——, 1963: The big freeze-up. *The Economist*, 208(6255), July 13, p. 193.

——, 1966: Crop insurance. Fed. Res. Bank of Chicago, *Business Conditions*, May, pp. 7–16.

——, 1967: The year the bees got grounded. *Time*, Nov. 10.

——, 1969: *The Wall Street Journal*, July 8.

ADDITIONAL REFERENCES

BERRISFORD, H. G., 1965: The relation between gas demand and temperature. A study in statistical demand forecasting. *Operational Res. Quart.*, 16(2): 229–46.

BLAXTER, K. L., 1964: The effect of outdoor climate in Scotland on sheep and cattle. *Veterinary Record*, 76: 1335–1451.

BROCK, F. V., 1959: Engineering meteorology: the effects of weather and climate on concrete pavement. *Bull. Amer. Met. Soc.*, 40: 512–18.

BROSCHE, L. E., 1963: What's new in the weather. *Public Utilities Fortnightly*, 72(12): 42–9.

CHAMPION, D. L., 1947: Weather and railway operation in Britain. *Weather*, 2: 373–80.

CHANG, JEN-HU, 1968: *Climate and Agriculture.* Aldine Publishing Co., Chicago, 304 pp.

CHANG, JEN-HU, CAMPBELL, R. B. and ROBINSON, F. E., 1963: On the relationship between water and sugar cane yield in Hawaii. *Agron. Jour.*, 55: 450–3.

COOPER, C. F., 1968: Needs for research on ecological aspects of human uses of the atmosphere. In: SEWELL, W. R. D., *et al.*, 1968: *Human*

Dimensions of the Atmosphere. National Science Foundation, Washington, D.C., pp. 43–51.

CRADDOCK, J. M., 1965: Domestic fuel consumption and winter temperatures in London. *Weather*, 20: 257–8.

CROWELL, E. H. and DEES, C. S., 1965: The weather: its effect on government contracts. *Briefing Papers* (The Government Contractor), No. 65-4, Federal Publications, Inc., Washington, D.C.

DeWILJES, H. G. and ZAAT, J. C. A., 1968: Influence of climate upon the number of weather-working hours in combine harvesting in The Netherlands. *Archiv. für Meteorologie, Geophysik. und Bioclimatologie*, Ser. B. 16(1): 105–14.

DONALDSON, G. F., 1968: Allowing for weather risk in assessing harvest machinery capacity. *Amer. Jour. Agric. Econ.*, 50: 24–40.

DORE, R. P., 1966: Climate and agriculture. The intervening social variables. *UNESCO Symposium on Agroclimatological Methods.* Reading, England, pp. 201–8.

FRIEDMAN, D. G. and HENDRICK, R. L., 1960: The role of a meteorologist in an insurance company. *Weatherwise*, 13: 141–5.

HALTINER, G. J., BLEICK, W. E. and FAULKNER, F. D., 1968: A proposed method for ship routeing using long range weather forecasts. *Monthly Weather Rev.*, 96: 319–22.

HEINEMANN, G., NORDMAN, D. and PLANT, E., 1966: Relationship between summer weather and summer loads; a regression analysis. *I.E.E.E. Trans. Power Apparatus and Systems*, 85: 1144–51.

HOLLAND, D. J., 1967: The Cardington rainfall experiment. *Met. Mag.*, 96: 193–302.

KESTIN, M., 1967: Effect of weather on cosmetic sales. *Amer. Perfumer*, 82: 28.

KRICK, I. P., 1954: Technical and economic aspects of weather modification in relation to water supply. *Jour. Inst. Water Eng.*, 8: 336–52.

MUSGRAVE, J. C., 1968: Measuring the influence of weather on housing starts. *Construction Review*, 14(8): 4–7.

PAUL, D. E., 1958: What can the weather do to net income? *Public Utilities Fortnightly*, 61: 411–13.

PLANK, V. G., 1969: Clearing ground fog with helicopters. *Weatherwise*, 22: 91–8, 126.

RAINEY, R. C., 1969: Effects of atmospheric conditions on insect movement. *Quart. Jour. Roy. Met. Soc.*, 95: 424–34.

ROUSE, G. D., 1961: Some effects of rainfall on tree growth and forest fires. *Weather*, 16: 304–11.

SCHERHAG, R., WARNECKE, G. and WEHRY, W., 1967: Meteorological parameters affecting supersonic transport operations. *Inst. of Navigation Jour.*, 20: 53–63.

SHELLARD, H. C., 1967: Wind records and their application to structural design. *Met. Mag.*, 96: 235–43.

SMITH, L. P., 1968: *Seasonable Weather.* George Allen & Unwin, London, 146 pp.

SPILLANE, K. T., 1967: Clear air turbulence and supersonic transport. *Nature*, 214(5085): 237–9.

STALLINGS, J. L., 1961: A measure of the influence of weather on crop production. *Jour. Farm. Econ.*, 43: 1153–62.

STROUT, A. M., 1961: Weather and the demand for space heat. *Rev. Econ. Stat.*, 43: 185–92.

SUTHERLAND, R. H., 1964: Seasonal orientation in the Sydney meat market. *Quart. Rev. Agric. Econ.*, 17: 113–19.

TAYLOR, J. A. (Editor), 1967: *Weather and Agriculture*. Pergamon Press, Oxford, 225 pp.

TUCKER, G. B., 1960: Some meteorological factors affecting dam design and construction. *Weather*, 15: 3–13.

TURNER, D. B., 1968: The diurnal and day-to-day variations of fuel usage for space heating in St. Louis, Missouri. *Atmos. Environment*, 2: 339–51.

TURNOVSKY, S. J., 1969: The demand for water: some empirical evidence on consumers response to a commodity uncertain in supply. *Water Resources Res.*, 5: 350–61.

TWEEDIE, A. D., 1967: Water and the city: prospects and problems. *Aust. Geog. Studies*, 5: 1–14.

WANG, JEN-YU, 1963: *Agricultural Meteorology*. Pacemaker Press, Milwaukee, 693 pp.

WEBB, R. E. and SMITH, F. F., 1969: Effect of temperature on resistance in lima bean, tomato, and chrysanthemum to *Liriomyza munda*. *Jour. Econ. Entomology*, 62: 458–62.

WILLIAMS, G. D. V. and ROBERTSON, G. W., 1965: Estimating most probable prairie wheat production from precipitation data. *Canad. Jour. Plant Sci.*, 45: 34–47.

WILLIAMSON, K., 1969: Weather systems and bird movements. *Quart. Jour. Roy. Met. Soc.*, 95: 414–23.

WINTLE, B. J., 1960: Railways versus the weather. *Weather*, 15: 137–9.

WRIGHT, P. B., 1969: Effects of wind and precipitation on the spread of foot-and-mouth disease. *Weather*, 24: 204–13.

W.M.O., 1968: The World Weather Watch and meteorological service to agriculture. *World Weather Watch Planning Report*, No. 22, WMO, Geneva, Switzerland, 14 pp.

——, 1965: It all depends on weather. *New England Business Rev.* April, pp. 10–11.

——, 1966: Retailers' sales rise with the temperature. *Business Week*, July 23, p. 36.

——, 1967: Long, cold spring for retailers. *Business Week*, June 3, pp. 34–5.

IV Sociological and physiological aspects

A. THE SETTING

Though man has had a visible effect on his environment he has also been forced to react to it in certain ways. One of the most ubiquitous elements of this environment is weather, and as man's technology advances in the fields which enable him to cope with the weather, he is discovering effects which, until a while ago, were unknown to him. Old wives' tales and myths discuss some of these effects and the resultant behaviour of man, but until this century little had been done to investigate cause-and-effect relationships, partly because man was not interested, and partly because he did not have the technical means to study or control them. However, it is now generally agreed that many of our sociological and physiological reactions are associated with the state of the atmospheric environment. This is particularly true with reference to temperature and humidity. Parr (1968) states, for example: '. . . that we can bear a great deal of cold weather with equanimity and little ill effect so long as we do not feel cold, or we can suffer excessively from a rather slight exposure if our sense of misery is reinforced by features of the milieu that have nothing to do with temperature except in their psychological effects'. We do, in fact, speak of buildings or entire landscapes looking cold or warm, and we may be attracted or repelled by situations according to whether we ourselves feel chilly or hot at that moment.

The relationships between environmental design and human responses to weather and climate are therefore important. Parr has suggested that psychologists and sociologists, working in conjunction with architects and urban planners, should examine these relationships for, with the aid of such an approach, it may be possible to establish goals of environmental design for any location in any climate, and to implement and co-ordinate these aims with those arising from other demands upon the milieu.

The 'value of the atmosphere' from a sociological or physiological point of view embraces many more aspects of life than those mentioned above. These include the effect of weather on human behaviour, crime rate, human comfort and even race riots. The Santa Ana wind in California, for example, has been associated with crime rates in Los Angeles (Miller, 1968), and several investigators such as Maunder (1962) and Terjung (1966a, b) have endeavoured to evaluate the human climates of various parts of the earth. Other studies have shown that weather, climate and health are in some cases closely associated. An apparent causal relationship between high death-rates, areas of industrialization, air pollution and certain meteorological conditions is also evident in some areas of the United States and the United Kingdom.

The tourist trade is also influenced to an important degree by variations in the weather. For example, extended periods of precipitation in the summer often result in considerable variations in tourist traffic to and from holiday resorts, while lack of snow reduces the revenues of ski resorts, as well as the enjoyment of hundreds or even thousands of skiers. Few studies have been undertaken to assess the influence of weather on the tourist, or in the closely associated area of sport. Nevertheless, most sports-minded persons are well aware of the 'value' of weather to their game, especially those concerned with, for example, horse racing, football, tennis, yachting or cricket. The monetary value of the weather in such cases may perhaps be relatively small. Nevertheless, it is clear that in many sports where 'national' reputations are at stake, such as would be the case on the last day of a five-day cricket 'test' between England and Australia, sunshine or rain may be of immeasurable value.

B. HUMAN ASPECTS

1. Weather and human behaviour

Three levels of climatic environment are seen by Bates (1966) in looking at the role of weather in human behaviour. First, one may distinguish the conditions surrounding a given individual organism; second, one can identify the climate of the habitat, that is, in the case of man, of the buildings in which he lives and works, or of the fields that he cultivates; and third, climatic conditions measured by standard meteorological methods, or the geographical climate, may be distinguished. Bates suggests that these three distinctive climates may be called the 'microclimate', the 'ecoclimate' and the 'geoclimate', and using this terminology it is readily apparent that man has been modifying his microclimate for a long time by wearing clothing, and that he has modified the ecoclimate even longer through the use of fire and the building of shelters. However, only in recent years has he come to think of intentionally modifying the geo-climate, though he has long affected the 'local' climate by such activities as clearing forests, building cities and polluting the atmosphere.

Whatever the level of climatic environment, however, many environmental components are usually considered to form part of weather, and those likely to influence human behaviour include precipitation (rain, snow, dew), temperature, humidity, solar radiation, wind (especially destructive storms), lightning and air transparency (fog, mist, cloudiness). In addition, one should add air pollution, both the 'natural' pollution of dust storms and volcanic eruptions, and the 'unnatural' but increasingly important pollution resulting from human activities.

a. Values in human–weather relationships

Though it is difficult to place quantitative values on many aspects of weather there appear to be two approaches to it from the standpoint of human behaviour and reaction. First, the loss in productivity, earnings and taxes, as a result of a reduction in labour force or in work slow-down due to weather conditions, or the increased occurrence of weather affecting reactions and medical conditions, such as hay fever and asthma; and second, the cost of modifying the environment of the worker or the patient to provide an environment conducive to satisfactory working and living conditions.

Several other points are important when decision-making is considered. One of them is that weather may not be the only factor affecting the decision. Another is the effect of random occurrences which could have far-reaching effects such as if a government official at a crucial moment feels 'subnormal' due to the weather, and allows a point of great importance to slip by. Such an occurrence may be infrequent and very difficult to detect, let alone quantify, but it may nevertheless have far-reaching effects.

The physical nature of the weather and its effects on visibility, surface conditions and comfort will also affect a person's decision to travel, and if he does decide to travel, by what means, and what type of clothing to wear in order to remain comfortable. Depending on the importance of the trip, individuals, large organizations and even governments could be affected. Another direct effect of weather on productivity is the cost to the worker or the company for extra clothing, and protective materials, or the cost to the company of providing a completely controlled environment. In the latter case, definite savings could be made, but the cost of controlling the environment in certain areas may be very high.

There are also the side effects of weather which are not easily sensed, or, if they are, not readily recognized, and these can have wide-ranging effects on human beings and their behaviour. Dordick (1958), for example, comments on the correlation between post-operative failures in areas of Europe where föhn effects occur; and Tromp (1964a) points out that the electromagnetic and electro-static impulses of the mind and nervous systems may possibly be influenced by changes in the electromagnetic and electrostatic fields of the atmosphere. If this is in fact the case, the environmental con-trol possible for helping persons with mental and neurological prob-lems are wide.

The weather criteria which involve decisions are many, but the following examples (Table IV.1) show that they can, on occasions,

TABLE IV.1 : *Weather-related decisions and human behaviour*

Factor	Decision	Weather criteria	Adaptabilities
Arthritis	move	warm, dry	live with pain medical help
Travel			
(a) business	fastest means	visibility precipitation storms	change time
(b) pleasure	where to go route	various	change time change destination
Work or study	possible alternates degree of application	various	protection will-power
Recreation	desire	wet, cold	clothing, transportation
Voting	civic duty, desire	wet, cold	mobility
Shopping	urgency of purchases	various	transport

be important, and often of economic significance. Medical problems, for example, can in some cases be eased and even solved by a change of climate through local atmospheric modification, or an investment in travel to a suitable climate. In the latter case, personal savings in medical costs and the unaccountable value of less suffering may be countered by the loss of friends, a family home, a job, or restrictions imposed by the required climate.

The importance of man-produced weather on behaviour should also be noted, particularly in view of the fact that the behaviour of people is in many cases associated with 'smog' conditions, especially if such conditions continue for more than one or two days.

Though the literature indicates a fairly definite correlation between weather and behavioural factors, the value of reactions (or the cost, as the case may be) is very difficult to quantify. For any given situation, some kind of benefit–cost analysis would presumably have to be done for all possible alternatives to see what is the real economic effect of the weather. It is nevertheless apparent that individual qualities and adaptabilities do play an important part in any evaluation of an individual's behaviour, and the relative health and mental condition of an individual making a decision are often related to the effects of the prevailing weather conditions.

b. Climatic determinism

The theory of climatic determinism is a subject that has caused considerable controversy during the twentieth century, and although the theory is not widely held today (except for some of the obviously limiting environments, such as the Sahara, Central Australia or the Antarctic), the subject is still of considerable importance.

The most well-known advocate of a causal relation between stimulating climate and human inventiveness was Huntington, and his views are set forth most completely in his last book, *Mainsprings of Civilization* (Huntington, 1945). Another strong supporter of climatic determinism was Mills (1942),who wrote *Climate Makes the Man.* Huntington's theory has been commented upon by many. Brooks (1949), for example, noted that Huntington's contention is that a certain type of climate, now found mainly in Britain, France and neighbouring parts of Europe and in the eastern United States, is favourable to a high level of civilization, this climate being characterized by changeable, stimulating weather. In commenting on this, Bates (1966) states that a most important factor is that civilization could not develop in higher latitudes until methods of coping with the cold and darkness of winter had been developed, especially efficient hearths, chimneys and window-panes. Following Markham (1947), one could thus suggest, not that the climate of western Europe is stimulating, but that civilization was not possible there until methods had been found of overcoming climatic defects. That is, as Bates states, the geoclimate remained the same, but the ecoclimate was greatly modified.

Regardless of the interpretation of climatic determinism, it appears from the viewpoint of a person who has lived in 'disturbed westerly' types of climate, that it does have or at least has had, some merit. It is also reasonable to suggest that there *are* optimal climatic conditions for human activities. But, as is equally obvious, individual preference is wide, and what is best for one sort of activity or one group of people may not be best for another activity or group of people.

c. *Sociological aspects of weather variations*

The sociological aspects of the human dimensions of the atmosphere have been reviewed by Haas (1968), who says that while a few sociologists early in the twentieth century speculated on the way in which the environment may shape man's institutions, no meaningful research on the possible relationship was conducted. Moreover, during the period 1930 to 1950, there appears to have been a strong negative reaction to the suggestion that man's social behaviour may be determined significantly by his 'natural' physical environment. However, sociologists have now developed what Haas calls a more 'open' attitude towards the possibility of this relationship, and have suggested that perhaps the wide publicity given to air pollution problems, and the emergence of research on natural disasters, has produced some 'second thoughts' on the part of the sociologists. Nevertheless, little really relevant research on the sociological aspects of the atmospheric environment has been completed.

Several major areas of research are suggested by Haas, who says that at present we know very little about the relation between specific dimensions of 'normal' weather variations and the various human activities in a community or region. Research leading to a better understanding of the relation of weather variations to several factors is therefore required in the following areas: (1) incidence of illness and use of health facilities and personnel, (2) incidence of crime, and law enforcement efforts, (3) use of educational facilities and student performance, (4) flow of persons and goods through transportation networks, (5) nature of major recreational and leisure time activities, (6) disruption and use of basic community services (fire protection; electric, water and gas distribution; and mass communications), (7) participation in political activity (registration, voting, etc.), (8) incidence of racial disturbances, and (9) demand for and use of welfare services.

In addition, there are the related questions, which are considered in detail in Chapter VI, dealing with the collection, dissemination and use of weather forecasts, which include such questions as: what is the influence of weather forecasting on human activities; and how does the content, mode and speech of dissemination of forecasts relate to interpretation and use? Man's modification of atmospheric conditions raises still another group of questions, and these are considered in Chapter VII. For example, as Haas suggests, if man has developed a set of adjustment mechanisms to cope with the atmosphere as he has percieved and experienced it over a long period of time, and that set of atmospheric conditions is altered significantly by inadvertent actions or planned efforts, some kind of readjustment is bound to occur. What, for example, asks Haas, are the discernible consequences of weather modification, what kinds of individual and group readjustments can be anticipated, and do individual and community responses to *planned* weather modification differ significantly from those to inadvertent modification such as air pollution?

There are in fact many unanswered questions regarding man's perception of, and adaptation to, 'extreme' weather events, such as hurricanes, blizzards, tornadoes and floods. To quote Haas:

> Do citizens in such disaster-prone communities develop a heightened sensitivity to weather forecasts in general, or only to severe storm warnings? Or is apathy the dominant mode of adjustment? Does adaptation vary significantly among various socio-economic levels? Are communities which frequently experience such weather hazards more likely to have effective organizational mechanisms for minimizing disruption, damage and injury than those which seldom experience such extreme weather events? Can adequate 'disaster preparedness' be induced in communities which seldom experience severe storms or floods?

Many sociological aspects of modern living are in fact geared to weather conditions. The interest shown by the man in the street about the weather stems from the effect it will have on his daily personal comfort, including his choice of clothing and method of transport. The weather also influences his leisure time, whether to mow the lawn, to go visiting, to play sport or to paint the house. The housewife, too, is interested in the weather to plan the daily household chores, to select the family menu, and to decide whether or not to go shopping. And even attendance at church has both seasonal and weekly variations due to weather conditions, according to a survey in Princeton, New Jersey, by Mazzarella (1967).

d. Weather and riots

Although the weather can act either as a catalytic or dampening factor on race riots, it is obvious that the reasons for riots are far deeper than the fickle changes in our atmosphere. Nevertheless, any attempt to assess the effect that weather can have on man, both physiologically and psychologically, is of importance. Moreover, although the effect of weather and weather changes on human action and reaction is part of a complex interplay of man and his environment, it seems clear that with respect to riots, the weather can have an influence in those locations where there is already great tension between peoples.

It appears that the way in which weather can instigate a riot is by creating a stress situation, that is, a state in which the vital functioning of man is threatened. Reactions to this stress are of three distinct kinds, defensive, adaptive or disorganized, and if the adaptive or defensive activities fail, such as in the case of extreme heat or humidity, and the removal of excess clothing or the remaining in a cooler portion of a building is not possible, then disturbance of behaviour may occur. This disturbance, which may be expressed by rioting, is, according to Deutsch (1967), '. . . most likely to occur when stress is intense and long lived and when defensive and adaptive mechanisms are weak or their capacity is exceeded'. Thus, if various weather parameters, such as temperature or humidity, deviate from the mean to such an extent that they cause a stress, and they then last for some duration, disordered behaviour, anxiety or other emotional disturbances may result.

One of the by-products of stress is fatigue, and with fatigue there tends to be a loss of perspective where minor happenings become intensified, or when threats lose their true significance. For example, in the United States many race riots are said to start from some trivial incidents between a few Negroes and the police. The situation often expands rapidly and a mob may form. Then, through the application of some form of stress, tension is built up to the point where it explodes into a riot, and it is suggested that the development of this stress can be initiated or at least heightened by changes in various weather parameters such as temperature, rainfall, humidity and possibly atmospheric pressure.

In the United States such a tension often arises out of the squalor and poverty of Negro ghettoes, and here weather could be said to act as a spark to light the already-present fuel, but is not the fuel itself. In this kind of situation the Negro community is under a constant stress, and when the weather elements combine to increase the body temperature and the heat stress, irritability prevails and tempers flare. Further, since these changes are often associated with fatigue, they may cause the situation to explode. This appears to be the case in the riot of June 1943 in Detroit, which ended with 35 dead and 530 injured.

While heat and humidity appear to be associated with rioting, it is also reasonably clear that atmospheric cooling may have the effect of stopping the rioting. In addition rain, which would seem to remove the tensions incurred by monotonous hot and humid weather, has also frequently been reported as a factor in ending the rioting.

Although it would be presumptuous to assume that heat and humidity are always associated with great stress, which in turn is a warning that riots may occur, the following study of the East Harlem race riots of July 23–26, 1967, is noteworthy, when 'the value of the weather' is considered. This clash developed on Sunday evening, July 23, after an off-duty policeman killed a Puerto Rican who had been in a knife fight with another man. The relevant weather data for New York City, for the period July 21–29, 1967, is shown in Table IV.2.

TABLE IV.2: *Weather during East Harlem riots, July 23–26, 1967*

Date (July)	Max. temp. (°F)	Min. temp. (°F)	Difference of temp. from mean	Max. T.H.I.	Hours with T.H.I.>75	Wind (mph)	Precipitation (in.)
21	80	63	−4	76	2	NE SE 5	—
22	81	65	−5	81	4	SE 5	AM 0·13 PM 0·21
23	87	70	+1	79	8	SW 8	PM 0·60
24	90	75	+5	82	16	SW 9	—
25	88	70	+1	81	10	SW 8	PM 0·58
26	86	69	0	77	7	W 6	AM 0·78
27	85	72	+1	78	10	SE 7	—
28	80	69	−2	76	6	SE 7 NE	
29	83	67	−2	77	6	Variable 6	—

(Rows 23–26 are marked *Duration of rioting*)

AM denotes 12 hrs ending at 8 a.m. – PM denotes 12 hrs ending at 8 p.m.

This shows that the temperatures were several degrees below normal for the two days prior to the riot but on the day that the riot erupted the temperature rose to 1°F above normal, that is a 6°F increase from the previous day. In addition, the temperature–humidity index was above 77 for eight hours on July 23, and was 80 or above for six hours on July 24.

Throughout the riot period the wind was quite weak and hence did not aid in the dispersement of body heat. At the same time, rain occurred on three of the days of the rioting, but it only made the atmosphere more steamy and uncomfortable. Nevertheless, the *New York Times* reported on July 26 that 'many prayed for rain – a heavy rain that would discourage the crowds'. The following day, after the return of relative calm, it was stated that 'heavy intermittent showers from sunset to the early morning hours helped to prevent new outbreaks by the surging mobs'.

The small number of race riots considered from a weather point of view, are far too limited in number to be able to say that a significant association exists between weather and riots. Nevertheless, it is believed that they give a hint of a relationship which possibly could be validated by a more detailed analysis. A study by Haas (1968) has indicated that little if any systematic research has been done on weather and race riots, but the evidence that is available seems to indicate that weather could very well have an important influence on rioting.

e. Weather and crime

The effect of weather conditions on crime rates has not been investigated in any detail. Nevertheless, there is some evidence available to suggest that the state of the atmosphere is associated with certain kinds of crime, a factor which, if substantiated, could lead to a greater appreciation by the law enforcement authorities and the public of the value of the weather.

The seasonality of crime is reasonably well established. For example, the Federal Bureau of Investigation (1965) indicated what appears to be a direct and significant relationship between seasons and the incidence of at least some types of crime. Eleven conditions which affect the amount and type of crime that occurs from place to place are listed by the F.B.I. and one of these is 'climate, including seasonal weather conditions'. In addition, the Bureau states that murder follows a seasonable pattern, occurring more frequently in the summer months, and the seasonal variation for aggravated assault shows a high number of offences in the summer months with a tapering off to a low in the colder months of the year.

The 'long hot summer' particulary as it applies to certain areas of the United States, and its coincidence with national annual peaks in various crime categories, precipitated interest by Miller (1968) into the possibility that certain short-term but marked changes in local area weather conditions may also be reflected in crime statistics. Such short-term but natural changes in the weather include winds like the sirocco, chinook, föhn and Santa Ana, which have such identity that their names and natures are familiar both to meteorologists and to residents of areas which experience them. Such winds, during periods when they are dominant, create an area-wide meteorological environment which is quite different from normal. Miller comments that many authors have noted this difference and, in doing so, have generally described its effect on man as detrimental.

In discussing the Santa Ana winds, Miller says that if people in southern California are uncomfortable and irritable on days when there is a pronounced Santa Ana condition, it might be expected that an above-normal number of crimes would be committed during those days. A preliminary attempt to determine whether such a correlation existed in the Los Angeles area was therefore made by Miller.

The dominant identifying characteristic of the Santa Ana is the low humidity, and it was the criterion used to identify the days during 1964 and 1965 on which the Los Angeles Civic Center experienced such a condition. The Police Department and the Sheriff's Office provided Miller with the number of homicides reported in their jurisdictions by day for 1964 and 1965, and since this is the ultimate crime of violence, it should according to Miller 'constitute the most appropriate single index'.

These two sets of data were combined to produce a single total number of homicides by day, and from this combined list the number of homicides reported for each of the 53 days, during 1964 and 1965, when the Civic Center experienced a noon relative humidity of 15% or less was obtained. In addition, the average number of homicides reported during both years by *day* of the week for the eight months, September to April, during which Santa Anas typically occur were calculated, these figures (Monday 0·8, Tuesday 0·9, Wednesday 0·8, Thursday 0·7, Friday 1·0, Saturday 1·4, Sunday 1·2) being taken as the 'normal'. Each of the 53 actual daily number of homicides was then compared to determine the 'departure from normal'. This information together with the relevant weather data (for 1965 only) is given in Table IV.3.

This shows that of the 53 days during 1964 and 1965 with a pronounced Santa Ana condition at Los Angeles Civic Center, 34 had an above-normal number of homicides, three had a normal number and 16 had a below-normal number. The total number of homicides for all 53 days was 58, compared with a normal of 50·8. The departure from normal was therefore 7·2 or 14·2%. The period October 20, 1965, to October 26, 1965, was the longest sustained Santa Ana with a noon relative humidity of 15% experienced at Los Angeles Civic Center during the two years under consideration, and of these seven days, four had reported homicides and three did not. However, Miller notes that 'the *total* of 10 reported homicides was 47% above the 6·8 for a normal week'.

TABLE IV.3: *Weather and homicide data for the Los Angeles area on days in 1965 with pronounced Santa Ana condition* (after Miller, 1968)*

Date		Weather data			Homicide data		
		Departure from normal					
	Prevailing wind direction	Relative humidity (%)	Max. temp. (°F)	Wind speed (mph)	Number reported for day	Normal for same day of week	Departure from normal
Feb. 16	NE	−33	+ 4·0	−1·5	1	0·9	+0·1
24	NE	−34	+10·0	−1·5	1	0·8	+0·2
25	NE	−33	+12·0	−1·9	1	0·7	+0·3
Mar. 2	SW	−32	+ 4·4	−1·1	1	0·9	+0·1
4	E	−33	+ 3·4	−0·1	1	0·7	+0·3
Sept. 30	W	−34	+ 7·6	−0·9	2	0·7	+1·3
Oct. 1	NE	−28	+20·7	−1·4	2	1·0	+1·0
2	NE	−28	+20·7	−1·7	2	1·4	+0·6
3	NE	−26	+15·7	−1·7	2	1·2	+0·8
20	N	−26	+17·7	−2·3	0	0·8	−0·8
21	NE	−27	+20·7	−1·4	0	0·7	−0·7
22	NE	−27	+22·7	0·0	2	1·0	+1·0
23	NE	−29	+20·7	−1·8	5	1·4	+3·6
24	NE	−28	+19·7	−1·4	2	1·2	+0·8
25	NE	−28	+17·7	−1·8	1	0·8	+0·2
26	NE	−26	+14·7	−2·7	0	0·9	−0·9
28	NE	−28	+15·7	−2·4	1	0·7	+0·3
Dec. 3	NE	−27	+15·5	−2·2	0	1·0	−1·0
4	N	−28	+19·5	−1·8	1	1·4	−0·4
				Totals:	58·8	50·8	+7·2

* Only days with a noon relative humidity of 15% or less at the Los Angeles Civic Center are listed. Weather data are derived from the monthly Local Climatological Data – Los Angeles, California, Civic Center reports for 1965, and the 1965 issue of Local Climatological Data with Comparative Data – Los Angeles, California, Civic Center both published by the U.S. Weather Bureau, or from unpublished records in that Bureau's Los Angeles office. Homicide data are derived from unpublished records of the Los Angeles City Police Department, and the Los Angeles County Sheriff's Office.

Note: The original table listed data for 1964 and 1965, but only those for 1965 are shown in Table IV.3.

Miller says, in conclusion, that it should be recognized that the time span for which data have been assembled is only two years, and there is no positive assurance that findings are not in part attributable to other currently unidentified factors. Accordingly, results of the study are believed to be indicative but not necessarily conclusive of the association between weather conditions (specifically the Santa Ana winds) and crime (specifically homicides) in the Los Angeles area.

f. Weather and comfort

In considering the meteorological aspects of human comfort, it is clear that any comprehensive index should include both the meteorological and non-meteorological factors, the latter group including clothing characteristics (such as albedo, thermal conductivity and opportunity for convective heat exchange near the skin) and respiratory and metabolic rates. However, the representative meteorological observations, and virtually all non-meteorological measurements are expensive to make and few are available. Net radiation measures, for example, are complex because they involve measurement of incoming short-wave and both incoming and outgoing long-wave radiation; furthermore, as Hounam (1967) points out, it is difficult to use this data in any general formula for estimating comfort. Nevertheless, the importance of the net radiation balance remains, and several biometeorological indices, amenable to practical application by the climatologist, have been evolved. These include the effective temperature (Houghton and Yagloglou, 1923), discomfort index (Thom, 1959), thermal strain (Lee, 1958) and heat stress index (Belding and Hatch, 1955). In addition, the temperature humidity index (T.H.I.) is used by the U.S. Weather Bureau to give a numerical indication of human comfort. The resulting index is based on the fact that about 10% of the population are uncomfortable at a T.H.I. of 70, more than half are uncomfortable at 75 and almost all uncomfortable at 80 or above.

Any comprehensive study of the effects of weather on comfort must of course include many factors not usually assessed in any comfort index. Hounam mentions some of them, and they include: (1) sequences of hot and humid days, since a concentration of consecutive extreme days may result in an accumulated physiological strain; (2) wind speed, for although wind reduces the effective temperature a higher wind speed may result in discomfort if dust or fine sand is lifted, such as on the foreshore in a strong sea breeze;

(3) the persistence of a weather type, since there may be a psychological reaction to extended periods of the same weather even within the limits of comfortable effective temperature; (4) insects such as mosquitoes and sandflies, since they affect comfort, and their life cycle is dependant on meteorological conditions; and (5) the effect of air pollution. This latter factor is particulary important, since meteorological conditions in the first few hundred feet control the degree of pollution of the atmosphere by industrial plants, and man's health can suffer significantly in this environment.

The relationship between weather and comfort in the hot dry and hot moist areas of the world has also attracted a good deal of attention this century. Papers by Hounam (1967), Stephenson (1963), McLeod (1965) and Watt (1967), for example, give indices of comfort for Alice Springs, Singapore, Gan and Bahrain respectively. Details are available in these papers, but a typical comment is illustrated by one of the conclusions reached by Hounam, who shows that although Alice Springs, in the centre of Australia, is uncomfortable for a substantial part of the warm season, the climate is more comfortable than the tropical north of Australia and some overseas arid areas. One of the main factors is the larger diurnal range at Alice Springs, which results in more comfortable nights than in areas near the tropical coast.

In view of these factors and many others not mentioned, it is clear that the value of the weather in tropical parts of the world is particularly related to the 'comfort' aspects of the atmospheric environment.

The effect of weather on comfort is not, however, restricted to the hotter and more humid areas of this world; indeed, the comfort in the temperate areas of the world is probably more significant to the 'natives' in temperate areas than it is to those in the tropical areas. Various optimum weather indices for temperate climates have been proposed including those of Davis (1968), Poulter (1962), Fergusson (1964), Rackliff (1965) and Hughes (1967).

Two further aspects of weather and comfort as applied to the value that can be placed on atmospheric variability are wind chill and air conditioning.

With respect to wind chill, various formulas have been proposed to express the dry convective cooling power of the atmosphere. However, very few are based on actual observations in a very cold climate such as that in the Antarctic (Wilson, 1963). Of the formulae available, Wilson (1967) says that the 'wind chill' of Siple and Passel (1945) is the one that has more closely correlated atmospheric cooling to stages of relative human comfort sensations and estimated approximate limits for the onset of freezing of human skin. Wilson investigated the correlation between wind chill index and the occurrence of frostbite in the Antarctic. The results are shown in Fig. IV.1.

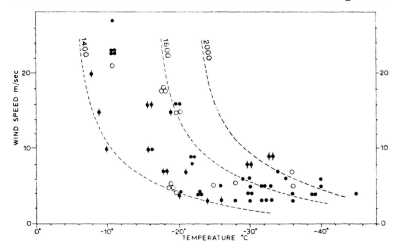

FIG. IV.1. *Recorded cases of frostbite in relation to temperature and wind speed. Open circles mark cases of frostbite that have occurred during shifting meteorological conditions (the mean of the extremes in temperature and wind is plotted). Dots with a vertical bar indicate cases of frostbite observed during sledding (wind speed estimated). Black dots are frostbite with established data for temperature and wind speed. The curved parameter lines correspond to a wind chill index of 1,400, 1,800 and 2,000 respectively (after Wilson, 1967).*

At the opposite extreme of wind chill and frostbite is the artificial cooling of the air such as air conditioning. The basic purpose of air conditioning is to cool, to ventilate, to purify, to dehumidify and to circulate the air. Thus, the air conditioner can be a very valuable appliance during the hot and humid weather, and the efficiency of the people in the house, office or factory can be increased. For example, if the peak efficiency of a worker occurs with a temperature/humidity ratio of 72°F/40%, it is generally estimated that with the same relative humidity and a temperature of 85°F, the efficiency drops to 80%.

Air conditioning is vital to many key areas, such as hospitals, industry, business and houses. For example, it is used extensively for patients who suffer from asthma, bronchitis, sinus and hay fever, and in restaurants, theatres and department stores, where competition necessitates the installation of air conditioning in order to attract customers.

It can be seen, therefore, that air conditioning is a vital factor in our society, not only for its importance in hospitals, industry and houses, but also in the 'value' of hot and humid weather to the manufacturer, wholesaler, seller and buyer of air-conditioning units.

2. Human classification of climate

During the past 100 years many classifications of climate have been published. Gentilli (1958), for instance, refers to no less than seventy-five classifications or part classifications of climate that have been suggested since 1880. These include the classifications by W. Koppen (Koppen, 1931, 1936), the two major classifications of C. W. Thornthwaite (Thornthwaite, 1931, 1948), together with those of Flohn (1950), Trewartha (1954) and many others.

Most of these climatic classifications involve the basic elements of annual or seasonal rainfall, and annual or seasonal temperature, although some use precipitation effectiveness and thermal efficiency. Almost all classifications or part classifications of climate relate climate to vegetation or in a few cases agriculture, although some such as Flohn (1950) have classified climate according to dynamical considerations. Very few, however, have classified climate from a human point of view. Brazol (1954) with his system of 'comfort months', and the U.S. Quartermaster General with his maps of climate used for assessing clothing needs of armed forces, are two of a small number of classifications published to 1960 that have considered man rather than plants as the climatic base, and even these two systems refer to only a few aspects of climate in relation to human comfort.

Climate and weather do, however, play a large part in our daily living. It is, therefore, rather surprising that although much work has been done on the effect of climate on human beings (such as that by Tromp, 1963a, b, 1964a, b), little thought has been given to analysing or classifying the climate with which we have to live. The tourist poster and travel guide will without hesitation praise the lovely climate of a particular area, but how does one or – probably what is more important – how do the 'authorities' know how lovely their climate really is? Furthermore, the same questions may be asked about the climates of the places in which we have to work and live. Is, for example, the climate of London better or worse than the climate of New York, Moscow or Sydney, and, if it is considered better or worse, how do we know?

Interest in climatic classifications based on man rather than on

1 *Hurricane Camille, photographed by ESSA 9, at 3.57 p.m., E.D.T., on 17 August 1969 as it was approaching the gulf coast of Louisiana and Mississippi* (Photo: ESSA)

YR MO DY HR MIN SC TK ZO S ESSA   M C LAT SP LONG SP ORBIT FR SUN GLINT,
69 8 17 19 57 28 3 60   9   7 2 35N 5 90N 5 2154 5   31N   95W

2 *Hurricane Camille's effect on a trailer court in Biloxi, Missouri* (Photo: ESSA)

3 *Tornado funnel near Enid, Oklahoma, 5 June 1966* (Photo: ESSA)

4 *After a heavy snowfall in downtown Columbia, Missouri – 3 March 1960*
(Photo: Columbia Missourian)

5 *Industrial smoke over Middlesbrough* (Photo: Aerofilms)

6 *Flood waters over farmland in the American Mid-West* (Photo: ESSA)

7 *Marine weather*
(Photo: ESSA)

8 *A computer at the National Meteorological Center, Washington, D.C.*
(Photo: ESSA)

9 *Computer printout showing atmospheric circulation* (Photo: ESSA)

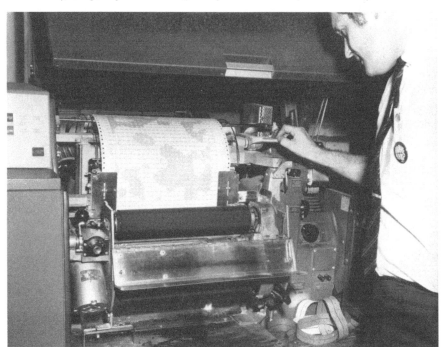

10 Satellite photography reveals eight major storms in Northern Hemisphere on 14 September 1967 (Photo: ESSA).

11 *Meteorologists analysing the Southern Hemisphere circulation use electronically rectified satellite photographs to guide them in sparse data areas* (Photo: ESSA)

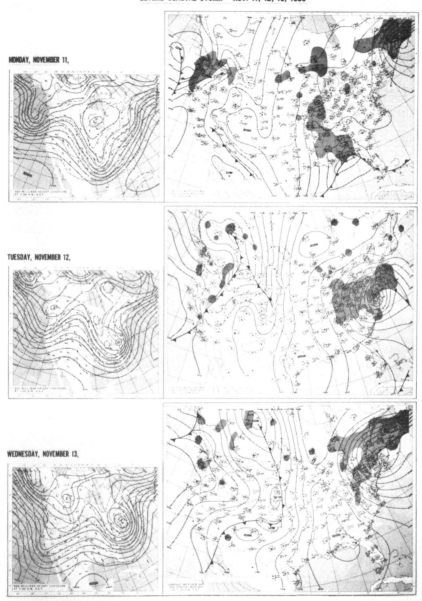

12 *Typical surface and upper level weather maps* (Photo: ESSA)

vegetation has, however, increased in recent years, and during the 1960's several human-oriented climatic classifications were proposed. These include those of Maunder (1962) for a human classification of climate in New Zealand, Terjung (1966a) on the physiological climates of California, Gregorczuk (1968), who mapped the bio-climates of the world related to air enthalpy, and Terjung (1966b) on a bioclimatic classification of the United States based on man. In addition, Prohaska (1967) points out some of the problems associated with terminology in bioclimatic classifications.

The human classification of New Zealand climates (Maunder, 1962) was formulated in part because classifications based on Thornthwaite and Koppen ignored such aspects of rainfall as dura-tion, diurnal variation and intensity; aspects of temperature such as degree-days, days with frost and days with extreme temperatures; and, aspects of the climate concerning sunshine, humidity and wind. It may be noted that whatever the value of the classifications of Koppen and Thornthwaite, they are classifications of only part of the whole climate. The inadequacy of these classifications is shown in Table IV.4, which compares some aspects of the climate of two New Zealand stations, Wellington and Hamilton. The first has on

TABLE IV.4: *Comparison of the climates of three New Zealand stations (after Maunder, 1962)*

Classification	Hamilton	Wellington
Koppen	Cfb	Cfb
Thornthwaite, 1931*	BB'ra	BB'ra
Thornthwaite, 1948†	$B_3B_1'a'r$	$B_3B_1'a'r$
Climatic element		
Mean annual rainfall (in.)	47	49
Mean annual duration of rain (hr)	512	676
Per cent rain 9 p.m.–9 a.m.	45·7	52·8
Mean annual sunshine (hr)	2,070	2,010
Mean days with screen frost	29	0
Mean daily maximum temperature July (°F)	56	51
Mean days with wind gusts ≥40 mph	23	151

* Quoted from Garnier (1950).   † Quoted from Garnier (1951).

the average 151 days per year with wind gusts greater than or equal to 40 mph compared with only 23 days at Hamilton, while Hamil-ton has on the average 29 days with screen frost per year compared with none at Wellington. It is clear, therefore, that the human climates of Hamilton and Wellington are different.

H

Three major difficulties impede the human classification of climates. First, the climatic elements which portray the human climate must be selected (i.e. one must decide what to include and what to omit); second, not all the desired climatic elements are available for classification (e.g. some may not be available, some may not be calculated, and some may be impossible to measure); and, third, once a selection of the climatic elements is made, they must then be 'weighted'. That is, it must be decided how important a particular climatic element is in relation to the total human climate. One must assess, for example, the importance of rain, and, if the rainfall element is further subdivided into amount, duration, intensity, etc., one must decide how to weight these individual items.

To formulate the human climatic index in the New Zealand classification, thirteen climatic elements (or aspects of climate) were considered, and, for each station, the climatic elements are rated 1, 2, 3, 4 and 5. The rating for each element was then multiplied by a factor – the weighting given to the particular element – and the sum of the thirteen weighted ratings is an index (X) of the human climate, X being defined as

$$X = (3R_1 + 3R_2 + 2R_3) + (4S_1 + 3S_2)$$
$$+ (2T_1 + T_2 + T_3 + T_4 + T_5) + (5H_1) + (2W_1 + 2W_2)$$

where: X = Human climatic index
R_1 = Mean annual rainfall
R_2 = Mean annual duration of rain
R_3 = Percentage of rainfall 9 p.m.–9 a.m.
S_1 = Mean annual duration of bright sunshine
S_2 = Mean winter duration of bright sunshine
T_1 = Mean annual degree-days
T_2 = Mean number of days with screen frost per year
T_3 = Mean daily maximum temperature of coldest month
T_4 = Mean annual maximum temperature
T_5 = Mean number of days with ground frost per year
H_1 = 'Humidity index'
W_1 = Mean number of days with wind gusts 40 mph and over
W_2 = Mean number of days with wind gusts 60 mph and over.

Since each climatic element is rated 1, 2, 3, 4 or 5 according to the 'favourability' of the climate, the most favourable total climate has an index of 30 (30 × 1), and the least favourable climate an index of

150 (30 × 5). In the case of the 22 New Zealand stations considered, the range is from 60 to 101. The basis for choosing the various elements is discussed in detail in the original paper (Maunder, 1962).

A different kind of human classification was put forward by Terjung (1966b), who states that the concept of human comfort and its distribution is an old one and has acquired many cultural and regional connotations through historical evolution. Because of this, in addition to the physiological facts of human metabolism brought to light by modern science, the physio-cultural complex of climatic sensations is riddled with customs, preconceived ideas, hearsay and irrational notions, each of which differs historically and regionally.

The social significance of these reactions is of importance, however, for it is reasonably clear that the interactions between climate and culture have had historical repercussions, in that they have influenced, although not necessarily determined, population movement, occupance, military strategy and cultural achievements.

The study by Terjung tested a classification of climatic environments based upon measurable human physiological and psychological reactions. The classification is applied to maps of the United States for purposes of illustration and evaluation, the resultant maps portraying what the author describes as '. . . how man reacts to his climatic environment and to what degree he feels comfortable or uncomfortable'. This, Terjung points out, could be of help in many different kinds of situations. For example, the mapped human climates could be of help to the retired, and to the victims of certain diseases intensified by climatic conditions. Terjung also points out the advantages of such a classification in the field of education, in that '. . . many of the existing cobwebs concerning climates could be eliminated'. By using such a system, the student will develop a better appreciation of the atmospheric environment. The maps could also be used to determine clothing requirements, nutritional needs and physiological and psychological repercussions of an area. In this regard, it is of interest to note that Soviet geographers and the U.S. Army Natick Laboratories have been active in this field of climate and clothing requirements. Liopo (1968), for example, says that an important problem in medical geography, bioclimatology, physiology, and hygiene associated with the opening up of the vast territory of Siberia and the Soviet Far East is the provision of optimal conditions for the life and work of man. A further suggestion made by Terjung is that bioclimatic maps would simplify knowledge regarding housing needs, building materials and heating or cooling requirements. The maps could also be useful in a regional analysis in gauging the potential of areas or places.

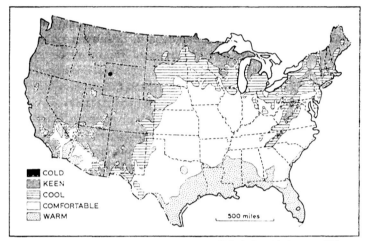

FIG. IV.2. *Comfort index for night-time in July (after Terjung, 1966b).**

It is appropriate to comment that Terjung's classification is a refreshing change from those classifications which leave man standing along the sidelines. Indeed, maps showing such things as comfort indices for night-time (Fig. IV.2) could be of significance when assessing the value of the weather.

C. HEALTH ASPECTS

1. Human biometeorology

Human biometeorology comprises the study of the influence of weather and climate on both healthy and diseased man. According to Tromp (1963b), many specialists in the field of medicine were inclined to consider the physico-chemical composition and the structure of the human body to be so constant that it was little affected by the meteorological forces in the surrounding atmosphere. However, Tromp points out that one of the great merits of the modern development of human biometeorology is that it emphasizes the continuous interaction between the internal physiological processes in man and the perpetually changing weather and climatic conditions in the atmosphere. Research into the field of human biometeorology shows a wide field of activities covered, and for a full discussion the reader is referred to any of the standard texts in the field such as Tromp (1963a).

* Reproduced by permission from the *Annals* of the Association of American Geographers, Volume 56, 1966.

It is relevant to note, however, that the factors of weather which appear to influence human health may be considered under four headings: sunlight, heat, wind and cold. Sunlight, for example, is one of the most important factors in man's environment, for included in its roles are the control of terrestrial temperature and the supply of energy both as food, and as power for the manifold activities of man. Some climates seem to be more healthful than others, but this would not appear to be closely related to the direct effects of sunlight on man. Two factors associated with sunlight, however, do appear to be significant from a 'value' point of view. First, exposure to the 'mid-day' summer sun may result in reddening of skin, or if exposure is long enough, there may be blistering and peeling of most of the superficial layers. Second, ultraviolet radiation may in some cases induce cancer of the skin. Licht (1964) in commenting on this says that it may be argued that a single dose never produces a cancer within the lifetime of the individual, but that all of us who have been exposed to sunlight have cancer of the skin developed to a varying extent on the exposed areas. In a similar manner the importance of heat (heat cramps, heat stress), wind (anginal attacks) and cooling (rheumatic pains, chilblains) should also be considered.

The science of biometeorology, however, according to Tromp (1964a) emphasizes the *continuous* interaction between the internal physiological processes in man and the perpetually changing weather and climatic conditions in the atmosphere. Tromp suggests that the weather probably affects the body in five principally different ways. First, it may stimulate the skin due to thermal stresses created by heat conduction, convection and radiation; second, it may stimulate the eyes and head, mainly as a result of solar radiation; third, the internal mucous membranes of the nose may be stimulated by changes in the humidity of the air; fourth, the lungs may be stimulated by changes in the ion content of the air, by trace substances in the air such as ozone and by gaseous or particulate, organic or inorganic air pollutants; and fifth, weather may stimulate directly the peripheral nerves possibly by electrostatic and electromagnetic fields. Each of these 'stimulants' has been described in greater detail in a number of publications, such as the influence of atmospheric pressure upon the physiological and psychological reactions of human beings (Dordick, 1958).

2. Effects of weather on certain ailments

Many studies have been made on the effect of weather and climatic conditions on specific diseases, but space will allow only brief mention of some of these studies here. It is important to note, however, that the economic effect of disease is in some cases significant, and that some kinds of weather modification, if continued, could have adverse effects on the prevalence of certain kinds of disease. On the other hand, it is quite possible that the right kind of weather modification could decrease the incidence of certain diseases, and it is reasonably clear that accurate medium- to short-range weather forecasts could have beneficial effects, in that precautions could be taken before the onset of weather conditions adverse to health.

A summary of the principal effects of weather and climatic conditions is given in Table IV.5. A detailed analysis of the influence of weather on asthma, rheumatic diseases and heart diseases, is given in the survey paper by Tromp (1964b), and many additional references are given in Tromp (1963a).

Additional studies include those of Davis (1958) and Derrick (1965), on the effect of weather on ulcers and asthma respectively. In the first paper a comparison of the number of duodenal ulcer cases admitted to the Philadelphia General Hospital over a five-year period with the temperature-change curve over the same period was made by Davis (1958). This showed close agreement, indicating that haemorrhages from duodenal ulcers may be brought on by marked variations in temperature. A warm climate with relatively little daily and seasonal change in temperature would therefore appear to afford the most suitable residence for those prone to suffer from duodenal ulcers. Accordingly Davis suggests that those northerners who are afflicted with duodenal ulcers might well be on the lookout for cold wave warnings in the autumn. He notes that such warnings cause construction workers to stop pouring concrete and serve to make fuel oil companies prepare for extra deliveries. They might therefore according to Davis '. . . be used to equal advantage by ulcer patients'.

The variation of asthma in Brisbane, Australia, and its relation to weather conditions is examined by Derrick (1965), who says there is a close relationship between seasonal variations and temperature. The relationship is twofold. First, when the mean temperature is under about 70°F (21·1°c) (which includes the six months May to October) it is linear, positive and delayed, and asthma variations

TABLE IV.5: *Reported effects of the influence of weather and climate on diseases* (abridged from Tromp, 1964b)*

Short periodical effects	Long periodical effects (seasonal or pseudo-seasonal)
Lung diseases	
Tuberculosis: Haemoptysis suddenly increases in clinics after oppressive warm weather before thunderstorms, after föhn, humid cold foggy weather or sudden heat waves.	Increased sensitivity to tuberculin test in March and April; low during autumn.
Asthma (bronchial): Increases with sudden cooling (particularly if accompanied by falling barometric pressure and rising wind speed); during high barometric pressure and fog (in W. Europe) very low asthma frequency.	Low in winter, suddenly increasing after June, max. in late autumn (W. Europe).
Bronchitis: Increasing complaints during fog (particularly in air-polluted areas) and specially if accompanied by atmospheric cooling.	High in winter, low in summer (in W. Europe).
Hay fever (and various forms of rhinitis): Allergic reactions often increase during atmospheric cooling.	Hay fever is related to flowering of certain plants or grasses, different for different countries. In W. Europe usually max. complaints in May–June.
Cancer	
Skin cancer: More common with increasing number of sun-hours and increased exposure of the skin to the sun.	—
Rheumatic diseases	
Most forms of arthritis react on strong cooling (falling temp.; strong wind). Humidity seems to have no direct effect, only indirect through cooling.	Arthritic complaints particularly common in autumn and early winter (W. Europe).
Heart diseases	
Coronary thrombosis, Myocardial infarction and Angina pectoris: Occur more frequently shortly after a period of strong cooling.	Highest mortality in Jan.–Feb. (in W. Europe and northern U.S.A.), lowest July–Aug. In hot countries (e.g. southern U.S.A.) highest mortality in summer, lowest in winter.
Infectious diseases	
Common cold: Weather changes affecting thermoregulation mechanism, membrane permeability and growth and transmission of common cold virus seem to initiate the diseases (e.g. very cold period followed by sudden warming up).	Max. in Feb.–March; increasing from Sept.–March (in W. Europe).
Influenza: Rel. humidity below 50% and low wind speeds seem to favour the development and transmission of influenza virus.	Max. in Dec.–Feb.; increasing from Sept.–March.

* For authors and references see Tromp (1963a), pp. 374–8, 575–84.

follow temperature variations after a lag of one to two months below
70°F; second, when the mean temperature exceeds about 70°
(usually from early November to mid-April) asthma decreases, the
relation being negative and without a lag. In the light of these
findings, Derrick suggested that the positive relation to temperature
under 70° depends on the growth of grasses and other vegetation
from which, after they grow and decay, allergens are derived, and
that allergen production is inhibited when the temperature exceeds
70°.

It should be noted that most biometeorologists do not appear to
place any great emphasis on the economic aspects of their findings.
It appears particularly appropriate to point out, therefore, that
since health, disease and economic activity are closely associated, and
that some association is also apparent between weather and health
and disease, it is reasonable to suggest that variations in the man's
atmospheric resources have, through health and disease, a direct
bearing on economic activity. The growing air pollution problem,
and the creation of larger and more efficient man-made climates,
and their association with the health of man, should therefore be
given close attention.

3. Weather and mortality

The association that exists between heavily polluted atmospheres
(e.g. London in 1952, and Donora in 1948) and excessive mortality
is mentioned in Chapter VI, and it is reasonably clear that under
certain atmospheric conditions, excessive pollutants can cause
many 'additional' deaths. By contrast, in normal atmospheric conditions, and with normal amounts of pollutants, mortality rates appear
to show few significant variations. However, there is usually a
reasonably significant variation in death-rates when associated with
seasonal weather conditions. For example, Momiyama and Kata-
yama (1967) report that in the United States, although the seasonal
fluctuation of the death index of man is found to be rather moderate
for all age groups and little variation is seen, particularly for infants,
in the cold season, the death index curve rises for old age groups. This
rise is much less pronounced in the United States than in Japan, and
the authors suggest that the moderate curve of the death index in
the United States may be ascribed to the wider use of effective room-
heating systems. Thus, 'protected' by an artificial climate, the
American people do not appear to be affected by seasonal changes
as markedly as those of other nations.

The authors point out, however, that the death index curve behaves differently for the American whites and non-whites. Two peaks appear for the latter, one in summer and the other in winter, whereas only one peak appears in winter for the former. This summer peak is ascribed to the lower standard of living of the non-whites in the United States, according to Momiyama and Kata-yama, a factor which is also probably related to the relatively fewer number of air-conditioning facilities in the non-white areas of the United States.

The seasonal variation of deaths in the United States has also been examined by Rosenwaike (1966), who showed that in the decade 1951–1960 the outstanding seasonal patterns were those of the two respiratory groups: influenza and pneumonia (except pneumonia of newborn, and bronchitis). The death rates in both categories in the peak winter month (January) were twice those in the low summer month (August). In addition, in virtually all of the major cardio-vascular–renal diseases (which account for slightly more than half of all deaths), the seasonal swings correspond to those of the res-piratory diseases (with which they are generally associated), al-though the amplitudes of the curves are much reduced. Rosenwaike notes also that the seasonal curve of deaths from suicide is ex-ceptional, a striking feature being a peak in the spring (usually April) and a trough at the close of the year (December). Kanellakis (1958) has also noted that the suicide curve rises to a peak in April in England.

Although a considerable amount of research is still to be done, it seems certain that weather and climatic conditions are significantly important in many mortality rates. It could also be suggested that man's artificial control of his environment has effectively reduced his mortality. As with most biometeorological problems, however, it is important to note that the weather and climatic conditions are only one factor in a highly complex matrix.

D. RECREATION AND SPORT

1. Recreation

The effect of weather on outdoor recreation is probably the greatest and the most influential variable on the flux in numbers and the cost of the various types of recreation. Moreover, because man erects buildings, transports himself in enclosed vehicles and attempts to change the atmospheric conditions which affect him, it may be assumed that, at least in part, this 'atmospheric modification' will be carried on when he uses his free time for his own enjoyment. In addition, social customs seem to have developed which require a certain type of weather for a particular activity, and in Anglo-America, in particular, an increase in time and money, as well as greater mobility, appears to 'permit' a narrower range of acceptable conditions. Therefore in order to maximize on a capital outlay for recreation facilities one may wish to lengthen the period of acceptable conditions through regional weather modification in order to increase returns.

Few papers have been published on the influence of weather on outdoor recreation, and Clawson (1966) points out that 'solid professional discussion of outdoor recreation is always difficult', since nearly everyone has experienced outdoor recreation personally, and he naturally feels he knows something about it. Even a modest acquaintance with outdoor recreation suggests, however, that it is weather-sensitive, and anyone who has had a picnic spoiled by a sudden downpour can testify to that. Nevertheless, Clawson suggests that we need to explore in just what ways, and at what stages in the recreation experience, weather and climate are most influential.

Clawson goes on to describe some basic considerations about the nature of outdoor recreation in order to better understand the basic relationships which weather is likely to influence. These include five interrelated sequential stages: first, anticipation or planning, when the person or family considers what to do, where and when to go, what to take, and how much time and money to spend; second, travel from home to the site; third, on-site experiences; fourth, travel home again; and fifth, recollection, when one relives the earlier experiences, and perhaps reinterprets them differently than one did at the time.

Most forms of outdoor recreation are also dependent upon a certain range of temperature, sunshine, humidity, wind velocity and other climatic factors, if they are to be enjoyed. But climate also affects the kind of environment in which outdoor recreation takes place, the most obvious effects being upon water supply, vegetation, and snow for skiing. In addition, the fact that leisure or free time for outdoor recreation is rather rigidly controlled by time is important, in that one must go when one has free time, not when weather is favourable. Moreover, one could suggest that the elements of available time and money form a ratio which, in recreation terms, may be termed 'desirability'. Consider, for example, the case where either time or money is limited. The alternatives to a person in this situation are twofold. If the time is short, then money will probably be spent on rapid transportation to a place that assures a high probability of weather conditions favourable to his intended form of outdoor recreation. On the other hand, if money is in short supply, then a slower and cheaper form of transportation may be taken to the same area he finds favourable for a shorter recreation time. Alternatively, limited time and funds may necessitate visiting an area with poorer weather conditions.

The possible value to recreation of weather modification has already been pointed out, but it is clear that more accurate and longer-range weather forecasts would also have considerable effect. Nevertheless, if everyone knew that the Saturdays and Sundays *next* month were going to be favourable for recreation, then problems would inevitably arise through crowded recreational facilities. Perhaps, therefore, we could do well to just wish for the ideal recreational climate, as viewed by Clawson as:

> . . . one where it never rained, was always pleasantly warm but not hot, was always mildly sunny, was never too humid, had only gentle breezes, etc.; and yet which had luxuriant forest and other vegetation, flowing clear pure streams, cold enough for trout but warm enough for swimming, and possessed various other desirable attributes.

Nevertheless, whatever the weather, one may still prefer to watch the football, hockey, cricket or tennis match on television in one's own environmentally conditioned living room, rather than to recreate at the nearby beach, forest or golf course, no matter what the weather is, is going to be, or how it is going to be modified.

2. Sport

In general, the effect of weather on sport can be considered in two ways. First, the effect of weather on the sport itself, whether this be cricket, tennis, golf, football or horse racing; and second, the effect of weather on the attendance and related activities at the sports arena or ground.

Baseball, for example, is a dry-weather game, a few of the weather-related problems in wet conditions being poor footing, wet equipment and soft fields. Football, of whatever code, is also affected by weather conditions, the three principal factors being the condition of the ground, sun and the wind. Similar weather-related problems apply to golf, tennis, horse racing and numerous other sports. The horse-racing industry, for example, relies on the spectator participation (money) to support it, and rain, of course, affects attendance. But it also affects the strategy of the betting, since some horses run better on a sloppy or muddy track than they do on a fast track; betting patterns change, therefore, in direct response to the spectators' view of how the prevailing track conditions will affect the various horses in the race. It is worth noting, however, that there appears to be a difference between the influence of weather on the amateur player and the professional, in that in amateur sports the chief concern is often protecting the field or grounds, whereas in professional sports it is the player, or in the case of horse racing, the horse, that is more often protected against adverse weather conditions.

The second major effect of weather on sport is the influence it has on the spectators. It can be generalized that most adverse conditions tend to keep spectators at home, away from outdoor as well as indoor sports, and this has a great effect on the management. For example, promoters normally prepare for a certain number of spectators by producing a comparable number of programmes, by ordering adequate concession supplies (a good deal of them perishable) and by hiring adequate help to handle concessions, gates and parking.

Some sports can, of course, draw crowds no matter what the weather conditions are, and important games, particularly professional sports, or the world-class sports like soccer, hockey, rugby and cricket, normally 'sell out' very quickly. But even here management has found it profitable to install underground electric cables in the Green Bay Stadium, Wisconsin, home of the National Football League 'Green Bay Packers', and at Murrayfield, Edinburgh, scene of many international rugby matches, in an effort to keep the field clear of snow and frost. In only one place, however, has there been an attempt to change the weather. This is at Houston, Texas, where a stadium was built to hold 60,000 people. What made it different was that they put a roof over it and thereby created a new microclimate. It has all the features of an indoor arena including upholstered theatre seats, and air conditioning, and not unexpectedly the effect on attendance of the regularly scheduled baseball games was tremendous. In 1964, for example, the year before the stadium was opened, the attendance was 725,000, compared with 2,539,000 in 1965, after the Astrodome was completed. Other features of this building include the Astroturf, a new 'grass' product, as well as the associated $7 million worth of roads and sewers. The building of the Astrodome has also pushed the price of land up from $10,000/acre to $100,000/acre, and all of this to insure against adverse weather.

Very few specific papers on the effect of weather on sports have been published, and much of the above can be found on the sports pages of any daily newspaper. However, both Shaw (1963) and Watts (1968) have published papers dealing with the effect of weather on baseball and Olympic yachting respectively.

Other cases could be given of the effect of the weather on the performances at the 1968 Olympics held in the rarefied atmosphere of Mexico City, and how athletes from Kenya and Ethiopia outran their counterparts from the United States and the Soviet Union. Mention could also be made of a certain August Saturday afternoon at Athletic Park in Wellington, New Zealand, when the weather conditions were so adverse that some 10,000 meat pies went unsold as a result of an attendance of only half the expected 60,000 at an international rugby match. But, that is another story.

BIBLIOGRAPHY

BATES, M., 1966: The role of weather in human behavior. In: SEWELL, W. R. D., ed., 1966: *Human Dimensions of Weather Modification.* University of Chicago, Dept. of Geography, Research Paper No. 105, pp. 393–407.

BELDING, H. S. and HATCH, T. F., 1955: Index for evaluating heat stress in forms of resulting physiological strains. *Heat, Piping Air Condit.*, 27: 129–36.

BRAZOL, D., 1954: Bosquejo bioclimático de la República Argentina. *Meteoros.*, 4: 381–94.

BROOKS, C. E. P., 1949: *Climate Through the Ages: A Study of the Climatic Factors and Their Variations.* McGraw-Hill, New York, 395 pp.

CLAWSON, M., 1966: The influence of weather on outdoor recreation. In: SEWELL, W. R. D., ed., 1966: *Human Dimensions of Weather Modification.* University of Chicago, Dept. of Geography, Research Paper No. 105, pp. 183–193.

DAVIS, F. K., JR, 1958: Ulcers and temperature changes. *Bull. Amer. Met. Soc.*, 39: 652–4.

DAVIS, N. E., 1968: An optimum summer weather index. *Weather*, 23: 305–17.

DERRICK, E. H., 1965: The seasonal variation of asthma in Brisbane: its relation to temperature and humidity. *Inter. Jour. Biomet.*, 9: 239–51.

DEUTSCH, A. (ed.), 1967: *The Encyclopedia of Mental Health*, vol. 6, F. Watts, New York.

DORDICK, I., 1958: The influence of variations in atmospheric pressure upon human beings. *Weather*, 13: 359–64.

FEDERAL BUREAU OF INVESTIGATION, 1965: *Uniform Crime Reports for the United States.* United States Department of Justice.

FERGUSSON, P., 1964: Summer weather at the English seaside. *Weather,* 19: 144–6.

FLOHN, H., 1950: Neue Anschauungen über die allgemeine Zirkulation der Atmosphäre und ihre Klimatische Bedeutung. *Erdkunde,* 4: 141–62.

GARNIER, B. J., 1950: The climate of New Zealand: according to Thornthwaite's classification. In: GARNIER, B. J., ed., 1950: *New Zealand Weather and Climate.* Whitcombe & Tombs, Christchurch, pp. 84–104.

GARNIER, B. J., 1951: Thornthwaite's new system of climatic classification in its application to New Zealand. *Trans. Roy. Soc. N.Z.,* 79: 87–103.

GENTILLI, J., 1958: *A Geography of Climate.* University of Western Australia Press, Perth, pp. 120–66.

GREGORCZUK, M., 1968: Bioclimates of the world related to air enthalpy. *Inter. Jour. Biomet.,* 12: 35–9.

HAAS, J. E., 1968: Sociological aspects of human dimensions of the atmosphere. In: SEWELL, W. R. D., et al., 1968: *Human Dimensions of the Atmosphere.* National Science Foundation, Washington, D.C., pp. 53–7.

HOUGHTON, F. C. and YAGLOGLOU, C. P., 1923: Determining lines of equal comfort. *Jour. Amer. Soc. Heat. Vent. Eng.,* 29: 165–76.

HOUNAM, C. E., 1967: Meteorological factors affecting physical comfort (with special reference to Alice Springs, Aust.). *Inter. Jour. Biomet.,* 11: 151–62.

HUGHES, G. H., 1967: Summers in Manchester. *Weather,* 22: 199–200.

HUNTINGTON, E., 1945: *Mainsprings of Civilization.* Wiley, New York, 660 pp.

KANELLAKIS, A. P., 1958: Seasonal swing in mortality in England and Wales. *Brit. Jour. Prev. Soc. Med.,* 12(4):

KÖPPEN, W., 1931: *Grundriss der Klimakunde.* Walter de Gruyter Company, Berlin.

KÖPPEN, W., 1936: Das Geographische System de Klimate, being Vol. 1, Part C. of *Handbuch der Klimatologies,* W. KÖPPEN and R. GEIGER, ed., Gebrüder Borntraeger, Berlin.

LEE, D. H. K., 1958: Proprioclimates of man and domestic animals. *Arid Zone Res.* (UNESCO), 10: 102–25.

LICHT, S., 1964: *Medical Climatology.* Waverly Press, Baltimore, 753 pp.

LIOPO, T. N., 1968: A method for computing optimal heat-regulating properties of clothing on the basis of the probability of variation of meteorological factors. *Soviet Geog.,* 9: 95–105.

McLEOD, C. N., 1965: An index of comfort for Gan. *Met. Mag.,* 94: 166–71.

MARKHAM, S. F., 1947: *Climate and the Energy of Nations.* Oxford University Press, New York, 236 pp.

MAUNDER, W. J., 1962: A human classification of climate. *Weather,* 17: 3–12.

MAZZARELLA, D. A., 1967: Sometimes on Sunday – it rains. *Weatherwise,* 20: 259–63.

MILLER, W. H., 1968: Santa Ana winds and crime. *Prof. Geog.,* 20: 23–7.

MILLS, C. A., 1942: *Climate Makes the Man.* Harper, New York, 320 pp.

MOMIYAMA, M. and KATAYAMA, K., 1967: Seasonal fluctuations of mortality in the U.S.A. *Inter. Jour. Biomet.*, 11: 223 (Abstract only).

PARR, A. E., 1968: Remarks on climate environmental designs. In: SEWELL, W. R. D., *et al.*, 1968: *Human Dimensions of the Atmosphere.* National Science Foundation, Washington, D.C., pp. 143–6.

POULTER, R. M., 1962: The next few summers in London. *Weather*, 17: 253–7.

PROHASKA, F., 1967: Climatic classifications and their terminology. *Inter. Jour. Biomet.*, 11: 1–3.

RACKLIFF, P. G., 1965: Summer and winter indices at Armagh. *Weather*, 20: 38–44.

ROSENWAIKE, I., 1966: Seasonal variation of deaths in the United States 1951–60. *Jour. Amer. Stat. Assn.*, 61: 706–19.

SHAW, E. B., 1963: Geography and baseball. *Jour. Geog.*, 62: 74–6.

SIPLE, P. A. and PASSEL, C. F., 1945: Measurements of dry atmospheric cooling in subfreezing temperatures. *Proc. Amer. Phil. Soc.*, 89: 117–99.

STEPHENSON, P. M., 1963: An index of comfort for Singapore. *Met. Mag.*, 92: 338–45.

TERJUNG, W., 1966a: Physiological climates of California. *Yearbook of Pacific Coast Geographers*, pp. 55–73.

TERJUNG, W., 1966b: Physiologic climates of the conterminous United States: a bioclimatic classification based on man. *Annals Assn. Amer. Geog.*, 56: 141–79.

THOM, E. C., 1959: The discomfort index. *Weatherwise*, 12: 57–60.

THORNTHWAITE, C. W., 1931: The climates of North America according to a new classification. *Geog. Rev.*, 21: 633–55.

THORNTHWAITE, C. W., 1948: An approach toward a rational classification of climate. *Geog. Rev.*, 38: 55–94.

TREWARTHA, G. T., 1954: *An Introduction to Climate.* McGraw-Hill, New York, pp. 234–7.

TROMP, S. W., 1963a: *Medical Biometeorology.* Elsevier, Amsterdam, 991 pp.

TROMP, S. W., 1963b: Human Biometeorology. *Inter. Jour. Biomet.*, 7: 145–58.

TROMP, S. W., 1964a: Weather, climate and man. In: *Handbook of Physiology.* American Physiological Society, Section 4, Chapter 16.

TROMP, S. W., 1964b: Influence of weather and climate on human diseases. In: A survey of human biometeorology. *W.M.O. Tech. Note Series*, No. 65, Chapter 5, Section 2.

WATT, G. A., 1967: An index of comfort for Bahrain. *Met. Mag.*, 96: 321–7.

WATTS, A., 1968: Winds for the Olympics. *Weather*, 23: 9–23.

WILSON, O., 1963: Cooling effect of an antarctic climate on man. With some observations on the occurrence of frostbite. *Norsk Polarinstitutt Skrifter*, Nr. 128.

WILSON, O., 1967: Objective evaluation of wind chill index by records of frostbite in the Antarctica. *Inter. Jour. Biomet.*, 11: 29–32.

ADDITIONAL REFERENCES

BRIDGER, C. A. and HELFAND, L. A., 1968: Mortality from heat during July 1966 in Illinois. *Inter. Jour. Biomet.*, 12: 51–70.

BROOKS, C. E. P., 1951: *Climate in Everyday Life*. Philosophical Library, New York, 314 pp.

DINGLE, A. N., 1957: Hayfever pollen counts and some weather effects. *Bull. Amer. Met. Soc.*, 38: 465–9.

DRISCOLL, D. M. and LANDSBERG, H. E., 1967: Synoptic aspects of mortality. A case study. *Inter. Jour. Biomet.*, 11: 323–8.

FALCONER, R., 1968: Wind-chill, a useful wintertime weather variable. *Weatherwise*, 21: 227–9, 255.

LANDSBERG, H. E., 1969: *Weather and Health – An Introduction to Biometeorology*. Doubleday & Co., Inc., Garden City, 148 pp. (American Meteorological Society: Science Study Series).

LEE, D. H. K., 1968: Culture and climatology. *Inter. J. Biomet.*, 12: 317–19.

LOWRY, W. P., 1969: *Weather and Life*. Academic Press, London, 305 pp.

MARSHALL, ANN, 1963: The prediction of indoor heat discomfort. *Aust. Geog. Studies*, 1: 115–23.

MOMIYAMA, M., 1968: Biometeorological study of the seasonal variation of mortality in Japan and other countries on the seasonal disease calendar. *Inter. Jour. Biomet.*, 12: 377–93.

RICE, R. W., 1969: Weather and pigeon racing in Malta. *Weather*, 24: 281–2.

TERJUNG, W. H., 1968: World patterns of the distribution of the monthly comfort index. *Inter. Jour. Biomet.*, 12: 119–51.

V Economic analysis of weather

A. THE SETTING

The annual global expenditure for meteorological services is currently of the order of one thousand million U.S. dollars (W.M.O., 1968). This amount represents what could be termed a recognition by the various governments of the present 'value' of meteorology. Improved techniques for monitoring the atmospheric circulation will require even larger expenditures, and the costs involved in the provision of satellites, electronic computers and high-speed communications systems, raise pertinent questions concerning the potential value of atmospheric resources to each country and to the world community.

It is not a simple matter to identify the benefits and costs involved in the use of atmospheric resources, for the dynamic nature and the common property characteristics of the atmosphere lead inevitably to complicated interactions. Perhaps because of these, few attempts have been made to evaluate atmospheric resources, and as a result there is no completely integrated study that attempts to give net effects, or even describe the complex relationships between weather elements and regional or national economies.

There are many theoretical models of the physical atmosphere, but few are concerned with the economic and social aspects of the atmosphere. An appraisal of the relevant literature indicates that little consideration has been given to what may be described as the end product in the meteorological chain. This chain has many parts, but the 'value of the product' is of basic importance. Simply, what is one inch of rain, or for that matter any rain, worth to a consumer, an area or a nation? Is it possible to place a value on a wetter-than-normal summer to the farmers in the U.S. Middle West, or a cold snowy winter to the people of Moscow, London or New York?

In developing econoclimatic models to answer these and similar questions it should be noted that the weather can provide an input into the productive process in one of two ways. It can provide an ingredient or *quantity* which is incorporated directly into the product. Rainfall, for example, can provide the water required for a parti- cular product, in which case the water is used only for the production of that produce. Second, the weather can provide the *conditions* in which a productive process can take place, such as the temperature required for the growing of corn. In such a case its use for this purpose does not preclude its use for other purposes. This characteristic of weather differentiates it, therefore, from other factors of produc- tion, and must be considered in the formulation of econometric models which include weather as an input.

Various techniques have been suggested for the evaluation of the weather, one of which is the assessment of the percentage *reduction* in 'weather damage' if weather services (including forecasts) are improved. For example, in the United States it is estimated that the weather losses in all classes of air transportation in 1965 amounted to $350 million; it has been further estimated that improved avia- tion weather forecasts could have reduced this by $30 million (W.M.O., 1967). In other sectors such as agriculture, particularly in developing countries, it may also be possible to estimate the improvement in yield or output following the *introduction* of meteoro- logical services, and thus place a value on the totality of weather information and forecasts, and not only the incremental benefit/ cost ratio.

The evaluation of modified atmospheric resources must also be considered (see Sewell, 1966). A complete evaluation of weather modification requires the identification of activities that are affected directly or indirectly by weather changes, and an analysis of the manner in which a given change results in gains or losses to such activities. In addition, account must be taken of the fact that weather modification is only one means of attaining the objectives that are specified. The merits of weather modification must also be compared with those of other alternatives, in order to determine the most efficient means of achieving the desired objective. Several possible approaches to the study of economic and social aspects of weather modification have been suggested (see White, 1966). One method would be to specify weather modification in the form of a 10% increase in rainfall, and then trace its impact on economic and social relationships. An alternative approach would be to assess the extent to which various activities are affected by the weather and then specify what aspects of weather modification would be required to improve the efficiency of the activity.

Most of the research has utilized the first approach, determining the effects of particular weather modification projects on particular activities. Little is known, however, about the extent to which various activities are affected by the natural weather, and consequently, the extent to which man-made changes might make any difference to their efficiency. Research into an evaluation of the natural atmosphere is therefore necessary before any true evaluation of the effects of the modified atmosphere can be made.

B. EVALUATION OF WEATHER AND CLIMATE

1. Problems of identification and measurement

During the 1960's an increasing demand for improved techniques for measuring the impacts of weather and climate on economic activities was apparent. This demand stems in part from mounting losses of property and income which result from extreme weather events, in part from man's increasing use of the atmosphere as a transportation route, and in part from his increasing ability to modify the weather.

Allied with this growing awareness of the economic, social and political aspects of meteorology is the programme of World Weather Watch, which provides for a considerable increase in weather services and information available to the world. Many of the results of this programme will contribute to man's basic understanding of the atmospheric circulation; it is more important to realize, however, that this understanding is much more than a purely scientific endeavour, for it also involves many practical aspects which are expected to bring significant economic, social and other benefits to the world as a whole (Davies, 1966).

To assess the nature of these economic and related benefits, a study of the subject was initiated by the W.M.O. Secretariat early in 1966. The report by J. C. Thompson (1966), concerning the economic and other gains which may be expected to accrue from the implementation of the World Weather Watch programme, revealed that little attention had been given by meteorologists to the non-scientific gains of their profession.

The report represents a study of the economic and other benefits which would accrue to the global economy through improved weather services associated with the World Weather Watch. Many of the benefits, although evidently very large, are for the present only qualitatively apparent. Of those benefits that are capable of quantitative assessment, some are expressed in non-monetary terms (saving in human life and health, for example), while others are complicated by the probable existence of secondary side effects which may partially or wholly offset the primary gains. Nevertheless, it seems clear that the attainment of significant benefits greatly exceeds the estimated cost of implementing the World Weather Watch.

The benefits of weather services are difficult to evaluate, and as Mason (1966) correctly points out, one cannot apply the criteria of the price the customer is willing to pay, because most if not all of the basic meteorological services are free. Nevertheless, an assessment of the economic value of meteorological services to the United Kingdom indicates a probable overall benefit/cost ratio of about 20 to 1, although neither the data nor the inevitable assumptions appear to support a very vigorous or refined analysis.

Economists unfortunately appear to have shown only a minor interest in the economic aspects of the human use of the atmosphere. Several reasons account for this according to Crutchfield and Sewell (1968). First, it is only recently that major problems have arisen in the management of atmospheric resources or, more accurately, that they have become a matter of real public concern; second, not until the 1950's was enough known about the atmosphere to specify accurately the physical dimensions of weather variations or the possibilities of modifying the weather and thus to provide alternatives for evaluation; and third, the systematic development and application of economic principles in the whole field of resource management are relatively new.

The human use of the atmosphere is not, however, limited to the use of the natural atmosphere *per se*, but is also concerned with man's adjustment to the atmosphere. Three major alternative adjustments to the weather are possible: (1) reducing uncertainty through improvement of weather information (including forecasts), (2) devising techniques to offset the impact of weather, and (3) developing means to modify the weather. Given these alternatives, several factors need to be taken into account to determine which approach is economically most efficient. Crutchfield and Sewell (1968) have indicated that three questions are pertinent: (1) which activities are sensitive to weather variations and in what degree, (2) to what extent are the various adjustments technically feasible, and (3) what are the gains and costs of each of the alternatives?

The specific relationship between the weather and various economic activities have, however, been explored only to a limited extent, most of the research having been given to the impact of weather on agriculture, and to a lesser extent on transportation, and the construction industry. The impact of weather on such economic activities as manufacturing or the service trades is less well known. Further, even in those activities in which research has been done, knowledge of overall impacts is imperfect. In agriculture, for example, some indication is available of the effects of a specific weather variable on a crop in a given region. But knowledge is still lacking of how the weather affects all crops in a region. It is possible, for example, to estimate how a dry summer affects butterfat production in the Waikato area of New Zealand (Maunder, 1968a), but estimates of the overall impact on that area's economy are fairly crude, and estimates of impacts of particular weather events on a national economy border on guesswork.

Variations in weather and climate affect economic activities in at least three major ways (Crutchfield and Sewell, 1968). First, temperature, precipitation and wind are important elements in certain production processes, variations in temperature and precipitation (or more specifically variations in the water and heat balances) often marking the boundary between success and failure in agriculture. The amount and timing of precipitation also have important effects on the operation of hydro-electric power systems. Second, the weather sometimes causes considerable losses of property and income, as in the case of strong winds, heavy snow, lightning or hail. Third, the constantly changing atmosphere is the most dynamic element of the natural environment. The timing and the magnitude of these changes are difficult to predict; accordingly, uncertainty is a major factor affecting economic use of the atmosphere. This is particularly the case in regard to the use of the resources of the atmosphere at some future time, when decisions are based on weather forecasts.

Whatever the economic activities, the analysis of meteorological costs and the related benefits is dependent to a considerable extent on the nature and accuracy of the economic data upon which it is based. In an ideal case, the data may be capable of expression in identical or similar units and thus costs and benefits may be compared directly. In practice, however, while costs may frequently be obtained or converted to monetary values, the benefits will generally be collected in a heterogeneous series of units. These range from numerical quantities such as money, or lives saved, to qualitative and intangible values such as personal satisfaction or an increased sense of security. In many cases it is difficult to combine this latter group into a mutually comparable system of units. Nevertheless, several techniques which can be described as econoclimatic models have been devised and these are now examined.

2. Econoclimatic models

In many sciences, we seek to understand the past in order to be able to predict and control the future. To do this, it is necessary to marshal whatever data, statistical or otherwise, and whatever theory, intuition or hypotheses that may be at hand into a consistent and manageable whole. In economic climatology, where adequate data are frequently lacking, it is especially important, therefore, that the structure be such as to allow theory and intuition to play a major role. In this regard the growing importance of mathematical and economic tools in applied meteorology is noteworthy.

In a study of the techniques available, McQuigg (1965) noted that it *is* possible to think of the atmosphere in purely physical mathematical terms, and to conduct investigations that are completely meteorological without concern about how meteorological events might affect life. It is also possible to study biological or physical processes without thought for meteorological events. It is equally possible to consider buying, selling and the management of production and distribution of agricultural wealth with little consideration of meteorology and biology, for an investigator in one of these specialities could justly claim that his particular discipline was complex enough without these additional variables.

Nevertheless, mathematical and economic tools do exist which allow analysis of a system that includes meteorological, biological and economic processes, but as McQuigg points out: 'In too many instances, the lack of specific economic data is the chief cause of inability to apply these modern management tools. Climatological data are readily available; weather forecasts are given wide distribution. The processing of large amounts of data is no longer a bottleneck.'

At the Symposium on the Economic and Social Aspects of Weather Modification, held at the National Center for Atmospheric Research in 1965, the problem of identifying and measuring the economic impacts of weather variations was considered. A proposal for the development of an Ideal Weather Model was suggested by Ackerman (1966). Such a model would identify and measure variations of actual weather conditions from conditions considered ideal for the pursuit of different economic activities, and would express the importance of these variations in economic terms. The development of such a model would require, according to Ackerman: (1) the determination of the requirements of each activity for temperature, precipitation and wind in terms of volume and timing; (2) the assessment of the economic costs of variations in different weather parameters to particular economic activities; and (3) the development of some means to trace the impact of a given weather variation through the economic system. Ackerman noted that considerable research is required before such a model can be applied, and in particular he emphasized the need to ascertain the weather requirements of various activities and the costs of weather variations to them.

Several techniques have been employed by economists and others to devise a model of the total economy of an area. These include among others (1) input–output, (2) simulation, (3) regression, (4) linear programming, and (5) benefit–cost analysis. In most cases these techniques have been applied without considering the climatic factors which may affect an economy, most econometric models being based on the 'non-meteorological aspects' of an economy. Since, however, even a casual look at the economy of most areas will readily show that meteorological factors cannot be ignored, existing techniques for economic analyses must be assessed with caution. A critical appraisal of the variation techniques available was included in the Report of the Task Group on the Human Dimensions of the Atmosphere (Sewell, Kates and Maunder, 1968) and the summary which follows is basically from this publication.

a. Input–output analysis

One technique that can be used to assess the real economic impacts of weather and climate is input–output analysis. This technique, reviewed by Chenery and Clark (1959), is concerned with the quantitative analysis of the interdependence of producing and consuming units in a modern economy. In particular, it studies the interrelations among producers as buyers of each other's outputs, as users of scarce resources, and as sellers to final consumers. Once the linkage among various industries are known, it is possible to trace the impacts of a change in output of one industry on other industries in an economy.

The technique of input–output analysis was conceived by W. W.

Leontief in 1951, and excellent summaries of the concepts and applications of the techniques are available (Leontief, 1951; Isard *et al.*, 1960). A number of sophisticated models based on input–output concepts have been developed for analysing economic changes in particular regions (see, for example, Isard, Langford and Romanoff, 1966; Moore and Paterson, 1955; and Tieart, 1962). Attempts have also been made to develop a model of the United States economy using this technique (Fromm and Klein, 1965; Goldman *et al.*, 1964), but in most input–output studies the factors involved usually exclude all climatic factors. The question may well be asked why this is so, and further, if weather factors can be included in a meaningful manner.

Typical of the non-climatic econometric models is the Arizona input–output model (Martin and Bower, 1967). In considering the agriculture processing sector of Arizona, this study shows that in order to produce a dollar's worth of agricultural products, it takes roughly 31 cents' worth of input from the livestock sector, five cents of input from the crop sector, eight cents from trade and transportation and so on. Inputs from rainfall, temperature, sunshine, wind, etc., however, are not considered.

The same authors in another study, do, however, attempt to assess the value of water to the Arizona economy (Martin and Bower, 1966). This report indicates that of the 5·0 million acre-feet of water consumed in 1958, 1·5 million acre-feet were consumed by cotton, 1·4 million by food and feed grains and 1·3 million by forage crops, etc. A subsequent analysis of this data indicated that the direct and indirect requirement to fill an order for an additional $1,000 of final demand varies from 65·0 acre-feet for the food and feed grains sector, to 0·03 acre-feet for the trade, transportation and services sector. The considerable indirect economic effect of water is demonstrated in the poultry and eggs sector, which ranks twenty-second out of 24 in its direct water coefficient but third out of 24 in its indirect coefficient. In this regard, Martin and Bower state that it takes nearly 21 acre-feet of water usage in the state's agricultural industries to produce $1,000 of poultry products for the consumer, the explanation for the apparently high total being that poultry and eggs require large amounts of agricultural inputs, particularly food and feed grains which are themselves large consumers of water. Here again the values of rainfall, temperature and sunshine are not considered. If, however, these climatic factors were or could be incorporated into econometric models of the type developed for Arizona, the impacts of changes in these weather parameters on the state's economy could be studied.

Variations in the weather can obviously result in changes in the production functions of various activities. Given a change in the production function of one activity, input–output analysis can make it possible to determine the impacts of that change on other activities as well. An increase of precipitation at a particular time, for example, may permit a significant increase in agricultural output, which in turn will increase the demand in the agricultural sector for inputs from other sectors. Increased output in the agricultural sector, however, does not necessarily imply that there will be increased outputs in all other sectors, for some economic activities may experience substantial gains, but others may suffer declines in the demand for their outputs. Input–output analysis, therefore, can provide a useful means of identifying which activities may gain and which may lose as a result of given weather variations, whether these be natural or man-made.

b. Simulation analysis

Simulation techniques also offer useful possibilities for examining an economic system and the effects of weather and climate on its activities and on the system as a whole. The simulation technique has been developed so that investigations can be conducted using computer programmes instead of physical models to represent the parts of the system and the interactions between the parts. Such models incorporate the variables that are believed to be the most important, and can then be used to determine the effects of a change in a particular variable on the system as a whole. The use of such models has increased considerably with advances in computer science and technology.

One attempt to assess the economic impacts of weather by simulation was made by McQuigg and Thompson (1966), who suggested that relations between weather events, non-weather events, man's function as a decision-maker and the economic outcome of an enterprise may be represented by an equation of the form:

$$E = f(W, N, A) + u$$

where E = economic outcome, $W = (w_1, w_2, \ldots, w_n)$ some actual weather events, $N = (n_1, n_2, \ldots, n_k)$ some actual non-weather events, and $A = (a_1, a_2, \ldots, a_i)$ a subset of alternatives based on information supplied to the decision-maker. The term u represents a 'disturbance' factor, which is random and normally distributed.

Using this equation a simulation model was developed based on real economic data from an important weather-sensitive enterprise, the management of the flow of natural gas to a city in the winter. The results of the simulation model analysis were then studied, and a basic concept began to emerge: that improvements in the accuracy of weather information may allow the manager of a weather-sensitive process to make 'better' (hence more valuable) decisions, provided he has a sufficiently precise rational method for translating weather information into operational terms. Hence, as also reported by Russo (1966) in his study of the U.S. construction industry, the value of weather information and the use to which it is put depend on a much more detailed knowledge of weather impacts on the weather-sensitive activities than is normally available.

Simulation techniques may also be usefully applied to the evaluation of weather modification programmes which have characteristics similar to activities that have already been successfully subjected to analysis by simulation, such as transportation systems and environmental health. Simulation techniques for analysing water resource programmes, particularly those which have evolved from the Harvard Water Program (Hufschmidt, 1966), also show promise for application to the analysis of weather modification programmes.

c. Regression analysis

Regression analysis is often a valuable tool in many kinds of investigations, and in spite of some problems in interpretation, it is a useful technique in studies of the effects of climatic variations on aspects of consumption and production. In the past multiple regression techniques could only be applied when the number of independent variables was relatively small. Today, however, with electronic computers regression analyses with 10, 20, 30 or more independent variables can be processed very quickly. Consequently, the 'choice' of independent (climatic) variables is not as critical as it once was.

Of the many climatological gains which should be considered in any analysis of the areal economic impact of weather and climate, those affecting the agricultural community are perhaps of primary importance, the gain or loss of income from agriculture through the effect of climatic variations on agricultural production being of real concern in both developed and developing countries.

Towards this end an attempt to assess the effect of significant climatic variations on agricultural production in New Zealand and its ultimate impact on agricultural incomes was made through the use of a regression model (Maunder, 1965). Regression techniques have of course been used for a considerable time, some classical studies in agroclimatology having used these methods (see, for example, Hooker, 1922; Kincer and Mattice, 1928; and Rose, 1936). Nevertheless, most studies published have considered only the effect of various aspects of climate on specific crops, animal or fruit production, and do not attempt to encompass the effect of the increased or decreased production on agricultural incomes, and even more important on the total economy of an area.

The New Zealand model mentioned above, however, does include agricultural incomes as an essential part of the study (Maunder, 1968a), and with modifications could be applied to the total economy of the area. With other changes, this model could also be applied to the study of climatic impacts on agricultural production in other countries as well. Regression techniques may also be used to trace impacts of weather changes on the non-agricultural aspects of the economy such as retail trade.

d. Linear programming

A relatively new contribution to the field of inter-industry economics is the technique of activity analysis or linear programming, developed first by Dantzig and Koopmans (see Koopmans, 1951). Most of the applications of this technique have been to problems of single plants or firms but the method itself is also useful for industry-wide and inter-industry analysis.

Basically, activity analysis is a method of analysing any economic transformation in terms of elementary units called activities. It provides the conceptual framework for the mathematical technique of linear programming, which can be used to determine optimal solutions to various kinds of allocation problems. This technique requires that the relationships between the relevant variables be expressed in linear mathematical form. These expressions, together with the appropriate constraints, make up a set of equations that have many solutions, the purpose of the linear programming procedure being to identify a particular combination of variables that will produce optimum results.

One example of linear programming is the PARM system (Wood, 1965), which is a model built up of basic variables called activities, which are represented in the computation system by time-phased vector groups in tables. The maximum size of these tables is limited by the available computer memory, the Prototype Model having 983 such activity vectors. Wood suggests that regional applications would involve the use of a supplementary model adjoined to the national model, and application of the regional model in any time-period would assume the prior application of the national model to that time-period. Further, he suggests that like the national model, the regional model would be applied first to the period immediately following the historical data, and then successively to subsequent periods. In this way the data computed for each time-period would be added to the historical time-series and used in the computation of the next period.

The application of this technique to problems in weather economics, or weather modification economics, has still to be developed, but as with input–output analysis the outlook is promising.

e. Benefit–cost analysis

Benefit–cost analysis can be a very useful tool in weighing the merits of alternative courses of action, particularly where the public sector is involved. The relationship between the benefits and the costs, however, have been studied primarily within a framework of programmes limited in functional and geographic scope. Seldom has benefit–cost analysis been undertaken for large comprehensive programmes such as weather and climate modification. As a result, according to Gutmanis and Goldner (1966), the many problems associated with such undertakings are relatively obscure. At the same time, however, there is growing demand for rigorous justification and evaluation of public policy, especially where public expenditures are involved, and this includes many aspects of weather information, weather forecasting and weather modification. Benefit–cost analysis, therefore, in spite of its perplexing operational problems, should be considered as an approach in the analysis, development and improvement of public policy decisions.

Much work has been done in recent years to refine its theoretical basis (see, for example, Prest and Turvey, 1965) and to broaden its field of potential applications (Dorfman, 1965). Progress has also been made, according to Crutchfield and Sewell (1968), with some of the more difficult problems of applying the analysis – notably in the treatment of risk and uncertainty, the treatment of time, and the identification and measurement of externalities. A few attempts have also been made to apply this technique to weather-related problems, such as air pollution (Wolozin, 1966; Ridker, 1967; Ogden, 1966) and weather modification (Gutmanis and Goldner, 1966).

In the latter paper, an attempt is made to illustrate the usefulness and limitations of the benefit–cost approach, through examination of the operational applicability of such analysis to the field of weather and climate modification. Based upon this examination, areas are suggested where such an analysis can be readily undertaken, and those where considerable operational limitations exist. The authors indicate that a number of difficulties in applicability of benefit–cost analysis to such large and complex programmes as weather and climate modification arise, the four more important being: (1) the extensive geographic and functional scope of such programmes, (2) the difficulties in obtaining the necessary qualitative and quantitative data, (3) the difficulty resulting from the availability of several possible technological approaches which may be employed in varying degrees either singly or in combination, and (4) the difficulty in integrating and supporting benefit–cost analysis with welfare economic theory.

However, if weather modification is limited in geographic and functional scope, undertaken with predetermined and known technology, and scheduled for completion within a specified time-period, then many of these problems are less significant. Examples of such limited projects are fog dispersion at airports (see Beckwith, 1966), or cloud seeding to artificially induce rain in a limited geographic area such as described by Battan and Kassander (1962).

For such limited weather modification projects Gutmanis and Goldner suggest that benefit–cost analysis can be readily undertaken. In particular, the limited geographic and functional scope allows collection of pertinent data; the use of known technology and the immediate application of this technology preclude the need to analyse the impact of future time and technology 'trade-offs'; and the economic impact of such programmes upon public welfare is relatively limited, and for practical purposes can be ignored. On the other hand, many difficulties arise in weather and climate modification programmes which are comprehensive in scope, very large or nation-wide in geographic coverage, or undertaken with limited technological knowledge. These, therefore, require a continuous research and development effort.

Despite the inherent difficulties, benefit–cost analysis has been proposed by most economists working with weather and climate modification programmes. In addition, most of the economists who have completed economic analyses of weather and climate have recognized a specific need for such analyses (see, for example, Nelson and Winter, 1964).

The role of benefit–cost analysis is to determine comparative social costs and benefits of several possible approaches. It is clear, therefore, that the benefit–cost technique of the evaluation of the atmosphere is a useful one, and one which should be carried out if meaningful decisions with regard to weather forecasting and weather modification are to be made. Especially important is the use of benefit–cost analysis in long-range weather forecasting, towards which many meteorologists are striving without any real appreciation of the economic and social benefits and costs, and in the more fundamental aspects of weather modification such as the steering of hurricanes.

f. Limitations of models

Applications of the various analytical techniques just described in the evaluation of weather, weather forecasting and weather modification have generally been more a matter of discussion than of accomplishment. For example, the advantages and disadvantages of using these analytical tools for such evaluation have not been reviewed in detail, neither has sufficient attention been given to the problem of what types of data or research are required before these techniques can be used successfully. Kuh (1965) states that 'the quality of most available statistics is weak. . . . In many areas of research, computing capability and theory of estimation and behaviour have clearly outstripped the ability of our statistical agencies to produce pertinent data for testing and estimation, even though matters are constantly improving.'

The inadequacy of our knowledge of the impact of weather is also demonstrated by McQuigg (1967), who states that anyone trying to write a set of equations, or a set of logical constraints, suitable for inclusion in a model to be used in a study of the impact of weather events on human activities, will find very quickly that we know precious little in *quantified terms* that can be used in this way. Research is needed, therefore, as indicated by Sewell, Kates and Maunder (1968), to select useful data which can be expressed in quantifiable terms, to determine relationships between weather variations and weather-sensitive activities, and to direct imaginative efforts towards developing and adapting techniques for measuring impacts of weather and climate on various economic activities.

C. APPLICATIONS

1. Area applications

Weather affects the profitability of many agricultural enterprises. Its effects might be immediate, as demonstrated by a rain shower that halts harvest operations, or long run, such as a long drought that dries up shallow wells and small streams. In either case, most weather variables affecting agricultural production are random and unpredictable in nature, thereby surrounding the farmer with an abundance of uncertainty (McQuigg and Doll, 1961).

In a review of the pertinent literature McQuigg and Doll state that research in production economics is usually directed towards the maximization of profits, or the minimization of costs. The usual treatment in analyses that involve a weather input is to assume that weather conditions will be 'average', such as the paper by Peterson and Swanson (1956), who analysed estimated annual yield, fertilizer and labour requirements for eight crop rotations and two farm sizes, using linear programming methods. In another paper Orazem and Herring (1958) related grain sorghum yields in south-west Kansas to soil moisture at planting time, rainfall during the growing season, and various levels of nitrogen application, the analysis showing that all three factors are important in determining final yields, but that the strongest relationships exist between soil moisture at seeding time and yields.

Other relevant papers, according to the review by McQuigg and Doll, consider the problem of evaluating weather forecasts in an economic sense (Gringorten, 1959) and the economic gains and losses due to weather events as described in a paper by Gleeson (1960). In this paper, Gleeson presents two techniques, one of which is to be used when the probabilities of weather events can be forecast or computed from weather records, and the other, based on game-theory, to be used when the probabilities of weather events are not known. A method of relating the economic factors involved in the use of weather forecasts, and the economic utility of weather forecasts, was also made by Thompson and Brier (1955). This method was based on a comparison of the ratio of the cost of protection to the loss that would occur if the event took place and no protection were provided, with the forecast probability that the adverse weather would occur.

Many other studies, of course, have been done and a review of some of those that are appropriate in the field of agroclimatology is available (Maunder, 1968b). Of the papers reviewed in this paper, those of Bean (1967) on 'Crops, weather and the agricultural revolution'; Thompson *et al.* (1964) on 'Weather and our food supply'; Hogg (1964) on 'Weather and agriculture'; Sanderson (1954) on 'Methods of crop forecasting'; Watson (1963) on 'Climate, weather and plant yield'; and L. M. Thompson *et al.* (1966) on 'Weather variability and the need for a food reserve', are of particular significance in relationship to a study of the 'value of the weather'.

The major conclusion obtained from a survey of the literature, however, is that, despite much progress, only a beginning has been made into the analysis of econoclimatic decision problems as they relate to areas. The pertinent study by McQuigg and Doll (1961) indicates, however, that on a sunny *day* in June in central Missouri, 100 acres of land receives about 8,820,900,000 B.T.U.s of heat from the sun, and that the photosynthesis process makes use of about 1% of this energy. This amount, when it is considered that it is equivalent to a year's fuel supply for the tractors used on 100 acres of corn, is considerable. An even larger percentage of the energy from the sun would be involved in maintaining soil temperatures, evaporating water and heating the atmosphere. It is apparent, therefore, that a farmer has control of only a very small fraction of the total energy required to produce a crop, for he has in fact 'control' over only the energy used in the form of fuel for tractors, electricity and manual labour.

These ideas were further developed, and in particular a study of the weather-related question 'Should grain drying equipment be used?' was examined in detail by McQuigg and Doll. They state that weather affects profits from grain drying in two ways; first, weather conditions during the growing season affect yields and consequently the number of grain-drying bins needed; second, weather conditions after harvest affect the cost of drying. Specific results are given in the paper, and under the assumptions used, the conclusion is that grain drying in central Missouri is profitable, although the expected profits naturally depend upon the price and cost data and the frequency distributions of the relevant weather variables.

The use of the simulation technique in area econoclimatic studies is demonstrated by Covert, Goldhamer and Lewis (1967) in a paper on the estimation of the effects of precipitation and scheduling of extended outdoor activities. They state that scheduling of outdoor activities such as road construction, farming and military training is, to a very large extent, dependent on the occurrence or non-occurrence of precipitation greater than some specified level. Probability and simulation models are developed and applied to a hypothetical construction company. The example given is based on the assumption that the XYZ Co. proposes to bid on Phase A of a stadium construction project in Midway, U.S.A., commencing work on June 5, 1967. From past experience, the company knows that 30 working days are required for completion of the job and that rain of 0·2 inch or more will effectively halt all work. One drying day is also required after each wet day or sequence of wet days before work may resume. The company is required to specify a realistic completion data to be awarded the contract and penalties will be assessed if the project is not completed on schedule. Two questions arise: first, what completion date should be specified by the contractor in the bid to be 90% confident of meeting that date; second, how many days would be saved by working on Saturdays?

The solution suggested, by Covert, Goldhamer and Lewis, using a simulation model, is that at the 0·90 level of confidence, 64 days are required to yield 30 working days, placing the completion date on August 8, 1967. If Saturday were also a workday, only 55 days would be required, placing the completion date on July 29, 1967. Consequently, if the XYZ Co. proposes a completion date of August 8, 1967, based on 65 calendar days and on five workdays per week, adding Saturday as a workday raises the confidence level for completion to something greater than 99%.

The authors of this investigation suggest that the simulation model might easily be extended to provide solutions for other weather phenomena such as frost and fog which are of vital concern to fruit growers and the air transport industry, respectively. In addition, other factors, such as the probability of a freeze after precipitation, could be included as limiting parameters.

Another example of a use of a simulation model in econoclimatic studies is that of McQuigg (1968) in a study of surface temperature modification at Columbia, Missouri. In this model, McQuigg asks as an initial question: 'Is there a need for modification?', this in turn being followed by two further questions: (1) Is the maximum temperature today expected to exceed a certain critical value?, and (2) Is the available soil moisture today estimated to be below a certain critical value? If the answer to both questions is 'No', then the decision for 'Today' is to make no attempt to modify.

The kind of atmospheric modification being simulated in the
I

model is the creation of contrail cirrus clouds. If, as McQuigg states, natural cloudiness is already abundant on the day in question, little effect can be gained from adding more cloudiness; thus it is pertinent to ask: 'Is the per cent possible sunshine less than some critical value?' Further, given the need for modification and the lack of sufficient natural cloudiness, the next question is: 'Are upper air temperatures suitable for contrail formation?' This question in turn is answered by using an array of probability values, the method of obtaining them being detailed in McQuigg's paper.

The model was applied using different estimates of important parameters, and in brief the results indicated that the model was producing a modified temperature series with components of variance that could reasonably be attributed to changes in the amount of cloudiness claimed.

In a follow-up of this paper, Nicodemus and McQuigg (1969) say that on the average there were about 55 days during the summer season (May to September) when modification was claimed to be 'successful', that is when the model produced estimates of the effect on surface maximum temperature through decreases in per cent sunshine by 'creation' of contrail clouds. On the other hand, in an average season, there were 40 to 50 other days when modification was needed, but where there was a 'failure' to modify. The effect on the application of the model to 20 years of temperature records at Columbia, Missouri, data is also discussed. On the average, the change in seasonal temperatures is less than 1°F, but on days when modification is a 'success', the average decrease in daily maximum temperature is from 3·0 to 5·0°F.

The authors conclude that if one is able to accept the basic assumption made in the simulation model, that is, that it *is* possible to create enough contrail cloudiness at times to decrease the per cent of possible sunshine by from 15 to 35%, then it appears possible to modify (i.e. decrease) afternoon maximum temperatures in central Missouri by 3 to 5°F on about half of the days when soil moisture levels are below 'desirable' levels, or when temperatures are expected to be above 'desirable' levels.

The application of this model to simulation experiments involving the effect of temperature modification on electric power demand, dairy production, crop production and other weather-sensitive operations is proceeding. In addition, the use of simulation techniques in estimating soil moisture conditions and in turn a 'workability index' for road construction in specific areas is being studied.

2. Regional Applications

Although some studies have been made on the value of weather, weather information and weather forecasting to specific activities in specific areas no study appears to have been made of the weather on a regional economy. Nevertheless, a few econoclimatic studies that could be termed semi-regional have been made.

A model for the evaluation of economic aspects of weather modification programmes for a system of regions, for example, was proposed by Langford (1968). In this paper a model is suggested for identifying and evaluating some of the possible economic effects which might result from alternative programmes of weather modification. A basic input–output model is used, and the effects of intermittent adverse weather conditions upon the construction industry are assessed. For example, if an economic loss estimate of 1·76% for the total construction business is assumed for the Philadelphia Standard Metropolitan Statistical Area, this loss amounts to approximately $14,188,000. Further, using data from the Philadelphia Region Input–Output Table, 89 industries would be affected by changes in the aggregate construction demand, the changes ranging from approximately $53 for electric lamps to $2,158,542 for wages and salaries. Each of the 89 sectors would in turn change its demand, and the effect would spread with differential intensity throughout the economy. The effect of the intermittent adverse weather conditions on an explicit production function was not examined by Langford (1968) to determine detailed probable expenditure changes. However, he suggests that it is reasonable to believe that substantial changes would be observed in labour and in those industries producing protective equipment and perishable goods.

An important modification to the basic model according to Langford is to make it capable of spatial or inter-regional differentiation. With respect to weather, this is particularly important as modification may not occur throughout the nation, but rather to spatially differentiated segments of the economy, and the impact of changes in a given region may be totally unlike those in any other region. Investigations are being conducted in regional input–output analysis to quantify regional effects, and this spatial extension will permit consideration of the impacts upon industries in region S resulting from a weather modification in region R because of the economic interdependence of the regions. Langford in concluding says:

The work in developing a basic input–output model for evaluating economic aspects of weather modification programs has been a challenging and expensive task. There is a growing belief among some regional economists that with modifications to basic parameters, such a highly detailed model can be adapted successfully to various regions at relatively low cost.

Another study which could, with modifications, probably be used to assess the value of inputs of weather into a regional economy is the input–output analysis for Arizona (Martin and Bower, 1967). To construct the Arizona model the economy was divided into 26 endogenous sectors consisting of ten agricultural sectors, five agricultural processing sectors, seven manufacturing and mining sectors, three service sectors and one other. Endogenous sectors are both producers and purchasers in the economy and it is assumed that the quantity and type of inputs required by these sectors is related to output through constant proportions. In addition, seven exogenous sectors: scrap and by-product, maintenance and new construction, state and local government, federal government, households and net state imports and exports were considered. These sectors are also purchasers or suppliers of products in the Arizona economy.

Regrettably, no climatic inputs are included in the analysis, but the model does show that to produce a dollar's worth of livestock it takes roughly 21 cents' worth of inputs from the crops sector, two cents from agricultural processing, four cents from service and utilities, 41 cents of net imports and so on. The authors state that 'total purchases' come to one dollar, but in reality they do not, for what is missing is the value (in cents) of the input of rainfall, temperature, sunshine, etc. to the production of one dollar's worth of livestock, and as indicated with regard to central Missouri (see p. 232) such inputs of the sun's energy into an agricultural economy are considerable and impressive.

Simulation models for water-resource systems and their utility in measuring the physical and economic effects of weather forecasting and weather modification have been summarized by Hufschmidt, Fiering and Sherwani (1968). The authors describe a digital computer simulation programme for a water-resource system which accepts streamflow as an input. However, the deriving of streamflows from precipitation records is not provided for, nor does the model effectively assess the gains (and losses?) associated with weather forecasting and weather modification.

In order to adapt such a simulation programme so that it can handle weather modification and forecasting situations, the authors suggested that several additions and changes are required. Among these are the historic precipitation and climatological record, information on the nature of the modification of precipitation, a method of translating precipitation data into runoff data, and a revised economic evaluation sub-model, adapted to handle deficits under conditions of natural and modified precipitation, and conditions of streamflow forecasting and absence of forecasting. The system subroutines required to adapt existing water-resource simulation programmes to deal effectively with weather modification and forecasting are examined in some detail in the paper.

Simulation techniques, when combined with rational-decision models, appear to show promise in helping to assess the effectiveness of weather modification and forecasting. Hufschmidt, Fiering and Sherwani point out, however, that the problems of using and adapting these techniques and models to this purpose should not be underestimated.

Although most econoclimatic studies emphasize the lack of real economic data, a problem which is equally important is the evaluation of an adequate measure of weather. The model proposed by Doll (1967) is of interest, for it represents a model that may be used to derive a weather index based upon meteorological measurements, such as rainfall and temperature. Weather is suggested as being characterized by a function, representing all of the meteorological variables influencing the final crop yield in a given year. The relationship between final crop yield and the weather during the growing season is complex, but Doll says that in general every possible meteorological function occurring during the growing season is associated with a unique final yield.

The formulation of a weather index using several meteorological variables is obviously a difficult problem, and Doll suggests that for purposes of simplicity weather can be considered as a function of only one meteorological variable. He then goes on to develop a model to estimate a rainfall index for corn in Missouri, 1930–1963. Thirty-seven weather stations throughout Missouri were selected. Corn yield and acreage data, available by counties, were assigned to the closest weather station, and weekly rainfall variables for the state were obtained by weighing each of the 37 rainfall series by the percentage of the state corn acreage assigned to that series. Corn yields have increased substantially since 1930, and to allow for this trend a cubic time function was introduced. Weather indices were then computed as the ratio of the yield predicted for the actual weather that occurred during the year, to the predicted yield had average weather occurred. The weather indices calculated for Missouri from 1930 to 1963 varied from 32 in 1930 to 142 in 1948 and 1961.

The rainfall index 'explained' two-thirds of the variation in yield deviation from trend, but as Doll indicates it is clear that further variables, especially temperature, should be included in the model. Further, the distribution of meteorological variables throughout the growing season and the occurrence of extremes are also important factors.

The final econoclimatic study on a regional basis to be considered is that of the effect of significant climatic factors on agricultural production and income in New Zealand (Maunder, 1965, 1966a, 1968a). In this study an 'agroclimatological model' (Maunder 1966b) was formulated for 18 different agricultural factors and for 27 different areas (mostly counties) in New Zealand, based on variations in agricultural production and climate, mainly since the 1930's.

The climate variables used were the seasonal and monthly data for rainfall, temperature and sunshine, the climatic data available being used in conjunction with the agricultural data for the county within which the climatic station was located or the nearest appropriate county. In all, 27 climatic station–county pairs were available, and the appropriate agricultural unit production data were assessed for each. Analyses were made for each of the 18 agricultural factors, these factors being divided into three main divisions – crops and butterfat production, wool and meat production, and apple and pear production. Separate econoclimatic evaluations were made for each relevant agricultural factor in each county or area; thus, in the case of wheat production, 16 analyses were made since wheat production was considered to be important in 16 (of the possible 27) counties (see Maunder, 1966a for a summary paper).

In order to estimate the 'effect' of variations in several aspects of the climate (specifically, rainfall, mean temperature and sunshine), the measure *specific climatic variation* was formulated, and is defined as a variation from the average of one standard deviation. A month was described as 'wet', 'warm' and 'sunny' if the departure from the average rainfall, mean temperature and sunshine was at least one standard deviation above average, respectively, whereas the terms 'dry', 'cool' and 'cloudy' were applied to months having a similar negative departure from the average.

The coefficients in the appropriate multiple regression equations associating agricultural production with climatic variations were used to estimate the 'effect' on production of those specific climatic variations found to be significant, and of the many results given in the original survey the following are relevant to this regional econo-climatic study. Specifically, an attempt to show the economic importance of the more significant agroclimatological associations in New Zealand was made by comparing the partial correlations significant at the 1% level by areas and agricultural factors. The 'economic' effects of these significant agroclimatological associations are summarized in Table V.1.

Analysis of Table V.1 shows that of the 24 partial correlations, 10 were associated with butterfat production, five with oats, three with wheat, two with barley, two with pear production and one with wool production per acre, and apple production per tree. The major feature of the analysis is the significance of climatic factors in their influence on butterfat production, and considering all 10 correlations, the effect of a significant standard deviation monthly departure from the average varied from $2·2 to $5·6 per cow.

The value of these variations per cow when related to particular areas fluctuates considerably, and depends (in this case) on the number of dairy cows in milk, this being clearly demonstrated when one area is compared with another. For example, in the Waikato County the 'county value' of the variations in butterfat production discussed above were over $200,000, whereas in Masterton the equivalent 'county value' was in the range $9,000–$23,000. These values are of course only an index of the total value of the variations in butterfat production in the two areas. A more realistic estimate of the total effect of a significant specific climatic variation on butterfat production and the subsequent variations in the income from butterfat can be obtained by assessing the number of dairy cows in milk located in the county and surrounding areas. In the case of the Waikato County, a first approximation of this could be taken as the dairy cow population of the South Auckland Land District which totals approximately 750,000, this representing nearly 40% of all dairy cows in New Zealand. Further, if the 'values per cow' in the Waikato County are taken as an index of the variations in butterfat income per cow for the South Auckland area, it can be suggested that a significant climatic variation such as 'wet' January is 'worth' about $2 million to the area (750,000 cows × $2·6 per cow). The probability of a 'wet' January (based on the period 1936/37–1959/

TABLE V.1: *Economic effects of significant* agroclimatological associations† *(from Maunder, 1968a)*

Agricultural factor	County	Relative climate	Month/season	Effect of climatic variations‡	
				Value/unit ($N.Z.)	County value ($N.Z.)
Wheat	Southland	Dry	Oct.	4·0 ± 1·0/ac	52,000 ± 14,000
	Southland	Cloudy	Jan.	7·2 ± 1·4/ac	92,000 ± 19,000
	Southland	Sunny	Feb.	7·6 ± 1·4/ac	100,000 ± 19,000
Oats	Waimea	Cool	Feb	5·8 ± 1·6/ac	400 ± 120
	Marlborough	Cloudy	Nov.	3·4 ± 0·8/ac	1,300 ± 200
	Marlborough	Warm	Jan.	1·8 ± 0·6/ac	680 ± 200
	Marlborough	Cool	Feb.	2·0 ± 0·6/ac	840 ± 260
	Levels	Cool	Dec.	4·4 ± 1·4/ac	2,500 ± 800
Barley	Southland	Warm	Oct.	4·0 ± 1·0/ac	780 ± 200
	Southland	Cool	Jan.	7·6 ± 2·4/ac	1,500 ± 460
Butterfat§	Ohinemuri	Wet	Nov.	4·6 ± 1·4/cow	90,000 ± 30,000
	Waikato	Wet	Oct.	2·6 ± 0·6/cow	260,000 ± 60,000
	Waikato	Wet	Jan.	2·8 ± 0·6/cow	290,000 ± 60,000
	Waikato	Sunny	Jan.	2·0 ± 0·6/cow	200,000 ± 60,000
	Waikato	Wet	Feb.	2·6 ± 0·2/cow	250,000 ± 12,000
	Masterton	Cloudy	Nov.	2·2 ± 0·6/cow	9,000 ± 2,000
	Masterton	Wet	Dec.	2·8 ± 0·8/cow	10,000 ± 3,000
	Masterton	Sunny	Jan.	5·6 ± 1·2/cow	23,000 ± 4,000
	Masterton	Wet	Feb.	5·0 ± 1·4/cow	21,000 ± 5,000
	Masterton	Cool	Feb.	3·4 ± 0·8/cow	14,000 ± 3,000
Wool/acre	Hawkes Bay	Wet	Previous autumn	1·0 ± 0·2/ac	799,000 ± 160,000
Apples/tree	Waimea‖	Wet	Summer	1·4 ± 0·2/tree	470,000 ± 90,000
Pear production	Waimea‖	Cloudy	Previous winter	—	36,000 ± 10,000
	Waimea‖	Dry	Summer	—	42,000 ± 10,000

* At the 1% level.
‡ Effect for month/season with relative climate as shown. All values are 'credits'. A similar 'debit' would be associated with the 'opposite' relative climate.
§ Herd testing group data in county.

† Based mainly on 1964 prices, and 1961/62 agricultural data.
‖ Nelson and Mapua fruit district data in county.

1960 at the Ruakura climatological station, the base station in the Waikato County) is one in six, compared with one in eight probability for a 'dry' January. The analysis therefore indicates that if these rainfall probabilities are taken to be representative of the climate of South Auckland, then once in about six seasons a 'wet' January will be associated with an 'increase' of about $2 million in the income of dairy farmers of South Auckland. Conversely, it might be expected that a comparable 'fall' in income would occur once in about eight seasons as a 'result' of a 'dry' January. These substantial variations in income thus provide a measure of the potential economic importance of significant climatic departures from the mean, as they affect butterfat production in South Auckland – New Zealand's foremost dairying area.

3. National applications

The effect of weather on a national economy has not been studied in any comprehensive manner. Some preliminary studies have been made, however, and the role of weather services and weather forecasts on various national economies have been assessed. The study by Mason (1966) on the 'value' of meteorology to the United Kingdom economy is particularly pertinent.

In this study, Mason says that an effective weather service seeks to guide the community in its use of weather services so that the great number of weather-sensitive decisions arising daily in all walks of life can be made in the best interests of the economic and social welfare of the country. Further, he correctly points out that although there may be no immediate prospect of controlling or significantly modifying the weather (at least to any great extent), its favourable aspects may be better exploited and its worst effects alleviated or avoided by acting upon meteorological advice. This is, of course, well realized in aviation, agriculture and some sectors of commerce and industry, but many areas are still unaware of the services and benefits that can be offered, often at trifling cost.

Many examples are given by Mason of the 'value' of meteorological advice to civil aviation, agriculture, commerce and the building industry. In regard to the latter, it is stated that the total annual output of the industry is valued at £2,900 million and is produced by a labour force of about one million. The value of the time lost because of bad weather has been estimated at £100 million per annum (3·5% of production), equivalent to about 10 working days. A weather service proposed by the U.K. Meteorological

Service, however, could be expected to allow builders to take protective measures to plan indoor work during bad weather, and to avoid its worst effects, thereby saving at least one working day per year, which would have a value of £10 million.

The demand for electricity is also particularly sensitive to temperature changes. A fall of 1°C increases the load by 1·3% in summer and 1·8% in winter, but an error of 2°C in the forecast temperature may, according to Mason, result in an underestimate of the demand by more than 1,000 megawatts on a winter's day. The Central Electricity Generating Board insures against such errors by running a spinning reserve capacity that costs about £2 million per year. The Board receives comprehensive meteorological advice, for which a charge of £12,000 is made. However, they estimate that without this advice the reserve capacity would have to be increased by at least 10% at an annual running cost of £200,000. In this case, the benefit/cost ratio is at least 20 to 1.

In another example given, Mason assesses the value of possible forecasts at twopence per family per day. Not a very large sum, but when totalled it represents £30 million per year – 15 times the cost in the United Kingdom of the basic service to industry and the general public. The benefit/cost ratio is therefore high.

Detailed studies by a professional economist, according to Mason, would no doubt provide firmer figures than those given in the analysis (and briefly mentioned above); nevertheless, as he correctly points out '. . . neither the data nor the inevitable assumptions will support a very rigorous or refined analysis'. Mason indicates that the total economic value of the Civil National Weather Service in the United Kingdom is at least £50 to £100 million per year, for a total cost of only £4 million. This it should be pointed out is the value of *weather services*, and not the value of weather *per se*. What that is, is unfortunately not known.

An evaluation of the weather and climate of the United States by socio-economic sensitivity indices (Hibbs, 1966) is also of interest. In this paper Hibbs suggests that 'weather and climate can be viewed as a resource which at various times and locations may be classified along a "favourable to unfavourable" spectrum'. Five not necessarily exclusive ways of man adapting his activities to this resource are suggested. These are passive acceptance, avoidance of areas and actions unfavourable to effective use of resource conditions, operational and defensive actions based on assessment of meteorological information, modification and direct control of weather and climate, and structural and mechanical defences such as air conditioning.

In developing an effective and economic programme based on these decisions, six closely related areas of needed activity have been identified by Hibbs. These include: (1) identification of potential weather service users; (2) determination of user requirements; (3) mapping of the climatological phenomena to which the user classes are sensitive, with an identification of area or zones where the phenomena may be expected to be favourable or unfavourable to the user classes; (4) determination of the geographic location of the user classes; (5) determination and evaluation of alternative feasible modes of user response to the weather–climate resource; and (6) devising and testing measures of benefit as an aid to E.S.S.A. in planning and policy decision-making.

As with most national meteorological services, the Environmental Science Services Administration through the U.S. Weather Bureau provides services aimed at satisfying the requirements of a number of external clients, and a useful distinction divides the clients or 'user groups' into two classes: consumers and producers. Hibbs defines the two groups as follows:

> *Consumers* use weather services to enhance values generated outside the market place. Personal enjoyment of leisure time and other non-income producing (in the dollar sense) activities are included. Also included are governmental actions using weather services to maintain or increase health, safety, general welfare, natural resource preservation etc.
> *Producers* use weather services for purposes of enhancing their professional, commercial or industrial activities. Thus, specialized services are used by professional meteorologists, agriculturists, commercial airlines and private business pilots, and marine transportation. Benefits accrue to members of this group as producers.

An analysis is described and a map of the distribution of the potential benefits of both information and modification services is given for eight regions of the United States. This information in tabular form is given in Table V.2.

These index values are according to Hibbs crude and untested, nevertheless, they appear to be useful for further development of the measurement of the sensitivity of various activities to the resources of the atmosphere.

One of the few national econoclimatic models devised so far appears to be the model for Canada (Maunder, 1968c).* The reason

* Work is proceeding in developing a model for the United States.

TABLE V.2: *National and regional potential benefits from weather in-formation and modification services ($ million) (from Hibbs, 1966)*

Area	Production activity benefit potential		Consumption activity (general public) benefit potential	
	From info. services	From mod. services	From info. services	From mod. services
United States	7,136	2,820	4,661	2,973
Far West	161	404	633	422
Rocky Mts	32	72	104	69
South-West	83	188	306	204
Plains	104	248	361	241
Great Lakes	238	634	946	630
South-East	196	446	714	476
Mid-East	267	687	1,106	737
New England	57	141	291	194

for designing this model was to establish a working econoclimatic model in order that a more quantitative appreciation of the effects of weather on the Canadian economy could be obtained. This involved four basic tasks:

(1) The evaluation of a national climatic index.
(2) The calculation of the deviations of the various economic indices from those 'expected' (the deviations from the 'trend line' are used as a means to take into account the 'trend for technology, population and prices').
(3) An assessment of the most important economic factors found to be most closely associated with national climatic variations.
(4) The evaluation in dollars of the worth of a specific variation in the national climatic index in so far as the index is associated with deviations from the economic trend line.

A climatic index for Canada was first evaluated. In this evaluation population concentration multiplied by an appropriate climatic factor was chosen as an initial 'indicator' of the impact of climate, particularly as it applies to aspects of the Canadian economy directly related to population density such as retail trade, home heating costs, ice cream and cold beverages sales, etc. There are, of course, many ways of obtaining a representative climatic index for an area, and it is obvious that the method of obtaining a national climatic index would vary according to the type of economic activity being considered.

In a preliminary study, twelve climatic stations were selected out of seventeen stations for which appropriate climatic data were available (selection was based on the availability of data on total precipitation, mean temperature and hours of bright sunshine for the period January 1961 to December 1967), and for each of these stations a representative population figure was tabulated based on the 1961 census. These varied from 16·4 for Montreal to less than 1·0 for urban areas such as Victoria, B.C., the representative population being obtained by expressing the metropolitan population in the area as a percentage of the total Canadian population.

A computer programme was then designed to calculate a weighted climatic index for each of three weather elements (rainfall, mean temperature, sunshine) on a monthly basis over the period January 1961 to December 1967. The three climatic elements used are, of course, not necessarily the most 'significant' ones, and it is clear that further research will aid in the filtering of the many weather variables so that the most economically sensitive weather elements may be known for each economic activity.

The simple formula used to obtain the weighted climatic index was:

Weighted index

$$= \frac{\sum_{1-12} \left(\frac{\text{actual observation}}{\text{average or normal}} \times \frac{\text{representative population}}{\text{for each station } (1-12)} \right)}{\text{total representative population}}$$

The weighted indices for Canada for 1965 are given in Table V.3. Similar indices were also obtained for other years.

Several important points arose out of this evaluation of a national climatic index.

(1) The indices measure (or at least are intended to measure) for Canada as a whole a relative amount above or below *normal* for the three climatic elements. However, an index of 1·00 could arise if one area of Canada experiences above average conditions while another area (of similar population) experiences similar below average conditions.

(2) There is probably a level where an extreme in one of the three factors will overstress the significance of the climatic indices. For example, three or four inches of precipitation above normal may represent the maximum effect that the weighted Montreal index should (and possibly does) have on the total Canadian index when taking into account specific economic activities.

That is, it may not matter whether it is 'wet' or 'very wet' to the economic consumer, in which case the percentage increase or decrease of sales under 'wet' or 'very wet' conditions would probably remain unchanged.

(3) Although the amount above or below the normal may be important, it may be the *relative difference* between the months that is more significant. For example, a rainfall of say 2·00 inches in one month following a month with a rainfall of 1·00 inch, may have a similar economic effect to a rainfall of say 4·00 inches in one month following a month with a rainfall of 2·00 inches. The importance of these *relative* differences (compared with the differences from the *normal*) suggested that an additional approach to the evaluation of a national climatic index was necessary. Accordingly, *differences* in each of the three climatic elements (weighted according to population) were also calculated, these differences compared with the actual index being shown in Table V.3.

The variations of the economic data from the 'trend line' were next considered, the economic data (for Canada) being first transformed so as to take into account the 'trend for technology, popula-

TABLE V.3: *Actual climatic indices and weighted climatic differences* for Canada: 1965 (from Maunder, 1968c)*

Month	Precipitation		Mean temperature		Sunshine	
	Difference	Actual	Difference	Actual	Difference	Actual
January	−0·06	0·95	+0·16	0·91	+0·17	1·15
February	+0·25	1·20	−0·05	0·86	−0·11	1·04
March	−0·72	0·49	+0·03	0·89	+0·21	1·24
April	+0·43	0·92	+0·08	0·96	−0·16	1·08
May	−0·06	0·86	+0·04	1·01	+0·02	1·10
June	−0·17	0·69	−0·02	0·99	+0·08	1·18
July	+0·18	0·87	−0·03	0·96	−0·14	1·05
August	+0·56	1·43	+0·02	0·98	−0·13	0·92
September	−0·38	1·06	−0·03	0·95	+0·03	0·95
October	+0·22	1·28	+0·05	1·00	−0·01	0·94
November	+0·07	1·35	−0·08	0·92	−0·06	0·88
December	−0·50	0·85	+0·32	1·24	+0·04	0·92

* Based on differences between the weighted climatic indices (e.g. the precipitation difference in August 1965 is the difference between the August index (1·43) and the July index (0·87) or +0·56 as shown).

tion and prices'. To try and account for this 'trend' a linear trend line (a non-linear trend could of course have been evaluated) was used as the simplest least-squares curve to best represent the economic data. The *deviations* from the expected economic trend line were then associated with the climatic variables using various correlation techniques.

Two important factors arise in any assessment of deviations from an economic trend line.

(1) It is difficult to establish what is the true trend line, and since econoclimatic studies supposedly start from the premise that some variations in the economic indices are related to weather and climate variations, it is important to be able to isolate those economic variations which are *not* associated with weather and climatic variations. The problem, essentially, is to filter out the 'economic noise' from the economic indices so that the variations remaining can be studied in relation to atmospheric variations.

(2) In dealing with monthly economic indices, it is difficult to combine successive months (as one would combine successive seasons in agroclimatological studies, such as the association of climatic variations with wheat yields per acre), since all months are not 'equal'. Indeed, it is clear that since the significance of climatic variations in their association with variations in various economic indices varies from month to month, the month-to-month (e.g. October to November) variations in economic conditions are probably not very useful in econoclimatic studies. It is, nevertheless, fairly clear that deviations of the economic indices from the 'trend for technology' are significant when the deviations for successive January months, or February months, etc. are compared.

In view of these considerations, the trend for technology for each economic index, for each of the 12 months January to December, were calculated (Table V.4), the deviations from the trend for successive January, February . . . December months being then used in various correlation analyses.

An attempt to ascertain the relationship between the national climatic indices (absolute and relative), and the variations of the economic factors from the 'trend' line, was made through the use of regression techniques. Specifically, the multiple and partial correlations were calculated for all January months, all February months,

TABLE V.4: *Economic indices* and deviations from the 'trend for technology, population and prices'. Canada: March 1961–1967 (from Maunder, 1968o)*

March	Retail sales ($m)					
	Women's clothing		Shoes		Fuel	
	Actual†	Deviation	Actual†	Deviation	Actual†	Deviation
1961	25·6	−0·3	15·9	−0·9	29·1	−0·7
1962	28·2	+1·0	18·7	+1·4	32·2	+1·8
1963	28·4	−0·1	17·8	0·0	33·0	+2·0
1964	28·7	−1·1	17·4	−0·8	28·9	−2·7
1965	30·7	−0·4	18·3	−0·4	30·9	−1·3
1966	33·5	+1·2	20·7	+1·5	30·6	−2·2
1967	33·3	−0·3	18·8	−0·8	36·5	+3·1

* Three examples only given in this table.
† Dominion Bureau of Statistics data.

. . . etc. for the period January 1961 to December 1967. Such analyses allow one to make two econoclimatic observations:

(1) The proportion of the variations in each economic variable. associated with the three climatic variations (rainfall, mean temperature, sunshine).

(2) The contribution and significance of each of the climatic variables in their association with variations in the economic variables, after allowance has been made for the contribution of the other two climatic variations.

In addition, it is possible through the use of coefficients in the multiple regression equation to evaluate the effect of 'a specific climatic variation' on the economic variables, the specific climate variations being taken to be the standard deviation of the variable. The evaluations so obtained are of course only approximations of reality, but until more relevant research is undertaken, such evaluations may be used as a guide as to the value of a 'natural' cold or wet month and therefore the value of a 'forecast' cold or wet month.

The apparent importance of the relative difference between months, compared with the difference from the normal, suggested two methods of evaluating the economic effect of a specific climatic variation (or difference). The first method utilized the actual weighted climatic indices, the second the differences between the actual climatic indices. In each case two approaches are available: first, the effect of *one* of the climatic indices on the economic index (such as the effect of a temperature variation on the retail sales of house heating oil), and second, the combined effect of all *three* climatic indices on the economic index (such as the effect of a 'dry, warm and cloudy' month on the retail sales of home-heating oil). Only the first approach, however, is considered here.

In the effect of temperature on home fuel sales, for example, the partial correlation analysis suggests that in January the sale of home-heating fuel is negatively correlated with mean temperature. Further, the coefficient in the regression equation for mean temperature is -10.81, and the standard deviation of the mean temperature index is 0.16. Accordingly, if it is assumed that the rainfall and sunshine indices remain unchanged then the effect of a month with a temperature index of 0.84 (i.e. the normal of 1.00 minus the standard deviation of 0.16) is to 'increase' the deviation of the economic index from the trend line by 1.7 (-10.81×-0.16), from what it would have been with 'normal' temperatures (-10.81×0.00). Conversely a 'warm' month (i.e. a month with a temperature index one standard deviation above average) would be associated with a 'decrease' of -1.7 ($-10.81 \times +0.16$). It may therefore be suggested that as an initial approximation a 'cold' January (defined as above) results in an 'increase' in the economic index of home fuel oil sales of 1.7, whereas a 'warm' January results in a decrease of 1.7. The 1.7 is an index of variability from the trend for technology line, and in this case is equal to home fuel sales (across Canada) of $\$1.7$ million. (The average total January home fuel oil sales across Canada, by way of comparison, is approximately $\$31.0$ million.)

Similar assessments were made with regard to the effect of the *differences* in the climatic indices. Table V.5 gives some results for the effect of climatic variations as described above.

One factor of considerable importance in any econoclimatic analysis is the non-availability of appropriate economic data for either small areas or for short time periods. For example, weekly, daily and hourly weather data are readily available, but economic data is usually available only on a quarterly or a monthly basis. One notable exception to this is the weekly data on retail sales in the United States. These data, however, apply only for the United States, and in order to use such data in any national econoclimatic analysis, it is necessary to compute a United States weather index for each week. This was done by considering the weekly mean temperature and weekly precipitation data for 148 places across the United States. Most of these places had a 1965 metropolitan popu-

TABLE V.5: *Effect of specific climatic variations on Canadian economic indices: 1961–1967* (from Maunder, 1968c)*

Month	Retail trade	Climatic element	Standard deviation	Reg. coeff.	'Economic effect' for ±S.D. ($m)	
March	Men's Clothing	Sunshine	0·08	15·87	+1·27 (sunny)	−1·27 (cloudy)
April	Furn. & App.	Precipitation	0·20	10·91	+2·18 (wet)	−2·18 (dry)
May	Variety	Sunshine	0·06	−26·84	−1·61 (sunny)	+1·61 (cloudy)
	Furn. & App.	Sunshine	0·06	−18·14	−1·09 (sunny)	+1·09 (cloudy)
June	General	Precipitation	0·19	9·88	+1·88 (wet)	−1·88 (dry)
	Variety	Precipitation	0·19	16·29	+3·10 (wet)	−3·10 (dry)
	Fam. Clothing	Precipitation	0·19	6·59	+1·25 (wet)	−1·25 (dry)
	Hardware	Precipitation	0·19	5·77	+1·10 (wet)	−1·10 (dry)
		Sunshine	0·05	9·67	+0·48 (sunny)	−0·48 (cloudy)
August	Shoes	Precipitation	0·23	−3·23	−0·74 (wet)	+0·74 (dry)
		Sunshine	0·08	−5·56	−0·44 (sunny)	+0·44 (cloudy)
September	Hardware	Temperature	0·04	31·46	+1·26 (warm)	−1·26 (cold)

* All partial correlations considered were significant at the 5% level.

lation of 300,000 or more, and they had a combined population of more than 120,000,000. A 'weight' was computed for each of the 148 places based on the 'buying power index'* as published in the *Marketing Magazine* (June 10, 1969), '1969 Survey and Buying Power'. These weights were then multiplied by the corresponding mean temperature and precipitation differences, from normal, for specific weeks. Using this method, the weekly weather indices† – weighted according to the 'buying power index' – were calculated for a two-year period from April 1967.

Following the method used for Canadian *monthly* retail sales already described, regression analyses were used to assess the significance of weighted *weekly* precipitation departures from normal, and weighted *weekly* mean temperature differences from normal, for 16 kinds of retail establishments, and for various 11-week periods.

TABLE V.6: *Suggested associations between weekly retail sales* in the United States and weather conditions*

Retail trade group	Period†	Weather conditions‡
Furniture and appliances	summer	wetter and warmer
,, ,, ,,	autumn	drier and warmer
,, ,, ,,	late spring/summer	wetter
Lumber, building and hardware	late winter/spring	wetter and cooler
,, ,, ,,	autumn	drier
,, ,, ,,	summer	cooler
,, ,, ,,	late summer/early autumn	cooler
Food (firms operating 11 stores or more)	late winter/spring	wetter and cooler
Apparel	winter/early spring	drier and warmer
,,	autumn	cooler
Drug stores	late spring/summer	wetter and warmer
Eating and drinking	late summer/early autumn	warmer
,, ,,	late spring/summer	warmer
Gasoline service stations	winter/early spring	drier

* Retail trade data source: U.S. Dept. of Commerce publication *Weekly Retail Sales*.

† 11-week periods within these seasons.

‡ Weather conditions associated with *above* average sales. (Wetter indicates wetter than normal for that particular period, etc.)

---

* Calculated by giving a weight of 5 to the U.S. 'effective buying income', 3 to the per cent of U.S. retail sales, and 2 to the per cent of U.S. population. The total is divided by 10 to obtain the 'buying power index'.

† For example, the weighted mean temperature departure for the 'cold' week January 8–15, 1968 was $-8.5°F$, and the weighted precipitation departure for the 'stormy' week November 11–18, 1968 was $+0.67$ inches.

Allowance was made for seasonal trends during the 11-week periods, and certain holiday weeks were excluded. Research in this area is continuing, but a summary of a few of the suggested associations obtained so far is given in Table V.6.

These results are, of course, only a start to a very complex problem, and it is clear that a great deal of research will be needed before it will be possible to identify and measure the impacts of weather on a national economy with any real accuracy.

4. International applications

Global estimates of the benefits of weather, weather modification and weather forecasting are extremely difficult to make. Revelle (1964) suggests, however, that in the United States alone where the annual flood damage was estimated to average, over the years 1946 to 1960, $280 million, better long-range forecasts would have enabled this amount to be reduced by $70–$140 million a year. He further comments that, although the value of advance warning of weather changes is not easy to assess accurately, if farmers, fuel producers, public utilities, builders and water managers *were* able to make economies of only 5%, the probable total saving would be at least $5,000 million per annum.

In another study of the proposed satellite weather observing and forecasting system (Stanford University, 1966), the conclusion was reached that the annual potential benefits of that system to the world would total $16·9 billion, about 50 times the estimated annual cost. This estimate is based on incomplete data, but whatever the precise figure, it is clear that large international benefits are to be obtained from significant improvements in weather services. In addition, the international value of the total atmospheric resources cannot be ignored.

The importance of inter-regional and international dependencies in any assessment of the value of weather also needs to be appreciated. Curry (1966) has noted, on the economic side, that it is clear that weather conditions in one area produce price changes in another. This is especially the case in the United States with its considerable areal specialization and regional interdependence where, for example, frost in Florida increases the price of citrus for growers in California and Arizona. In view of this, Curry rightly reminds us that it is most unlikely that any sizeable increase in production in the U.S. West as a result of rain augmentation will leave prices paid to eastern farmers unaffected. Similar effects must also be considered when relating droughts on the Canadian prairies to Canadian wheat exports to China, and a whole host of other international dependencies. The real value of the weather to the world may therefore never be known.

BIBLIOGRAPHY

ACKERMAN, E. A., 1966: Economic analysis of weather: An ideal weather pattern model. In: SEWELL, W. R. D., ed., 1966: *Human Dimensions of Weather Modification*. University of Chicago, Dept. of Geography, Research Paper No. 105, pp. 61–75.

BATTAN, L. J. and KASSANDER, A. R., JR, 1962: *Evaluation of Effects of Airborne Silver Iodide Seeding on Convective Clouds*. University of Arizona, Institute of Atmospheric Physics, Scientific Report No. 18.

BEAN, L. H., 1967: Crops, weather and the agricultural revolution. *Amer. Statist.*, 21(3): 10–14.

BECKWITH, W. B., 1966: Impacts of weather on the airline industry: The value of fog dispersal programs. In: SEWELL, W. R. D., ed., 1966: *Human Dimensions of Weather Modification*. University of Chicago, Dept. of Geography, Research Paper No. 105, pp. 195–207.

CHENERY, H. B. and CLARK, P. G., 1959: *Interindustry Economics*. Wiley, New York, 345 pp.

COVERT, R. P., GOLDHAMMER, M. M. and LEWIS, G. F., 1967: An estimation of the effect of precipitation on scheduling of extended outdoor activities. *Jour. App. Met.*, 6(4): 683–7.

CRUTCHFIELD, J. A. and SEWELL, W. R. D., 1968: Economic research aspects of human adjustment to weather and climate. In: SEWELL, W. R. D., et al., 1968: *Human Dimensions of the Atmosphere*. National Science Foundation, Washington, D.C., pp. 59–69.

CURRY, L., 1966: Seasonal programming and Baysian assessment of atmospheric resources. In: SEWELL, W. R. D., ed., 1966: *Human Dimensions of Weather Modification*. University of Chicago, Dept. of Geography, Research Paper No. 105, pp. 127–38.

DAVIES, D. A., 1966: Foreword. In: THOMPSON, J. C., 1966: The potential economic and associated values of World Weather Watch. *World Weather Watch Planning Report*. No. 4, W.M.O., Geneva, p. iii.

DOLL, J. P., 1967: An analytical technique for estimating weather indexes from meteorological measurements. *Jour. Farm. Econ.*, 49(1): 79–88.

DORFMAN, R. (Editor), 1965: *Measuring Benefits of Government Investments*. Brookings Institution, Washington, D.C.

FROMM, G. and KLEIN, L. R., 1965: The new national econometric model: Its application. *Amer. Econ. Rev.*, 55: 348–69.

GLEESON, T. A., 1960: A prediction and decision method for applied meteorology and climatology, based partly on the theory of games. *Jour. Met.*, 17: 116–21.

GOLDMAN, M. R., MARIMONT, M. L. and VACCARA, B. N., 1964: The interindustry structure of the United States – A report on the 1958 input–output study. *Survey of Current Business*, Nov., pp. 10–24.

GRINGORTEN, I. I., 1959: Probability estimates in relation to operational decisions. *Jour. Met.*, 16: 663–71.

GUTMANIS, I. and GOLDNER, L., 1966: Evaluation of benefit–cost analysis as applied to weather and climate modification. In: SEWELL, W. R. D., ed., 1966: *Human Dimensions of Weather Modification*. University of Chicago, Department of Geography, Research Paper No. 105, pp. 111–25.

HIBBS, J. R., 1966: Evaluation of weather and climate by socio-economic sensitivity indices. In: SEWELL, W. R. D., ed., *Human Dimensions of Weather Modification*. University of Chicago, Dept. of Geography, Research Paper No. 105, pp. 91–109.

HOGG, W. H., 1964: Meteorology and agriculture. *Weather*, 19: 34–43.

HOOKER, R. H., 1922: The weather and crops in Eastern England: 1885–1921. *Quart. Jour. Roy. Met. Soc.*, 48: 115–38.

HUFSCHMIDT, M. M., 1966: The Harvard program: a summing up. In: KNEESE, A. V. and SMITH, S. C., eds., 1966: *Water Research*. Johns Hopkins Press, Baltimore, Md., pp. 441–55.

HUFSCHMIDT, M. M., FIERING, M. B. and SHERWANI, J. K., 1968: Simulation models for water-resource systems: Their utility in measuring physical and economic effects of weather forecasting and weather modification: Summary report. In: SEWELL, W. R. D., et al., 1968: *Human Dimensions of the Atmosphere*. National Science Foundation, Washington, D.C., pp. 121–35.

ISARD, W., et al., 1960: *Methods of Regional Analysis*. Wiley, New York, 784 pp.

ISARD, W., LANGFORD, T. W., JR and ROMANOFF, E., 1966: *Working Papers, Philadelphia Regional Input–Output Study*. Regional Science Research Institute, Philadelphia, Pa., December.

KINCER, J. B., and MATTICE, W. A., 1928: Statistical correlations o weather influence on crop yields. *Monthly Weather Rev.*, 56: 53–7.

KOOPMANS, T. C., 1951: *Activity Analysis of Production and Allocation*. Cowles Commission Monograph, No. 13, Wiley, New York.

KUH, E., 1965: Econometric models: is a new age dawning? *Amer. Econ. Rev.*, 55(2): 362–9.

LANGFORD, T. W., 1968: A proposed model for the evaluation of economic aspects of weather modification programs for a system of regions. In: SEWELL, W. R. D., et al., 1968. In: *Human Dimensions of the Atmosphere*. National Science Foundation, Washington, D.C., pp. 113–20.

LEONTIEF, W. W., 1951: *The Structure of the American Economy*. Oxford University Press, Oxford, 264 pp.

McQUIGG, J. D., 1965: Foreseeing the future: forecasts and decisions. *Met. Monographs*, 6: 181–8.

McQUIGG, J. D., 1967: *Some Brief Comments on the Use of Simulation Models as a Tool to Study the Relationship of Weather and Human Activity*. Unpublished paper.

McQUIGG, J. D., 1968: A simulation model for the study of surface temperature modification. In: SEWELL, W. R. D., et al., 1968: *Human Dimensions of the Atmosphere*. National Science Foundation, Washington, D.C., pp. 136–42.

McQUIGG, J. D. and DOLL, J. P., 1961: *Weather Variability and Economic Analysis.* University of Missouri, College of Agriculture, Research Bulletin No. 771.

McQUIGG, J. D. and THOMPSON, R. G., 1966: Economic value of improved methods of translating weather information into operational costs. *Monthly Weather Rev.*, 94(2): 83–7.

MARTIN, W. E. and BOWER, L. G., 1966: Patterns of water use in the Arizona economy. *Arizona Rev.*, 15(12): 1–6.

MARTIN, W. E. and BOWER, L. G., 1967: Input–output analysis: An Arizona model. *Arizona Rev.*, 16(2): 1–5.

MASON, B. J., 1966: The role of meteorology in the National Economy. *Weather*, 21(11): 382–93.

MAUNDER, W. J., 1965: *The Effect of Climatic Conditions on Some Aspects of Agricultural Production in New Zealand, and an Assessment of their Significance in the National Agricultural Income.* Unpublished Ph.D. Thesis, University of Otago.

MAUNDER, W. J., 1966a: Climatic variations and agricultural production in New Zealand. *New Zealand Geog.*, 22: 55–69.

MAUNDER, W. J., 1966b: An agroclimatological model. *Sci. Record*, 16: 78–80.

MAUNDER, W. J., 1968a: The effect of significant climatic factors on agricultural production and incomes: A New Zealand example. *Monthly Weather Rev.*, 96: 39–46.

MAUNDER, W. J., 1968b: Agroclimatological relationships: A review and a New Zealand contribution. *Canad. Geog.*, 12: 74–84.

MAUNDER, W. J., 1968c: *An Econoclimatic Model for Canada: Problems and Prospects.* Paper presented at the Conference and Workshop on Applied Climatology of the American Meteorological Society, Asheville, North Carolina, Oct.

MOORE, F. T. and PATERSON, J., 1955: Regional analysis: an interindustry model of Utah. *Rev. Econ. Stat.*, 37: 368–88.

NELSON, R. R. and WINTER, S. G., 1964: A case study in the economics of information and coordination – the weather forecasting system. *Quart. Jour. Econ.*, 78(3), 420–41.

NICODEMUS, M. L. and McQUIGG, J. D., 1969: A simulation model for studying possible modification of surface temperature. *Jour. App. Met.*, 8: 199–204.

OGDEN, D. C., 1966: Economic analysis of air pollution. *Land Economics*, 42: 137–47.

ORAZEM, F. and HERRING, R. B., 1958: Economic aspects of the effects of fertilizer, soil moisture and rainfall on the yields of grain sorghum in the Sandy Lands of southwest Kansas. *Jour. Farm. Econ.*, 40: 697–703.

PETERSON, G. A. and SWANSON, E. R., 1956: *Highest Return Farm Systems.* University of Illinois, Agricultural Experimental Station, Bull. 602, Urbana, Illinois.

PREST, A. R. and TURVEY, R., 1965: Cost-benefit analysis: A survey. *Econ. Jour.*, 75: 683–735.

REVELLE, R., 1964: Oceans, science and men. *Impact of Science on Society*, 14: 145–78.

RIDKER, R. G., 1967: *Economic Costs of Air Pollution: Studies in Measurement*. Praeger, New York, 214 pp.

ROSE, J. K., 1936: Corn yield and climate in the Corn Belt. *Geog. Rev.*, 26: 88–102.

RUSSO, J. A., 1966: The economic impact of weather on the construction industry of the United States. *Bull Amer. Met. Soc.*, 47: 967–71.

SANDERSON, E. H., 1954: *Methods of Crop Forecasting*. Harvard University Press, Cambridge, Mass., 259 pp.

SEWELL, W. R. D. (Editor), 1966: *Human Dimensions of Weather Modification*. University of Chicago, Dept. of Geography, Research Paper No. 105, 423 pp.

SEWELL, W. R. D., KATES, R. W. and MAUNDER, W. J., 1968: Measuring the economic impact of weather and weather modification: A review of techniques of analysis. In: SEWELL, W. R. D., *et al.*, 1968: *Human Dimensions of the Atmosphere*. National Science Foundation, Washington, D.C., pp. 103–12.

STANFORD UNIVERSITY, 1966: SPINMAP: *Stanford Proposal for an International Network for Meteorological Analysis and Prediction – Summary Report*. Stanford University, Palo Alto, Calif., May 31.

THOMPSON, J. C., 1966: The potential economic and associated values of the World Weather Watch. *World Weather Watch Planning Report*, No. 4, W.M.O., Geneva.

THOMPSON, J. C. and BRIER, G., 1955: The economic utility of weather forecasts. *Monthly Weather Rev.*, 83: 249–54.

THOMPSON, L. M., *et al.*, 1964: *Weather and Our Food Supply*. Center for Agriculture and Economic Development, Report 20, Iowa State University, Ames, Iowa.

THOMPSON, L. M., *et al.*, 1966: *Weather Variability and the Need for a Food Reserve*. Center for Agricultural and Economic Development, Report 26: 3, Iowa State University, Ames, Iowa.

TIEART, C. M., 1962: *Markets for California Products*. California Development Agency, Sacramento, Calif.

WATSON, D. J., 1963: Climate, weather and plant yield. In: EVANS, L. T., ed., 1963: *Environmental Control of Plant Growth*. Academic Press, New York, pp. 337–49.

WHITE, G. F., 1966: Approaches to the study on human dimensions of weather modification. In: SEWELL, W. R. D., ed., 1966: *Human Dimensions of Weather Modification*. University of Chicago, Dept. of Geography, Research Paper No. 105, pp. 19–23.

WOLOZIN, H. (Editor), 1966: *The Economics of Air Pollution*. Norton, New York, 318 pp.

W.M.O., 1967: Assessing the economic value of a National Meteorological Service. *World Weather Watch Planning Report*, No. 17, W.M.O., Geneva.

W.M.O., 1968: Economic benefits of meteorology. *World Met. Organization Bull.*, 17(4): 181–6.

WOOD, M. K., 1965: PARM – An economic programming model. *Management Sci.*, 11: 619–80.

ADDITIONAL REFERENCES

BURLEY, T. M., 1965: Flood and drought in the Hunter Valley of New South Wales and their impact upon the agricultural community. *Tijd. Voor Econ. Soc. Geog.*, 56: 193–9.

BYRNE, G. F. and TOGNETTI, K., 1969: Simulation of a pasture–environment interaction. *Agric. Met.*, 6: 151–63.

EDWARDS, R. S., 1969: Economic measurement of weather hazards. Aberystwyth Symposium 1968. *Weather*, 24: 70–3.

HASHEMI, F. and DECKER, W. L., 1969: Using climatic information and weather forecasts for decisions in economizing irrigation water. *Agric. Met.*, 6: 245–57.

HELBUSH, R. E., 1968: Linear programming applied to operational decision making in weather risk decisions. *Monthly Weather Rev.*, 96: 876–82.

McQUIGG, J. D. and DECKER, W. L., 1962: The probability of completion of outdoor work. *Jour. App. Met.*, 1: 178–83.

OURY, B., 1965: Allowing for weather in crop production model building. *Jour. Farm Econ.*, 47: 270–83.

SCHICKEDANZ, P. T. and DECKER, W. L., 1969: A Monte Carlo technique for designing cloud seeding experiments. *Jour. App. Met.*, 8: 220–8.

SHAW, L. W., 1964: The effect of weather on agricultural output: a look at methodology. *Jour. Farm Econ.*, 46, 218–30.

W.M.O., 1966: Data processing in meteorology. *WMO Tech. Bull.*, No. 73, World Meteorological Organization, Geneva, 180.TP.90.

ZUSMAN, P. and AMIAD, A., 1965: Simulation: a tool for farm planning under conditions of weather uncertainty. *Jour. Farm Econ.*, 47: 574–94.

VI Weather knowledge: benefits and costs

A. THE SETTING

The mounting losses of property and income which result from extreme weather events, man's ability to modify his atmospheric environment (deliberately or inadvertently), and the ever-present control which the atmosphere (the rain, the snow, the wind, the humidity and the sunshine) exerts over our 'economy', are all significant in today's society. The predictions and knowledge of the atmospheric scientist, and in particular those of the weather forecaster, are therefore of considerable economic and social value.

The identification of the weather-sensitive aspects of any activity, the extent to which appropriate meteorological advice can be given and the form in which it should be conveyed, require, however, close collaboration between meteorologists, climatologists and the users of weather information in order to ensure complete understanding of both the operational and meteorological problems. In complex organizations such as large industries or construction works, normal analysis by the management may reveal the weather-sensitive points. However, as Crow (1965) points out, the practical ability of the meteorological profession to provide useful assistance must be considered, and joint discussion between the management and the atmospheric scientist will usually be found to be most beneficial.

Benefit/cost ratios of national meteorological services are not known in any great detail, but the advent of the World Weather Watch in the late 1960's prompted several nations to do some 'meteorological book-keeping'. It has usually been concluded that the savings to a national economy are many times greater than the money spent by that country on meteorology. For example, the role of meteorology in the U.K. national economy was examined by Mason (1966). He states that although there may be no immediate prospect of controlling or significantly modifying the weather, its favourable aspects may be better exploited and its worst effects alleviated or avoided by acting upon meteorological advice. This, he says, is realized in aviation, agriculture and some sectors of commerce and industry, but many areas are still unaware of the services and benefits that meteorology can offer, often at 'trifling cost'.

It is now generally accepted that the overall benefit/cost ratio of a National Meteorological Service is approximately 20 : 1. The additional benefits that will result from improvements associated with the World Weather Watch have also been assessed, but the evaluations reported in *World Weather Watch Planning Report* No. 4 (Thompson, 1966b) must be treated with reserve because of the numerous unsubstantiated assumptions involved.

Weather and climate affect human activities in pervasive ways. Studies have shown that profitability in certain economic activities such as agriculture and air transportation is associated with fluctuations in the weather. Other investigations have indicated that human behavioural patterns are also influenced by variations in the weather. It is natural, therefore, that man should try to find ways of adjusting to and even altering the weather, but it is important to realize that modification of tomorrow's (or next month's) weather may be very dependent on the weather that is forecast for tomorrow (or next month), particularly if decisions whether or not to modify are going to be based on the predicted weather. Moreover, it is clear but not always realized that to the client both accurate weather forecasting and successful weather modification may have similar effects. The retail store manager, for example, is not usually concerned with why it rains, but only that it does or does not rain. It is equally important to realize that longer-range (and accurate) weather forecasts, as well as the ability to modify tomorrow's weather, are coupled with many disadvantages – as well as obvious advantages – especially if *everyone* knows what next week's weather is going to be, or if everyone is going to have a hand in modifying what the weather forecaster promises.

The benefits *and* the costs to society of weather forecasting and weather modification must, therefore, be considered in any assessment of weather knowledge. In many parts of the United States, farmers are apparently convinced that attempts to increase rainfall or reduce hail are successful, and warrant the fees charged by commercial cloud seeders (Sewell, 1966). In an attempt to increase runoff where hydro-electric power facilities are located, several electrical power utilities in California have also engaged in cloud seeding experiments to increase runoff and thus make increased power generation possible. These utilities believe, according to Sewell, that if cloud seeding proves to be effective, the cost of obtaining the additional generation from these operations would be cheaper than obtaining power from alternative sources, such as thermal power. Several municipal water supply systems have also engaged commercial cloud seeders to stimulate rainfall to replenish declining reservoirs, and various government agencies in the United States and other countries are engaged in programmes to suppress lightning, dissipate fog, reduce hail damage and even modify hurricanes.

It is clear that important benefits are derived from weather modification but not everyone is convinced that it is a good thing. Modification has been vehemently opposed on the grounds that it involves 'tampering with Nature' and may result in unknown side effects. Others, while not opposing it, view it with some caution and suggest that the potential side effects and implications be thoroughly explored before any attempt is made to alter the weather.

One could also question whether some kinds of weather modification have already gone too far, for it seems illogical to produce (at some cost) 'man-made rain' in an area before the value of 'natural rain' to that area is known. Nevertheless, some aspects of weather modification should be encouraged, for, as President Johnson stated in transmitting to Congress the 6th Annual Report on Weather Modification (Johnson, 1965):

> We hope someday to acquire the knowledge permitting us to minimize the incidence and severity of hurricanes, tornadoes, and other violent storms and, also be able to improve the temperature and rainfall conditions in agricultural and industrial regions. The hope is not fanciful or unrealistic, but it would be misleading to suggest that such a day is near now.

B. WEATHER INFORMATION

1. Types of weather information

Weather information may constitute a number of separate items. It may, for example, consist of the weather existing at the present moment, weather that is expected to exist at a specified time in the future, or analysis and interpretation of the records of weather that has existed in the past. All three types can be of value.

The type of weather information required to arrive at a decision is determined by the character of that decision. For example, an analysis of past weather is useful in arriving at planning decisions involving the location and design of many types of industrial facilities, and the production and marketing of weather-sensitive consumer items, whereas present weather and short-period forecasts of future weather are most useful in making operating decisions such as the flight plan of a commercial airline flight. Similarly, architect-engineers are constantly faced with such problems as siting, building-orientation, heating, ventilating, insulation and drainage, which require the use of weather information. However, erroneous 'solutions' may easily be reached by the unqualified. For example, differences between the direction of the prevailing winds and 'storm' winds, differences between average temperatures and average maximum temperatures, and the meaning of terms such as 20-year return period for wind gusts or flood flow are often economically significant.

2. Using weather information

An interesting use of weather information reported by Hallanger (1963) is in the design of an advertising campaign for room air conditioners. It was determined that the motivation to buy was closely associated with the existence or anticipation of extended periods of temperatures and humidities above certain critical levels. Analysis of weather records from various parts of the country, and the times when these patterns would probably occur in each market area, were therefore determined, and a timely co-ordinated advertising campaign was scheduled. The result, according to Hallanger, was a much higher return for each advertising dollar spent, and as a bonus, production and shipping schedules benefited from the information developed. In addition, there are many less obvious applications of meteorological and climatological information. The commercial operations of large banks, for example, are weather-sensitive activities, as are travel agencies, laundries, drugstores and departmental stores. In the special symposium on 'The Business of Weather', Hallanger made some very pertinent remarks to the businessmen and others who were participants:

> . . . the key point in the realization of the potential value of weather information to your activity [is that] . . . the meteorologist, working as a team with your people, must become familiar with your operation. Only then is he in a position to identify the true weather problems. Only then can he provide the appropriate weather information in the most useful form. Only then will the return exceed the cost by the greatest amount.

Thus, the potential value of weather information to an economic activity and to a particular operation will be realized *only* when the qualified professional meteorologist, working with the co-operation and backing of management, develops the information most suited to the specific needs of the activity.

Such observations are not new, nor were they first suggested by Hallanger (1963), for as McQuigg and Thompson (1966) point out, several investigations, including Demsetz (1962) on 'the economic gains from storm warnings in Florida', McQuigg (1964) on 'the economic value of weather information', Murphy (1960) on 'meteorology and heating load requirements', Rapp and Huschke (1964) on 'weather information – its uses, actual and potential' and White (1964) on 'new weather discoveries will serve you', have effectively demonstrated the 'value' of weather information, and have advanced the seemingly reasonable thesis that weather information has (or accrues) economic value because it makes possible better management decisions during the operation of a weather-sensitive process.

3. The value of weather information

It is true to say that it is difficult to compute, or even make reasonable estimates of, the increased economic value that might result from improvement in the accruing of weather information. Two of the reasons for this, according to McQuigg and Thompson (1966), are the dearth of real economic data from weather-sensitive processes, and the expense, danger or undesirability of conducting actual experiments in which the accuracy of weather information used by a decision-maker is deliberately varied. The various techniques and models available for evaluating weather are briefly described in Chapter V, including the simulation model of the relationship between weather events, non-weather events, man's function as a decision-maker and the economic outcome of an enterprise as described by McQuigg and Thompson. A schematic outline of these relationships is shown in Fig. VI.1, which shows clearly that the economic outcome of the enterprise is related to some actual weather events, some actual non-weather events, and to the particular choice of alternatives that is made by the decision-maker, the term u being included because the relationship between π and $(W, O, A*)$ is almost never known exactly. At the time the choice of alternatives is being made, the decision-maker does not know the actual values of W, O that will occur, nor does he know the exact value of u. He therefore, according to the authors, evaluates

$$\pi^* = f(W^*, O^*, A) + w$$

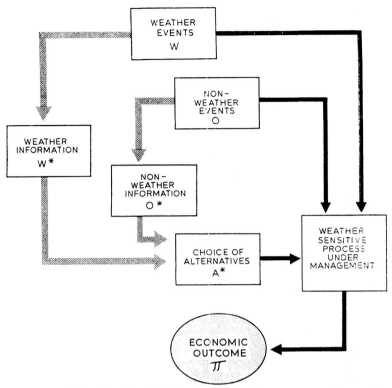

FIG. VI. I. *Schematic outline of relations between weather events, non-weather events, the choice of alternatives by management, and the economic outcome of an enterprise. Solid lines represent relationships between actual events and the process which is being managed. Dotted lines represent the production and the use of information (after McQuigg and Thompson, 1966).*

for each possible alternative, choosing A*, a subset of A, in some rational manner such that π* will come as close as possible to being consistent with his long-term strategy (one such rational strategy would be to attempt to maximize expected profit).

Management of the flow of natural gas through the distribution system of a company serving a city is a weather-sensitive process and was chosen by McQuigg and Thompson to serve as an example of simulation.

In brief, they assume that the pipeline serving a certain local gas company has many customers, and that the amount of gas that can be pumped through the pipeline under various pressures is limited. A simplifying assumption made by the authors is that the local gas distribution system in this simulation model has no storage facilities, and that it has an agreement with the pipeline serving it that calls for 'penalty' payments proportional to demand over a given amount on any given day. Further, in order to keep from paying the 'penalty' rate during periods of heavy demand for gas the local company has negotiated agreements with certain users which allow service to these users to be curtailed, with a few hours' notice. A further simplifying assumption is made that the manager has only one time (the beginning of each day) to choose one of the two possible alternatives open to him. These are:

a_1 = order no curtailment of service to local users.
a_2 = order curtailment of service to some local users so as to avoid paying the penalty rate.

Accordingly, one of two possible losses may occur on any particular day if the manager chooses the wrong alternative: he may order no curtailment of service when in fact he should have, which results in penalty payments, or alternatively he may order curtailment when he did not need to, thus losing sales that his company could have made.

In the application of the model the set of non-weather events is considered fixed, and in addition for simplicity the number of customers is assumed to be fixed. In the analysis, a series of daily observed values of the number of heating degree days (base 65°F) for December to March for 30 winters in Columbia, Missouri, was used in two ways. First, several series of 'forecasts' each with a given distribution of errors were generated; second, several series of 'observed' levels of demand Y_i were computed to simulate several levels of precision of the linear regression relationship. For a particular day, Y_i was computed as:

$$Y_i = A + BX_i - C_i + u_i$$

in which A and B are linear regression coefficient (A = +4157·0 in thousands of cubic feet of gas, B = 248·11 in thousands of cubic feet of gas per heating degree day), X_i is the observed number of heating degree days and C_i is the level of curtailment ordered. The term u_i can be thought of as being drawn at random from a normally distributed sample of values.

The above equation illustrates that the amount of gas demanded for a particular day is influenced by the weather that actually occurs (represented by X_i), by the decision made by the manager (represented by C_i) and by some random events (represented by u_i). Further, for a given pair of the standard deviation of the forecast error in °F, and the standard deviation of the disturbance term 'u' in thousands of cubic feet of gas, the amount of penalty sales is computed as $(Y_{Pen})_i = Y_i - Y_p$ where $(Y_{Pen})_i = 0$ in the case where $Y_i < Y_p$. However, if $C_i > 0$, and $Y_i < Y_p$, the amount of lost sales is computed as $(Y_L)_i = Y_p - Y_i$. McQuigg and Thompson assume that one unit of penalty sales cost 10 times the value of a unit of lost sales. A value of the 'loss function', L_i, is then computed as $L_i = (Y_L)_i + 10(Y_{Pen})_i$.

The average annual loss function was computed for various errors in temperature forecasts and variations in the disturbance term, as shown in Fig. VI.2, and as the authors point out, the values can be readily converted into terms of money by multiplying by the actual

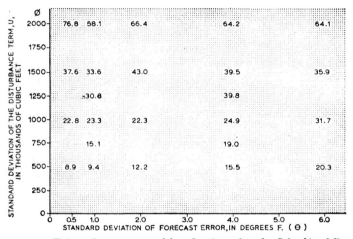

FIG. VI.2. *Estimated average annual loss function values for Columbia, Missouri (in millions of cubic feet of natural gas), for various errors in temperature forecasts and variations in the disturbance term (After McQuigg and Thompson, 1966).*

price. For comparison, it may be noted that the average daily gas demand for Columbia is close to 14·5 million cubic feet. McQuigg and Thompson, in commenting on the results shown in Fig. VI.2, state that there is a reasonably consistent increase in the values in the body of the figure from left to right in lower portion of the graph, where $\phi < 1000$ (this represents the situation in which the decision-maker is using a reasonably precise expression to translate weather information into operational terms). However, where $\phi > 1500$, it is apparent that there is no change in the average value of the loss function as θ increases or decreases. Thus, a manager using an imprecise relationship to translate weather estimates into estimates of

K

gas demand could not discriminate between degree-day forecasts that were almost perfect and those that were very poor.

The model discussed above shows that a simulation model approach can be used to estimate marginal changes in the value of improvements in weather information. Other problems similar to the flow of natural gas into a city can be envisaged, and the idea of a level of activity above or below which some 'penalty' is exacted, is common to many enterprises. For example, flooding does not occur until some particular water level is exceeded; freezing does not occur until some temperature is reached; a structure is not damaged until a certain wind speed is exceeded, etc.

Many other examples of the use of weather information could be given, and the summary provided by Reichelderfer (1964), a former Chief of the U.S. Weather Bureau, provides some insight into how 'weatherwise' some businesses are. He cites, for example, the case of February 1958, a particularly sunny and cold month in the U.S. mid-west, when temperatures were from 8 to 13°F lower than February 1957, and department store sales were 8 to 14% lower than in February 1957. Specific examples are given, such as the baker who uses professional weather services, to add some $250,000 a year to his profits, simply by gearing his baking and distribution to the daily weather forecast. In another case, small marketers can boost sales by using the rain, such as the New England company which sells paper umbrellas which repel rain for two or three hours. Its owners–managers, according to Reichelderfer, kept a weather eye on overcast skies and hawked their umbrellas at the ball park or football stadium when rain began sprinkling. Thus, being alert to rapidly changing weather enabled them to be on the spot with something that people needed and wanted. They therefore did not waste their efforts at times when the product would not be sold anyway.

C. WEATHER FORECASTING

1. Atmospheric predictability

The problem of weather prediction requires consideration of three fundamental questions: What is the weather? Why is there weather? What will the weather be? The first question (what is the weather?) is answered from the basic weather observations, including not only those in the local area at ground level, but also the observations of the different levels of the atmosphere over as large an area as is required. The second question (why is there weather?) involves the understanding of the reasons why the atmosphere behaves as it does, and numerous atmospheric models are in operational use to determine the nature of the general circulation of the atmosphere. The understanding of the atmosphere also requires an appreciation of the physical and dynamic properties of the 'small-scale' features such as sea breezes, tornadoes, thunderstorms, frontal systems, hurricanes, anticyclones and the monsoon circulation. If the answers to the first question (what is the weather?) is known in most, if not all, details, and the second question (why is there weather?) has solutions, then and usually *only* then will answers to the third question (what will the weather be?) be forthcoming.

In addition to the three questions mentioned above, a weather forecasting programme usually involves three distinct variables: area, space and time. Area in this context can be considered the area from which observations are taken, and the area for which a forecast is made, and in general the forecast area is much smaller than the observation area. For example, a weather forecast for eastern Canada normally requires observations from at least the whole of Canada and most of the United States. The second factor, space, can be considered to be the vertical dimensions of both the observational network and the prediction area, the extent of the space dimension being specifically related to the third factor, which is time. The time element is in turn related to two things, the period of validity for the forecast, and the time-period of 'historical' observations required before the issue of the forecast.

The importance of these factors is readily apparent if a two-hour forecast for a small area (e.g. New York City) and a long-range forecast (e.g. a month) for a large area such as Europe are compared. In the first case, observations would be needed for a relatively small area (probably only within 100 miles of New York), up to only relatively low levels and for probably only the previous 5 to 10 hours. By contrast, a month's forecast for Europe would involve observations of both present and past weather conditions, possibly for several months or years over the whole Northern Hemisphere, and probably even observations over the Southern Hemisphere, for various levels up to at least 100,000 feet.

a. Methods of prediction

The methods of weather forecasting which are currently in use have been summarized by Lorenz (1966) into three categories. First is *synoptic* forecasting. This is the subjective method in which the forecaster assembles all the information which can be analysed concerning the present and recent state of the atmosphere, and then formulates the data into meaningful interpretations of the weather situation in terms of systems, such as high- and low-pressure areas, air masses, fronts, cloud systems, and on a more local scale thunderstorms and tornadoes, etc. On the basis of the systems behaviour, the synoptic forecaster extrapolates their positions and configurations into the future (being careful to add any new systems whose genesis he feels is indicated) and then translates his prognosticated weather pattern into forecasts of weather at specified locations. This estimate of how a system will behave is based upon his knowledge or opinion of how similar systems have behaved in the past.

The requisites for a good synoptic forecaster are a keen analytical mind and a vast amount of experience, but even the best forecaster cannot assimilate all the available information concerning the current weather picture, and his experience cannot provide for all possible variations of the atmosphere. Incorrect forecasts are therefore inevitable; such as heavy snowstorms which have not been predicted even a few hours in advance, and the predicted heavy rain which fails to eventuate.

The disillusionment with synoptic methods of forecasting led many meteorologists to seek methods which do not rely upon human judgment and alertness. One procedure fulfilling these requirements entails *statistical* forecasting, in which the relevant features of the atmosphere are expressed as numbers. These numbers are submitted into mathematical formulas, based wholly upon the statistics of past weather behaviour, but as Lorenz (1966) points out: '. . . these statistics are really nothing more or less than experience expressed in the form of numbers'.

The third method of forecasting is *numerical weather prediction*. This method was envisioned early in the twentieth century by Richardson (1922), but in practice it is almost wholly dependent upon high-speed, large capacity computers, so that it did not really come into being until the 1950's with the advance of the computer. In numerical forecasting the computer is instructed to determine particular solutions of the mathematical equations which represent the physical laws governing the atmosphere and its environment. In principle, the computer solves the time-dependant dynamical equations of atmospheric motion, but in actuality a complete solution of the equations is beyond the scope of any existing computer, so that current forecasts are based upon rather crude approximations, reflecting limitations both in the understanding of the physical laws of the atmosphere and in the power of the computer.

The methods of weather forecasting allow considerable room for improvement, and it is clear that the technique of accurate weather forecasting is still far from perfection. The reason for this is that the mathematical solutions of the time-dependant equations of the atmosphere are basically unstable. This means that there is difficulty in constructing the mathematical models, and in the development of the techniques of numerical approximation to be used in computations. In addition, the accuracy and density of the synoptic data to be used as the initial state of the model atmosphere is far less than sufficient. However, the greater number of synoptic observations that will become available through the World Weather Watch, the better understanding of the physical laws of the atmosphere that is expected through the Global Atmospheric Research Programme, and advances in computer technology, will enable considerable improvements to be made in short, medium, and long range forecasting.

b. Problems of predictability

The actual 'predictability' of the atmosphere has been questioned by many observers. For example, the extent to which the weather is predictable has been raised by Robinson (1967), who indicated that the Global Atmospheric Research Project was planned on the assumption that if the initial state of the atmosphere is known with sufficient accuracy, large-scale motions are predictable '. . . as a determinate physical system for a period of approximately two weeks'. The proposals of the World Weather Watch have been similarly justified, and Robinson was concerned to test this claim, particularly the truth of the assumption that there is no fundamental limitation of the predictability of the atmosphere, but only practical limitation arising from the limitations of computing machinery and methods of measurements.

There are many meteorologists who disagree with these findings, including J. Namias, Chief of the Extended Forecast Division of the U.S. Weather Bureau, who in the Wexler Memorial Lecture given in 1968 made some very pertinent observations (Namias, 1968).

> Meteorologists have always dreamed of predicting weather in detail in both space and time. . . . Today, the number of eminent meteorologists who see detailed weather prediction (out to two or three weeks at any rate) as a realizable achievement over the next several years has substantially increased: indeed the optimism has become so universal that one finds references to the imminent

era of detailed predictions in most of the world's technical publications of meteorology and even in the lay press. Naturally, the opinions of the world's top meteorologists are varied . . . But among the enthusiasts are the most highly respected in our field, whose opinions are generally solidly based.

What has happened to bring about this wave of optimism asks Namias, and what is its foundation? There are of course many reasons, but advances in computer technology, and observational systems, along with what Namias calls 'a generation of top flight dynamicists able to employ these technological advances in this quest for objective weather predictions', are particularly important.

Accurate numerical predictions, not only of fluctuations around monthly or seasonal means but even of detailed weather forecast for a period of two or three weeks, are, according to Namias, foreseen by 'top-echelon scientists'. The rationale for this optimism was given in a report by a panel of experts (Panel on International Meteorological Cooperation, 1966), and included a predictability test describing the performance of three different iterative models of the general circulation. Each model was run twice for at least 30 days beyond an arbitrary interruption time, once undisturbed and once with a perturbation of the temperature added. When the fields developing from the unperturbed and perturbed initial conditions diverged enough to differ by as much as two randomly chosen fields, the limit of detailed predictability (about two weeks) was said to have been reached. However, Namias says that Robinson (1967) later remarked: 'This experiment may tell a lot about the models, but not necessarily anything about the real atmosphere.' Robinson also stressed a possible lack of a gap in the spectrum of atmospheric turbulence – a lack which Namias suggests could be fatal for long-range detailed prediction. However, Namias says that no one experienced in long-range forecasting has spoken out publicly or written on this matter, probably because of the complexities of numerical modelling (often baffling to the practitioner), because of the 'quick victories' with shorter-period prediction, and because of the unavailability of results on long-range forecasting with real atmospheric input.

c. Verification of predictions

In spite of these 'three good reasons to keep quiet' Namias suggests that 'cautious optimism' is the key phrase. He then describes some recent tests of the harsh yardstick of verification of weather forecasts.

One of the tests comprised a series of 14 forecasts utilizing

Shuman's 6-layer primitive equation model (Shuman and Hover-male, 1968), departures from the predicted and observed sea-level pressure being extracted for a grid covering North America. Cor-relation coefficients were computed between forecast and observed sea-level pressures, out to six days, and the results were compared to persistence, and to two partially subjective predictions by experi-enced forecasters. One forecaster used only standard baroclinic and barotropic prognostic charts, the other used the Shuman model prognostic charts as well. The results are shown in Fig. VI.3, and Namias considers that at least two conclusions can be drawn from

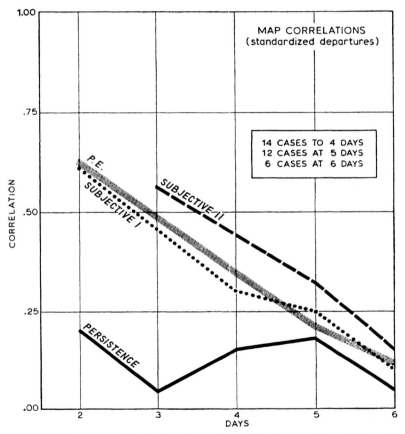

FIG. VI.3. *Statistical evaluation of 14 National Meteorological Center 6-layer primi-tive-equation (PE) model forecasts, and corresponding subjective forecasts, of sea-level pressure over North America made from December 1966 to June 1967. Subjective I was made without the aid of the PE predictions, subjective II was made with the aid of PE predictions (after Namias, 1968).*

the information given in the diagram: first, extension of numerical predictions to six days can assist in improving present methods, particularly when modified by experienced forecasters; and second, that the model will have to be greatly improved if skill out to two or three weeks is to become possible.

Despite the advance in numerical forecasting, most predictions of the atmosphere in the 1960's have been relatively restricted in nature, for in the majority of cases data from only the Northern Hemisphere has been used. It is clear, however, that the entire global atmosphere must be considered for extended range predictions, and this is given as one of the reasons for the World Weather Watch and the associated I.C.S.U.–I.U.G.G.–G.A.R.P. programmes. But, as Namias reminds us, even with the best implementation there arise a number of questions including: will the global atmosphere be observed in sufficient horizontal and vertical detail so that cumulative errors in iterative numerical predictions will not destroy the values of the prognoses after several days; is it economically feasible to set up observation platforms and systems to achieve the required resolution of the initial state; and, even with such an observational network will man ever understand the complex workings of the complex ocean, air and continent systems to such an extent so as to make appreciably skilful forecasts of weather at given places more than a week in advance? Or, as Namias asks: '. . . does the atmosphere possess a certain indeterminancy so that it hardly "knows" its own detailed behaviour beyond this time?'

d. Limits of predictability

The theoretical limits of predictability are examined by Gleeson (1967), who considered two questions raised by a panel on the feasibility of a global observation and analysis experiment (Panel on International Meteorological Cooperation, 1966). The two questions were: 'Given the inevitable small errors, will these remain small or will they grow, and, if they grow, how fast will they grow?' Gleeson only considered simple physical models in which a unit mass moves at constant and variable speeds, and he indicated that the distant-future location of the mass is unpredictable in both cases. This and other conclusions reached by Gleeson suggest that errors in deterministic predictions of a meteorological variable do not remain small but eventually become large enough for the predicted variable to be completely uncertain within its range of possible values.

A further indication of the problems associated with the instability of the atmosphere is afforded by numerical simulation, in which the initial conditions need not represent the present or any other actually observed weather situation, and the numerical solutions are extended over simulated periods of months or even years. Some sophisticated simulations have been made and Charney *et al.* (1966) concluded that a doubling of the amplitude of small perturbations in five days is a reasonable estimate. If this is correct, Lorenz (1966) says that there would be no reason why good forecasts should not be able to be made more than a week in advance, once the technique has been mastered. Forecasting a month or more in advance, however, is another matter, and it is hard to imagine that the errors in estimating the *current* state of the atmosphere will ever become so small that the forecasting errors that would occur in a monthly forecast would become small enough to be tolerated. Nevertheless, the conclusion that small perturbations double in five days, rather than in some other time interval, is highly tentative; and as Lorenz indicates: '. . . the situation may be much better or much worse'. Lorenz also points out that the existence of instability specifically implies that 'the configuration of the atmosphere at a specific time in the future will differ considerably from the predicted configuration at that same time. Nothing is said as to how the general behaviour over an extended period will compare with the predicted general behaviour over that period.'

The problems associated with prediction in the atmospheric sciences also have parallels in the social sciences, as pointed out by Thompson (1964), who suggests that outside the laboratory it is easy to *identify* situations potentially favourable for wars, revolutions and riots, but the *timing* of actual outbreaks is difficult. Similarly, in meteorology it is relatively easy to identify situations favourable for shower activity, but the timing of actual showers is difficult. However, depending ultimately on economics and the extent to which it was deemed worth while to apply time and effort to the task, and not on any inherent predictability of the weather, it may be possible to predict every shower. To carry out such a procedure to an infinite degree of refinement would, however, according to Thompson, be completely uneconomic and no one would do it, but '. . . given cheaply operating automatic weather stations and revolutionary developments in data collecting and computing, such as have in part occurred, there is no reason to think that the problem of comprehensive yet detailed weather forecasting cannot be solved in principle'.

The best prospects for immediate improvements of the technique of weather forecasting are in the field of dynamical or numerical forecasting, rather than in synoptic or statistical forecasting. Yet the ultimate in weather prediction will not be attained, according to Lorenz (1966), until some system has been developed for directly measuring the various weather elements at a distance, so that we may obtain a virtually continuous distribution of the conditions throughout the atmosphere.

The predictability of financing weather forecasting activities is also extremely important, and most relevant in any discussion on the value of the weather. White (1966a), for example, says that it is clear that the pace of development will depend greatly on the resources we can direct to these ends. But as White comments, the Federal (U.S.) budget process is as 'unpredictable as the weather and I can see no new technological developments on the horizon that will improve its predictability'.

2. Presentation of weather forecasts

Meteorological services aim at satisfying the requirements of a large number of clients who can be divided into consumers and producers. Hibbs (1966) states that consumers use weather services to maintain or increase health, safety, general welfare, natural resource preservation, etc. Producers, on the other hand, accrue benefits to themselves through the use of weather services in enhancing their professional, commercial or industrial activities. Accordingly, 'producer' weather forecasts are used by professional meteorologists, agriculturalists, commercial airlines, private business pilots, marine and transportation companies, and many other businesses.

During the past decade many advances have been made in the atmospheric sciences, including the improvement of techniques for weather forecasting, and a greater awareness of the demand for weather forecasts by society. Many National Meteorological Services, for example, have in recent years provided weather forecasts and services relevant for road maintenance (snow warnings, heavy rainfall forecasts), water control (forecasts associated with irrigation, flooding, droughts), forestry (low humidity, lightning suppression probabilities), agriculture (longer-range forecasts, frost warnings), scheduling of power production, oil deliveries, retail sales campaigns, as well as specialized marine and aviation forecasts. It is, nevertheless, fair to say that in general the public weather forecasts have not improved in presentation or impact. In addition, there is an inevitable change (through time) in the reaction of consumers to weather and weather forecasts. It is important, therefore, that atmospheric scientists are aware of society's changing desire for, and reaction to, both short-range and long-range weather forecasts.

The presentation of weather forecasts is a highly specialized and an important form of communication, and since the weather has a profound effect on our activities and businesses, most of us are called upon to make many 'weather-sensitive' decisions. These decisions are based, for the most part, on the weather forecast. The weather forecast can be viewed in three phases: the *preparation* of the forecast, the technical task of the meteorologist; the *presentation* (the 'weather package') and the communication of that forecast to the consumer; and the use or *application* of the forecast by the consumer.

a. The weather package

The weather 'package' has two important connotations. First, it describes the necessity of condensing the weather forecast into a few seconds or minutes of television or radio time, a few column inches of a daily newspaper, or in more specialized cases, a detailed technical weather forecast for industry, aviation, etc. Second, 'package' implies that the meteorologist is preparing it for presentation to the consumer in much the same way as any consumer good is packaged for efficient and effective marketing.

Meteorologists and the communications media began to realize the importance of the proper presentation of the weather package in the 1940's and there has been a revival of interest since 1960. The U.S. Department of Commerce, Weather Bureau (1964), identified 18 user groups of current and potential users of weather information: the general public, general science, health and safety, general welfare, resources utilization, land transportation, air transportation, water transportation, construction, water supply and control, energy production and distribution (light, heat and power), merchandising, recreation, manufacturing, fishing, agriculture, forestry and communications. All of these groups require special packaging, but probably of chief concern is the first group, the general public, which includes the needs of housewives, commuters, schoolchildren and vacationers.

The Weather Bureau used subjective methods to rank the influence of weather on the 18 user groups, and a high ranking implied that the user is very sensitive to weather. That is, that 'adverse conditions may involve life or very large monetary losses, *and* that the user can take appropriate corrective or evasive action on the basis of advance information'. Using this method and criteria the general public user group ranked first, followed by fishing, agriculture, general science and air transportation. The influence of the weather on the general public is therefore considerable, and the importance of the public weather forecast package is evident. It must in fact be a package appropriate to the needs of all the members of the general public, as well as being capable of being understood by all who wish to use it. The following examples from the Vancouver–Victoria area of British Columbia, Canada, illustrate a typical weather package:

> Cloudy skies with rain beginning tonight. Showers tomorrow morning with light winds. Thursday's outlook cloudy with rain. Vancouver 49, Victoria 47. (Radio CKLG, Vancouver, 9.40 p.m., Feb. 20, 1968.)

> Gale warning for Juan de Fuca. Cloudy, rain by noon. Mild. Winds southeast 25. Friday's precipitation ·29; sunshine 1 hour 42 minutes; recorded high and low at Victoria 55 and 45. Today's forecast high and low 48 and 42. Today's sunrise 7.07, sunset 5.48; moonrise 5.49, moonset 1.27. (*Victoria Daily Colonist*, Feb. 24, 1968.)

More detailed weather forecasts are available to the general public at certain times on other radio stations in the area. In addition, the television weather forecasts are more detailed than the examples given above. Nevertheless, many of the weather packages available to the 'general public' through mass communication media are unsatisfactory, and often confused.

b. Problems of weather forecast presentation

There are a number of reasons for the 'unsatisfactory' state of weather presentation. For example, Mason (1966) indicates that the goal of a national weather service is to improve the observation, understanding and prediction of the behaviour of the atmosphere, to disseminate this knowledge quickly and widely, and to guide the community in its use so that the great number of weather-sensitive decisions arising daily in all walks of life can be made in the best interests of the economic and social welfare of the country. However, it seems to be the case that the general public has not sufficient meteorological background to make use of the meteorological information that could be made available to them in a more detailed package. Further, it could also be suggested that most meteorologists do not have sufficient 'mass media' knowledge, despite the fact, as pointed out by Tenenbaum (1966), that the stories and forecasts from any major weather office are read and heard by millions daily.

A two-pronged educational attack is needed: education of the meteorologist in the techniques of journalism, and education of the consumer in meteorology in the use of the weather forecast. Both of these approaches have been used with considerable success, the first approach being given a trial at the Boston Forecast Center in 1964. The main theme of the programme, reports Tenenbaum, was that weather forecasts should have 'correctness, conciseness, coherence, clarity, and unity'. The weather forecast in fact should have all the elements of the complete news story: who, what, when, where, why and how; with the 'why' and 'how' being the least important elements as far as the majority of the consumers are concerned. However, it has often been suggested that the weather forecaster should always give a reason for, or explanation of, his forecast. This is usually done by TV weathercasters, who have thereby become popular, but in many cases the radio or newspaper forecast is given as a bare statement of fact without any explanation as to why. The use of actual weather maps with isobars, fronts, etc., a feature of nearly all daily newspapers in New Zealand and in some other countries, is however a newsworthy feature and it could find ready adoption in other parts of the world.

Few of these considerations are taken into account by many forecasters and broadcasters. For example, the forecast given previously from the *Victoria Daily Colonist* is typed on a standardized form which is sent directly to the typesetter and make-up man. No attention is paid to the important details of 'who', 'why', and 'how' and only the bare essentials of 'what', 'when' and 'where' are given, and always in the same format. In printing the weather forecast this newspaper, therefore, ignores many of the principles that are standard procedure in the rest of its operation, and it is clear that the newspaper cited is not alone in this respect. Nor are meteorological services blameless, for instead of preparing a package for the newspapers, they are willing to – or possibly have to – tailor a forecast to fit into the newspaper's blanks.

Another problem area in the weather package is the choice of words in describing expected weather, and as Shak *et al.* (1966) have pointed out:

> When we pass from the domain of conversation to scientific discussion and vice versa, it becomes evident that the use of many

words is altered or restricted. . . . Because of these differences it is essential that each locale devise a uniform set of terms to adequately describe temperature variations with regard to degrees of heat and cold. It is also essential that each forecaster be consistent in terminology to avoid confusion to the public from one forecaster to the next.

The authors provide charts for determining the use of words to describe temperatures in New York City forecasts, and in the proposed standard temperature terminology for New York City it is suggested that if the temperature deviated −20°F from the normal maximum on January 1 forecasters would use the term 'very cold', while a −20°F deviation on June 1 would call for 'unseasonably cool'. Similarly, a +20°F deviation from the normal maximum temperature would be described as 'unseasonably mild' on January 1, 'unseasonably warm' on May 15 and 'very hot' on August 1.

Similar problems also occur in communication of the other five elements commonly dealt with in the forecast: wind speed, wind direction, cloudiness, visibility and precipitation. Oddie (1964), for example, questions terms such as 'sunny periods', 'fine' and 'frost', and suggests the need for numerical definition of fog. It is common, for example, to hear weather forecasts '. . . with some fog in low areas', but the user has no way of knowing how extensive this fog will be or how it will affect his activities. Oddie suggests, therefore, that fog should be defined in terms of the visibility that it allows; in other words the weather forecast should state 'Fog in low areas, expected visibility 200 yards'.

Most of the words used in a public weather forecast aim at a purely objective description of the weather. But in everyday conversation many people use another class of words – like 'close', 'muggy', 'raw' and 'oppressive'. Whether or not the average weather forecaster should use such terms is open to debate, but as Oddie points out, 'In their favour it can be said that they tell the listener what he generally wants to know – how he will *feel*: moreover, most of them imply something about the humidity of the air, a matter which, though not unimportant, is traditionally neglected in forecasts.' Nevertheless, human reactions are both variable and uncertain, and it is possible that forecasters would be wise to keep to physical quantities.

c. Consumers' reaction to forecast presentation

A number of surveys on the consumers' reaction to weather forecasts have been conducted. In 1949, for example, a survey of University of Toronto (Toronto, Ontario) students on weather technology was published (Controller, Canadian Meteorological Service, 1949). In 1967 a similar study (Maunder, 1969a) was conducted at the University of Victoria (Victoria, British Columbia). Both studies used questionnaires, but in the earlier study they were completed by 200 second- and third-year psychology students, while in Victoria 400 first- and third-year geography students were involved. In each case the sample was small and not without bias. Despite this, it is believed that some meaningful comparisons can be made. A few of the questions and replies (expressed as a percentage of the total number of replies) were as follows:

(1) *How frequently do you normally read, listen (or watch) the daily weather forecasts?*

	U. Victoria	U. Toronto
(a) Almost always	15	20
(b) Fairly often	43	47
(c) Hardly ever	42	33

(2) *A forecast reading 'Tomorrow will be partly cloudy' means to you*

(a) Part of the sky will be covered by clouds all day	34	49
(b) Part of the sky will be clear, remainder cloudy	45	38
(c) Rain will fall intermittently during the day	1	2
(d) The forecaster is only partly sure what to expect	2	7
(e) Will be cloudy in some areas, clear in others	18	6

(3) *In a weather forecast which of these items is usually of most interest to you?* (rank position 1, 2, 3, 4)

(a) The *temperature* expected tomorrow	2·34	1·78
(b) The clearness or cloudiness of tomorrow's *sky*	2·77	2·82
(c) The expected *wind* direction and speed	3·51	3·64
(d) The *precipitation* (rain, snow) expected	1·37	1·84

The most significant question so far as the forecaster is concerned is probably question three, where students were asked to indicate the 'importance' to them of the items in a weather forecast. In the 1967 survey 48% indicated that the *precipitation* expected was the item they most wanted to hear about (with an average ranking of 1·37) compared with only 9% who thought that winds were of most interest (average ranking 3·51). The 1967 survey indicated quite clearly therefore that in Victoria, B.C., a weather forecast should emphasize the precipitation expected, temperature, state of the sky and winds, in that order. By comparison the 1948 University of Toronto survey indicated that the *temperatures* were the most important (average ranking 1·78), closely followed by *precipitation* (average ranking 1·84). These results could indicate a number of things, but probably of most importance is that the normal (climatic) conditions (i.e. just how important *are* temperatures, snow, rain, wind, cloud, sunshine, etc., to the consumer of the local weather forecasts) need to be taken into account, and that an ideal forecast (terminology-wise) in one area may not be ideal in another.

Opportunity was taken in the 1967 survey (Maunder, 1969a) of asking some additional questions. One question related to the use of the newspaper, the radio, the television as means of obtaining weather forecasts. The results are as follows:

How often do you normally:

	Listen to weather forecasts on the radio	*Read the weather forecasts in the paper*	*Look at weather forecasts presented on TV*
Every day	20	5	3
Almost every day	26	9	10
Sometimes	25	37	31
Rarely	8	30	36
Never	4	19	19
Listen but don't pay attention	16	—	—

These results indicate quite clearly that less than half the students surveyed took an 'active' interest in weather forecasts, and possibly surprisingly only 14% read the weather forecasts in the newspaper regularly, and only 13% looked at the TV weather forecast on a regular basis. Radio fared a little better, for 46% indicated that they listened to the weather forecast every day or almost every day. Perhaps the most significant conclusion drawn from the survey was that the great majority of students did not read, listen or watch the weather forecast. It may, therefore, be more appropriate for weather forecasters to concentrate a little more on the 20 to 25% who seem to be interested in their product than on the 75 to 80% who are not interested. The problem is to find out the type of person who *is* interested in the forecaster's product, and then give him the information required.

No claim for any degree of completeness is made for the survey reported briefly above. Nevertheless, as also reported by Sherrod and Neuberger (1958), such a survey indicates that the attitude of the 'general public' towards forecasting terms *is* important and that it may change from time to time and place to place. In particular, it appears that time may alter the consumer's understanding of weather forecasts and weather forecasting terminology, as well as the consumer's requirement for weather forecasts.

The feasibility of including a percentage-probability forecast in the weather package is often suggested, the following being an example of such a package:

> Cloudy today; rain tonight; clearing tomorrow. Rain probability: today 40 per cent, tonight 80 per cent, tomorrow 30 per cent.

The first part of this package is the usual categorical prediction which is presented for those who Thompson (1963) suggests wish '. . . to follow the economically unsound, but mentally less fatiguing policy of letting the forecaster make their decisions for them'. The second part, the percentage-probability forecast, allows the user to base his weather-sensitive decision on 'more accurate' information. Nevertheless, the probability forecast of precipitation does have its problems, for it is not at all clear in the United States (where such forecasts are used) if the term 40% probability means the same thing to consumer No. 1, consumer No. 2, forecaster No. 1, forecaster No. 2, area No. 1 or area No. 2.

d. Use of weather packages

The information provided by the public forecast can be used to great advantage by small business operators. Reichelderfer (1964), for example, in a study on 'using weather services in your business', indicates that a baking company decreases its shipment of rolls to suburban locations on rainy days (a decision based on the 24-hour forecast), but delivers more rolls to locations near subway entrances, railway stations and bus terminals. Why is this the case? Apparently on rainy days many housewives do not shop, but ask their husbands to buy the rolls on their way home. Hence, intelligent use of the daily weather forecast can pay dividends.

The specific weather forecast packages required by commercial and industrial users are many and varied. Nevertheless, the main function of a 'commercial or industrial weather forecaster' is to communicate his knowledge of weather factors in such a form that the overall profitability and safety of the client's operations are increased. Further, as pointed out by Borgman (1960), '. . . since the client's interest in the forecast service is in direct proportion to the possible profit, it is clearly in the ['industrial'] meteorologist's interest to study ways for maximizing the forecast profitability'.

The specialized weather forecaster must then be familiar with the requirements of his client and be able to translate his forecasts into an appropriate package. Consider, for example, two applications of the work of the private meteorologist in the United States. The first problem, as reported by Nelson and Winter (1963), concerned a large trucking firm which carries perishable goods on uncovered trucks. The trucks are loaded at night and leave the depot early in the morning, and a moderate or heavy rainfall while the trucks are on the road costs the trucker about $500 per truck in damages to the goods. The comparable cost of covering the trucks with a tarpaulin is $20 per truck. The firm hired a private meteorological consultant and asked him to predict moderate or heavy rain (rainfall in excess of 0·15 inches) each evening for the following morning. In this case, then, the package was simple: rain or no-rain in a given period. Because he was aware of the application of his forecasts the forecaster predicted rain whenever there was any doubt, thus tailoring his forecasts to his customer's needs. The second case was the hour-to-hour notification of weather developments provided by a private consultant (Allen, 1964) to the firm building the Chesapeake Bay bridge-tunnel. The usual forecast consisted of notification of impending weather and sea conditions, but when a dangerous storm threatened the meteorologist in this case actually advised the construction firm on a course of protective action. This sort of decision by a weather consultant is rare for it requires an intimate knowledge of the client's operation, but in a case such as this the explicit package saved thousands of dollars.

In these two examples, it can be seen that the specialist forecaster is able to present a weather package of great value to the commercial or industrial user by translating weather information into economically meaningful terms. The specialized package must, of course, normally be prepared by private consulting firms, for government weather services do not usually have the funds available to provide the thousands of individual packages which would be required by users daily.

e. Weather forecasting and the mass media

The use of the mass media – particularly radio and television – is of tremendous value in time of disaster, and in respect to severe weather warnings radio has proved invaluable on many many occasions. In particular, the importance of the broadcasting industry in the United States to the nation's weather warning system is well recognized, and White (1966b) shows just how important it can be in the case of

Hurricane Hilda, which moved from the Gulf of Mexico towards Louisiana in autumn 1964.

> The response to the Weather Bureau's first announcement was immediate. Local television stations in major cities pooled their resources to pick up and bring to the public the latest information on Hilda's strength, location, speed, and direction. Radio stations broadcast frequent bulletins. . . . The television and radio bulletins had an immediate effect. . . . Hilda had disastrous effects. But her effects would have been much more severe if television and radio had not been there to broadcast immediate and constant bulletins.

It is, nevertheless, apparent, as has already been pointed out, that in many cases the care and skill employed in producing the weather package is not in proportion to its potential value to the user. Examples of the packages available in most localities demonstrate that once the technical stage of the forecast has been completed there is little attention paid to its presentation. This obviously reduces the value of the forecast, for if it is poorly presented the forecast will be poorly received and much of its potential value to the decision-maker (whether he be John Citizen, a fisherman or a business manager) will be lost. But, as White (1966b) has pointed out, intelligibility and understandability are the joint responsibility of the weather forecaster and the mass media. Most weather forecasts are factual, sometimes dry and occasionally technical, and it is interesting to note the different interpretations given to the reading or televising of a weather forecast by a broadcaster and by a meteorologist. Some broadcasters, for example, try to inject some life into what appears to be a very dull weather bulletin, but on many occasions they do so at the expense of the weather information being given. On the other hand, the meteorologist, in giving a broadcast, may forget what kind of audience he has.

How *much* the public will listen to is often debated, but a survey of readers of the North American Newspaper Alliance, reported and discussed by Tobin (1966), indicates that 89% were in favour of the fullest possible presentation of meteorological information in preference to a brief official bulletin from the local weather bureau. He also comments on the fact that unlike the non-professional weathercasters (who often draw amusing cartoons or wear Arctic coats or raincoats to dramatize the weather forecast), professional weathercasters rely very little on novelties or gimmicks to hold audience interest, using instead the natural interest value of daily weather phenomena. Tobin as Communications Editor of *Saturday Review* summed up the whole position of the presentation of weather forecasts and information in these words:

> Our point is, of course, that nothing is as consistently fascinating, variable, unpredictable, and common to all as the day's weather and a hint of what will involve us tomorrow. . . . In a way, weather is about the only news that is totally new every twenty-four hours whether you're six or sixty, and when many papers insist as they do on a front-page weather piece every day, we couldn't agree with them more.

It is time then for an updating of weather forecast packages, and both the weather forecaster and the communications media must accept the responsibility of preparing and presenting forecasts which give not only the minimal data, but also sufficient background to make the weather forecast useful to as many users as possible and, at the same time, attractive to the consumer. Some suggestions have been made of the kind of improvement necessary such as clearly defined terms and percentage-probability forecasts. In addition, where applicable radio stations must have their broadcasters read weather forecasts at a rate slow enough to be intelligible, newspaper editors must learn to take the same care in preparing their weather forecast 'stories' as they do in the rest of their work, and television should accept more responsibility in educating its audience to the local weather situations. If these steps were taken the economic and social value of the familiar weather forecast would be improved immeasurably.

3. Reliability and probability of forecasts

It has been pointed out previously that the 'translation' from the weather forecaster to the weather consumer is not always done with the same care and attention as the 'translation' from weather map to weather forecaster. The translation procedure is in fact an important aspect – too often ignored – in the weather forecast matrix. Nevertheless, some attempts have been made to convey to the consumer some of the 'inner' thoughts of the weather forecaster, and one of the most useful types of forecast is the probability forecast. The importance of such forecasts is stressed by Nelson and Winter (1963), who state that while many decision problems can be adequately treated by distinguishing only 'rain' or 'no rain', for other problems the decision-maker will want to know a great deal more about the probability distribution of amounts of rain than the probability concentrated at the origin. Other problems depend on other weather factors – wind, temperature, type of precipitation, sunshine, humidity, etc. – and different decision-makers will in general be interested in events in different localities.

A convenient and widely used guide for translating probability forecasts into operational decisions is the cost/loss ratio. Following Thompson and Brier (1955), let $C =$ the cost of protection against an adverse event, $L =$ the loss suffered if protective measures are not taken and the event occurs, and $p =$ the probability that the event occurs. The quantity C/L is called the 'cost/loss ratio', and the decision criteria are: if $p < C/L$, do not protect; if $p > C/L$, protect; if $p = C/L$, take either course.

Shorr (1966) comments that when the cost/loss ratio criteria are put into operational use, dollar values quite often are used for C and L. This practice presumes that a decision-maker's dissatisfaction with a loss is a linear function of the dollar amount of the loss, e.g. that he will be twice as dissatisfied with a $2,000 loss as he will be with a $1,000 loss, and that he will be five times as dissatisfied with a $5,000 loss as he will be with a $1,000 loss, etc. This simple linear relationship, however, rarely exists in practice. A second difficulty concerns the fact that the dissatisfaction associated with a given dollar loss differs markedly among decision-makers. Thus, a $5,000 loss to an operator with limited resources has quite a different meaning from a $5,000 loss to an operator with substantial resources.

a. Probabilistic form of forecasts

Meteorologists today often express their predictions in probabilistic form, the precipitation probability statements in the U.S. Weather Bureau's public weather forecasts providing perhaps the best examples of such predictions. Several investigators have studied the problem of probability in weather forecasting, including Winkler and Murphy (1968), Murphy and Epstein (1967a, b), Curtiss (1968), Gringorten (1967) and Bryan and Enger (1967). These authors discuss in particular aspects of probability forecasting such as 'hedging', 'interpretation of precipitation probability forecasts', 'good probability assessors' and 'verification of forecasting skill'.

Winkler and Murphy (1968), for example, suggest that probabilistic predictions issued by meteorologists are subjective, i.e. that each prediction expresses the 'judgment' of a particular meteorologist on a particular occasion. Specifically, a subjective probability measures the confidence that an individual has in the truth of a particular proposition, e.g. a meteorologist's confidence in the statement 'precipitation tomorrow', and naturally this confidence may vary from individual to individual. The authors also formulate a framework for evaluating or determining the 'goodness' of the forecasters who assess the probabilities. These scoring rules are to encourage meteorologists to make 'honest' assessments, to evaluate meteorologists, and to help them become 'better' assessors.

In another paper it is suggested that a meteorologist who prepares probability forecasts should not 'hedge', i.e. the meteorologist's probabilities should express his true beliefs (Murphy and Epstein, 1967b). The essential difficulty arising from this problem of 'hedging' is that some weather forecasters (notably in the U.S. Weather Bureau) are evaluated according to a particular 'scoring system', and it has been suggested that a forecaster should (or even must) 'hedge' when preparing a probability forecast in order to obtain the best possible score. The forecast which he presents on a particular occasion will then depend upon both his true beliefs on that occasion and the scoring system which is employed to evaluate his forecast. Murphy and Epstein show, however, that a class of scoring systems exist which possess the property that the best 'hedge' is no 'hedge', that is, the meteorologist can expect to obtain the best score if, and only if, the forecasts he prepares express his true beliefs. Thus, the meteorologist is encouraged to produce the 'best probable' probability forecast based on his true convictions.

b. Accuracy and verification of forecasts

The need for verification to determine and measure forecasting skill is an important aspect of any weather forecasting programme, and since 1950 three distinct purposes for verification have emerged (Gringorten, 1967). One purpose is to test the accuracy of forecasts, another is to test the operational or economic value of a set of forecasts, and the third purpose is to test the skill of the forecaster. These differences would approach zero if forecasting could approach perfection, but perfect forecasts do not exist, and Gringorten attempts to determine if there exists any skill at all in a set of forecasts. In doing so, he uses a system of scoring developed in 1951 (Gringorten, 1951) to test the skill of a set of 1,098 forecasts of ceiling height at Duluth, Minnesota, given by Air Force Base forecasters during the years 1960, 1961 and 1962.

The Duluth forecasters, according to Gringorten, showed a skill of 0·19–0·23, which means that 'they used something beyond their knowledge of climatology and persistence of the weather to earn scores'. Unfortunately, the forecasting system developed by Gringorten has not enjoyed general acceptance since its publication in 1951 and, according to the author, one of the reasons for this is that it is a system yielding *skill* scores of only 20–30%, which is a sad commentary on the state-of-the-forecasting art, when the public has heard figures like 80 and 85%. Gringorten points out, however, that there are answers to such criticisms, and he states that high percentages of accuracy or skill, for their own sake, are meaningless.

Another fundamental part of applied research is the evaluation and testing of models. As Murphy and Epstein (1967a) indicate, predictions based on models properly assume an important role in the field of meteorology. Nevertheless, although meteorologists have been concerned with the practice of forecast verification for many years (Meglis, 1960), considerable controversy surrounds this practice (e.g. Brier and Allen, 1951), a fact which Murphy and Epstein believe is the result, in large measure, of the failure on the part of the meteorologists to give proper consideration to the nature of the process of evaluation itself.

The evaluation process is assessed in detail by Murphy and Epstein, and the authors also describe seven measures of probabilistic prediction: probability score (Brier, 1950), 'reliability' and 'resolution' scores (Sanders 1958), information quantity (Shannon and Weaver, 1949; Bross, 1953), information ratio (Holloway and Woodbury, 1955), 'validity' measure (Miller, 1962), 'skill' score (Gringorten, 1965) and distance measures (Epstein and Murphy, 1965). Two purposes for the evaluation of predictions are considered by Murphy and Epstein: first, the assessment of their absolute and/or relative 'skill' or 'value' to the meteorologist; and, second, the assessment of their absolute and/or relative 'utility' or 'value' to the decision-maker. These purposes, in turn, define two forms of the evaluation process, verification and operational evaluation. However, since the measure of 'utility' will be different for decision-makers and for different decision situations, the authors point out that a universal measure for operational evaluation does not exist. In fact, as Murphy and Epstein state '. . . let us not persist in arguments concerning which measure is "best", when the important considerations are the attribute it measures and the reason for making the measurement'.

c. Meaning of probability in forecasts

The meaning of probability in precipitation forecasts has been examined by Curtiss (1968), who says that in the United States where precipitation probability forecasts are now routine, two kinds of forecasts are given. First, a forecast may read 'shower probability, 40%', the probability referring to the entire forecast area which would generally include all parts lying within 20 to 30 miles of a centre. Second, sometimes an areal or temporal qualification is published along with the probability, as is exemplified by the following forecast from the *Miami Herald* for Sunday, June 4, 1967 (Curtiss, 1968):

Today's forecast

Miami and vicinity: Partly cloudy through Monday with scattered showers most likely during the night and morning hours except well inland during afternoons. High today in eighties. Variable mostly easterly winds 5 to 15 mph. Shower probability 50 per cent.

Curtiss says that it is natural for a scientist or mathematician who is familiar with quantitative probability theory, but who is not a meteorologist, to wonder whether such a numerical probability statement can be given a meaning within their theory. Further, in the absence of authoritative information as to the intent of the U.S. Weather Bureau, and under the assumption that meaningful quantitative *a priori* probabilities really can be assigned to precipitation events, there are quite a number of reasonable conjectures which can be made as to what a single-number precipitation probability might mean. These include: the probability of the event that some rain will fall somewhere in the forecast area sometime during the time-period covered by the forecast; the probability of the event that general rain will cover all of the area; the fraction of the forecast area which will receive rain in the forecast period; and the probability that a specific point in the forecast area will receive more than a trace of rain sometime during the forecast period. Of these possibilities, the 'correct' interpretation is the last one mentioned, and the official viewpoint of the U.S. Weather Bureau towards precipitation probability forecasts is given by Curtiss. The actual wording used follows very closely a formulation in a letter to

Curtiss dated March 13, 1967, from R. H. Simpson, Associate Director of the U.S. Weather Bureau.

> The event to which a precipitation probability applies is the occurrence of more than a trace of precipitation (water equivalent, if frozen) within a specific forecast period at a specific point in the forecast area. . . . For purposes of verification the specific point is taken to be the location of the rain gauge at the official Weather Bureau Station; for example, this is Midway Airport for Chicago and vicinity. When a single unqualified probability number is released in a local forecast, the assumption is made implicitly that local conditions will impose only one regime of events in the metropolitan area, and that the published 'point probability' is therefore at least approximately valid at each point in the forecast area.

Thus, according to Curtiss, a forecaster would be meeting 'government specifications' if he concentrated on estimating the relative frequency, given the forecast information of the event that more than a trace of rain will be recorded merely in the one gauge at the official station. If, therefore, a given forecast is based on the assumption that either general rain will cover the forecast areas completely during the forecast period or there will be no rain at all, then there can be no ambiguity in the meaning of the forecast probability for a user who is advised of the assumption. However, as Curtiss points out, if the alternatives to 'no rain' include various possibilities as to partial coverage of the forecast area by precipitation during the forecast period, then in order to interpret a single-number precipitation probability forecast some assumptions must be made as to the way in which the partial coverage will be distributed in the forecast area.

The importance of the meaning of probability forecasts cannot be underestimated, especially when the 'value' of weather forecasts is assessed. Indeed, it is clear that unless the weather forecaster, and the forecast consumer, who in most cases is a decision-maker, agree on what is meant by 'shower probability 50%', then pertinent and valuable decisions based on weather forecasts will not be made. The correct translation from weather forecaster to weather consumer is therefore of paramount importance.

d. Reliability and value of long-range forecasts

The accuracy of the weather forecasts of the various national meteorological services is a topic often discussed by the general public, and although accuracies of 70% to 85% are usually claimed, it is not at all clear what this means, and as already pointed out by Gringorten (1967), a 75% accuracy may in fact only give a 20% 'skill' score. But whatever the true accuracy, and this can probably only be assessed by the consumer of weather forecasts, it is evident that many improvements are still awaited. The value of forecasts can probably best be seen in long-range forecasts, and some idea of the accuracy involved and the possible value to the community of such forecasts are illustrated in the study by Freeman (1966) of the monthly forecasts of the U.K. Meteorological Office.

The twice-monthly publication of forecasts of mean temperature and total rainfall for a month ahead commenced in the United Kingdom in 1963. The expected mean monthly temperature is given as one of the five categories: much above average, above average, near average, below average or much below average, the limits for these categories having been chosen such that in the period 1931–1960 each category occurred equally frequently. A forecast of rainfall is given in a similar way as one of the three equally likely categories, above average, near average or below average, the limits being expressed as percentages of the average monthly rainfall. In Freeman's analysis, a 'score' was calculated, for each of the 10 regions in the United Kingdom, according to the values shown in Tables VI.1 and VI.2. In addition, the average scores for the United Kingdom were calculated and provided an assessment of the accuracy of the forecast as one of the five categories: A – no serious discrepancy; B – good agreement; C – moderate agreement; D – little agreement; and E – no real resemblance. In addition, an assessment was made on the accuracy of the additional information about weather such as thunderstorms, fog and frost, such phenomena being included in the U.K. forecast if they were expected to be notably more or less frequent than usual. The number of forecasts in the various categories for the 33-month period from December 1963 are given in Table VI.3.

As will be seen from Table VI.3, 25 out of the 66 forecasts were in good or better agreement. On the other hand 18 of the 66 forecasts in classes D and E could be classed as very poor. It was, however, rare for forecasts of temperature, rainfall and additional information simultaneously to be very good or very bad, so that overall marks of A or E were few. Perhaps the most important aspect of the evaluation was that the middle categories of both temperature and rainfall were forecast too often, and the extreme categories too rarely. In fact, temperatures much below average and much above

TABLE VI.1: *Scores for temperature forecasts*
(from Freeman, 1966)

Actual	Forecast				
	Much below	Below	Average	Above	Much above
Much below	4	1	−3	−4	−4
Below	2	4	1	−2	−2
Average	0	1	4	1	0
Above	−2	−2	1	4	2
Much above	−4	−4	−3	1	4

TABLE VI.2: *Scores for rainfall forecasts*
(from Freeman, 1966)

Actual	Forecast		
	Below	Average	Above
Below	4	−2	−4
Average	0	4	0
Above	−4	−2	4

TABLE VI.3: *Success of long-range forecasts in the U.K.*
(from Freeman, 1966)

Category	Mean temperature	Rainfall	Additional information	Overall marking
A = No serious discrepancy	10	10	12	1
B = Good agreement	24	10	17	24
C = Moderate agreement	12	20	19	23
D = Little agreement	8	17	14	15
E = No real resemblance	12	9	4	3
Totals	66	66	66	66

average were forecast on only 4% and 3% of occasions, whereas they actually occurred on 29% and 9% of occasions respectively.

The foregoing results, Freeman says, cannot engender complacency in our long-range forecasters, but they do represent a solid achievement. Nevertheless, the inability to forecast the 'extreme' categories (those occurring about 20% of the time) on most occasions is perhaps the most serious deficiency to be remedied, for it is clear that from an economic and social viewpoint, the ability to forecast much below normal conditions one month ahead will pay handsome dividends, whereas the ability to forecast 'average' conditions one month ahead will provide only average dividends. More research effort on forecasting the heavy snowfalls, the very warm summers and the very wet months would seem therefore to be well justified, even if it means that the forecaster has to 'go out on a limb'.*

4. Economic and social value of weather prediction

The mounting losses of property and income which result from extreme weather events, the increasing use of the atmosphere as a transportation route, man's ability to modify his atmospheric environment (deliberately or inadvertently), and the ever-present control which the atmosphere (the rain, the snow, the wind, the humidity and the sunshine) exerts over our 'economy', are all significant in today's society. The predictions of the atmospheric scientist, and in particular those of the weather forecaster, are therefore of considerable economic and social value.

a. Economic utility of weather forecasts

Several studies on the economic value of weather forecasts have been made, but it is appropriate to note that forecasts which show a high verification score (discussed in the previous section) are not necessarily economically useful predictions (see Thompson, 1952). In view of this, Thompson and Brier (1955) discuss a method of analysis designed to measure the economic utility of the forecast, and suggest a verification procedure based upon the operational risks involved in taking protective measures against adverse weather.

Consider the case of a potential user of a weather forecast faced with the problem of deciding whether or not to take protective measures against a certain adverse weather element: W. Thompson and Brier suggest that in general he should take such protective measures if, in the long run, some economic gain will be realized; otherwise no protective measures should be taken. In order to derive a criterion for making this decision, several terms are defined:

---

* During the period 1966–1969 the proportion of successful forecasts (issued by the Meteorological Office) has increased slightly. (M. H. Freeman, per. com.)

G_p = total expected gain for N = $(f_w + f_{nw})$ days of operation if protective measures are taken every day; G_{np} = total expected gain for N days if no protective measures are taken; C = cost of protection each day that protective measures are taken; L = loss suffered each day that adverse weather occurs and no protective measures have been taken; T = average daily net operating income exclusive of the cost of protection (C) which may have been taken, or the loss (L) which may have been suffered; f_w = frequency of adverse weather; and, f_{nw} = frequency of favourable weather. Using this terminology, the authors showed that if protective measures are taken every day, the total gain would be the daily net operating income minus the daily cost of protection, both times N, the number of days of operation. Thus, $G_p = (T-C)N$. On the other hand, if no protective measures are taken, the total expected gain would be: $G_{np} = (T-L)f_w + Tf_{nw}$. The total gain to maximize the profit on the entire operation should be as large as possible; thus protective measures should be taken whenever $G_p > G_{np}$ or when

$$(T-C)N > (T-L)f_w + Tf_{nw}.$$

That is, when $\dfrac{f_w}{N} > \dfrac{C}{L}$. Thompson and Brier state that the left-hand side of this inequality defines P, the 'probability' of adverse weather. As a result, protective measures should not be taken if $P < C/L$. The criterion for making a decision to protect or not to protect may therefore be expressed:

$$P \begin{Bmatrix} > \\ = \\ < \end{Bmatrix} \frac{C}{L} \quad \begin{matrix} \text{Protect} \\ \text{Either course} \\ \text{Not protect} \end{matrix}$$

Thus the value $P = C/L$ represents a critical ratio, above which protection should be provided, and below which it should not.

The economics of weather forecasts using the above expression are discussed in detail by Thompson and Brier. Thompson (1962) applies these concepts in another paper and Table VI.4 shows a series of N probabilities of adverse weather, where W and No W are the occurrence and non-occurrence, respectively, of an operationally adverse weather event, and a, b, c and d represent the frequencies of these events.

Using this table, Thompson shows that the total weather protection expense for an operation, E_f, will be due to the cost of protection whenever $P > C/L$, plus the loss suffered on those occasions when $P \leqslant C/L$ and W occurs. The equation is: $E_f = C(b + d) + Lc$.

TABLE VI.4: *Generalized contingency table showing results of probability predictions (from Thompson, 1962)*

Observed weather	Forecast probability		
	P < C/L	P > C/L	Totals
No W	a	b	a + b
W	c	d	c + d
Totals	a + c	b + d	$N = a + b + c + d$

On the other hand, if 'perfect' forecasting were to be attained, the total expense for the operation, E_p, would arise only from the necessity for protecting against adverse weather. Thus:

$$E_p = C(c + d).$$

The economic gain which might be achieved if perfect forecasts were attainable, is then assessed by Thompson as:

$$G_s = \frac{E_f - E_p}{NL} = \frac{1}{N}[(b - c)C/L + c]$$

In using these equations, the difficulty is that current weather predictions do not, in general, contain quantitative information concerning their uncertainty. Instead, a working assumption more or less equivalent to an 'average' value of the assumed economic risks for a large number of operations must be used to produce a categorical prediction of the future weather. Denoting the expense for these 'average' predictions by a subscript 'a', the equation becomes: $E_{fa} = C(b_a + d_a) + Lc_a$. Thompson suggests, therefore, that the economic gain which could be realized from an operations research study which makes optimum use of uncertainty information will be given by the difference between E_f and E_{fa}. If this improvement is G_0, then the unit improvement is:

$$G_0 = \frac{E_{fa} - E_f}{NL} = \frac{1}{N}[(b_a + d_a - b - d)C/L + c_a - c]$$

where N = the number of days of operation and L = daily loss if adverse weather occurs and no protective measures have been taken. The total economic gain G_t (that is the value exceeding that obtainable from currently issued 'average' predictions), that could be realized if perfect forecasts were obtainable, is therefore given by the summation of G_s and G_0. That is:

$$G_t = \frac{E_{fa} - E_p}{NL} = \frac{1}{N}[(b_a + d_a - c - d)C/L + c_a]$$

The application of the equations of G_s, G_0 and G_f to weather predictions is assessed by Thompson. Three series of probability weather predictions are analysed, and the results of one study which illustrates the economic gains for a series of probability forecasts of precipitation in San Francisco are shown in Fig. VI.4. It can be

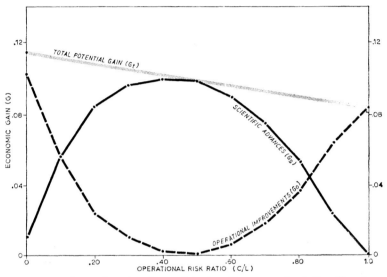

FIG. VI.4. *Economic gains for precipitation forecasts made at San Francisco, California, as described by Root (1962) (after Thompson, 1962)*

seen that for small and very large values of the operational risk ratio, C/L, the economic gains due to operational improvements (G_0) are greater than gains which might be achieved by scientific advances, G_s. For values of C/L centred near 0·50, however, a nearly optimum operating decision is presumably already provided by 'average' forecasts. Accordingly, as Thompson points out, the realizable operational gain is small, while the possible gain due to scientific advances is large.

Thompson illustrates two other studies, the economic gains of precipitation forecasting at Salt Lake City, Utah, and 'general weather' forecasts for several representative locations in the northeastern United States. The average economic gain from these analyses suggest that operational risks for the United States economy as a whole probably include all values of C/L, ranging from near zero to near unity. If it is assumed that all operations are equally likely and equally important, a simple arithmetic mean of the economic gain for each curve may be computed. The mean values are illustrated in Table VI.5, and this shows that the magnitude of

TABLE VI.5: *Mean values of the economic gains from scientific advances* (G_s), *operational improvements* (G_0) *and total potential gains* (G_t). *Figures are in units of potential loss* (L) *per forecast (from Thompson, 1962)*

	G_s	G_0	G_t
Precipitation forecasts, San Francisco, Calif.	0·06	0·04	0·10
Precipitation forecasts, Salt Lake City, Utah	0·05	0·04	0·09
General forecasts, north-eastern U.S.A.	0·05	0·02	0·07

these average values within each column is strikingly similar. For this sample at least, the economic gains from improvements in the operating decisions and in scientific knowledge appear to be of about the same order of magnitude, regardless of the weather element or section of the country considered.

The results of Thompson's study are presented in terms of the percentage of loss which would be suffered upon the occurrence of adverse weather. The total gains for an economic unit (e.g. a single company, an industry, or a homogeneous segment of the economy) may therefore be obtained by applying these percentages to weather losses associated with the appropriate weather element, location and forecast period. Such information, says Thompson, 'would be useful where planning must be based on the probable economic impact of future research and development in weather prediction'. The author, nevertheless, adds a note of caution and indicates that in assessing the economic value of weather prediction in this way, only protectable losses should be considered, since the total loss due to adverse weather often becomes a misleading statistic if all, or part, of the potential damage is in a form for which no protective measures are possible.

b. Industrial weather forecasting

The main function of an industrial meteorologist is to communicate his knowledge of weather factors in such a form that the overall profitability and safety of the client's operations are increased. Moreover, as Borgman (1960) points out, the client's interest in the forecast service is in direct proportion to the possible profit. Therefore, it is in the meteorologist's interest to study ways of maximizing the forecast profitability.

In most companies, the 'private' meteorologist (in contrast with many government meteorologists) works directly with the personnel carrying out the operation, and they naturally welcome any reasonable plan for increasing profits. However, since company personnel are normally unfamiliar with the sophistications of weather theory, the meteorologist must be very careful to state his forecasts so that they can easily be translated into profitable decisions. Borgman suggests, in fact, that this translation from forecast into action is too frequently the weak link in the forecast utility, and that the meteorologist's forecasts and recommendations should be stated in positive terms which suggest the most profitable protective action. Borgman continues:

> . . . In a very real sense, weather forecasting can be regarded as a game in which the meteorologist pits his imperfect prediction methods against nature. He makes a move by making a forecast or a recommendation of protective measures; then nature makes her move by responding with some weather condition. The cost of the protective measure and the damage suffered, if any, represent the score made on the move. The meteorologist is motivated to keep this score as low as possible while nature is ponderously disinterested in the whole game.

At the beginning of any industrial weather forecast 'game', there is, according to Borgman, a choice of two alternatives. First, if the meteorologist feels that some continuously instructed protective measure will cost less on the average than actions based on his forecasts, he may refuse to play the game and recommend instead that the client follow the routine measure. Second, if the forecaster feels that his skill is sufficient to make a profit, he may accept the challenge and proceed with the game. But, as the author comments, 'the meteorologist hurts his client as well as himself if he undertakes to play the game when he should not, for he may lose future contracts for other operations because of his indiscretion on the one operation'.

The important question of 'how should the meteorologist decide when to play the game and what forecasts to make in order to minimize weather damage' is then asked by Borgman. Several solutions are described, but the simplest situation involves a damaging weather event which either occurs or does not occur. For example, the industrial operation can be protected completely by a protective expenditure C, but cost D will be incurred if the event occurs without the protection. Four possible strategies are as follows:

L

c_p = cost of always protecting the operation regardless of the forecast,

c_d = cost of always taking the damage and repairing it later,

c_a = cost resulting from protecting it, only if the damaging event is forecast by a meteorologist who has a positive skill score, and

c_0 = cost resulting from the advice of a meteorologist with a zero skill score. [The skill score in c_a and c_0 is defined by Brier and Allen (1951) to be $S = (n - E)/(N - E)$ where n correct forecasts have been made in N forecasts and where E correct forecasts would be expected on the basis of pure chance.]

From this a certain minimum skill score must be attained by the forecaster before it is profitable for the client to follow the forecaster's advice in preference to another of the available strategies, and this is demonstrated by Borgman in a number of examples, using a 'Minimum profitable skill score' graph (Fig. VI.5). One example given by Borgman concerns a fuel-oil storage problem. Here the cost of storing the oil is divided by the profits that would

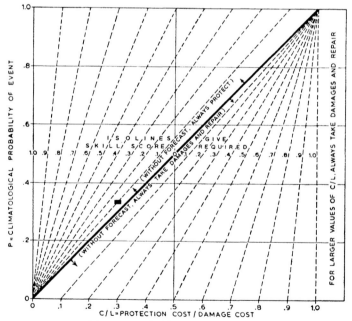

FIG. VI.5. *Minimum profitable skill score (after Borgman, 1960).*

be lost if the oil were not available for sale in a cold spell, and is about 0·3; that is $C/L = 0·3$. Now, if a period cold enough to bring about a particular sale of the oil occurs about once every three years, and a meteorologist is forecasting whether or not the cold period will occur in the next year, a problem arises as to what skill score he would have to have in order for it to be profitable to follow his forecast and store the oil in anticipation of the predicted cold winter. This point, $C/L = 0·3$, $P = 1/3$, is shown on Fig. VI.5 with a small square, and it shows that some profit will occur over a period of time if the forecaster's skill score exceeds 0·10. However, if no forecast is available the most profitable action, according to Borgman, would be to store the oil so that it is always available if a cold spell occurs.

Many forecast situations are of course much more complicated than this. For example, Borgman discusses the case of the forecast situation in which X_1, X_2, . . ., X_n represent a mutually exclusive and exhaustive classification of all possible weather events that have particular significance relative to the client's operations, and S_1, S_2, . . ., S_m represent all possible strategies which the client can take to protect himself against the various weather events X_1, . . ., X_n.

It should be emphasized that throughout Borgman's paper emphasis is placed on profit and safety criteria, rather than accuracy, as the main measurements for the success of an industrial meteorologist. This does not indicate, of course, that accurate forecasts are not necessary, but it is important to appreciate that accurate forecasts are not sufficient to guarantee a high utility unless the forecaster makes the required effort to word his forecasts so that they are usable in making decisions.

5. Decision-making

Most people, if they are to live in safety and comfort and to engage in profitable 'business' enterprises, must make many weather-related decisions. Such decisions are made by people for many purposes. Moreover, these decisions are based in part, if not in whole, on the type of weather information available. Nevertheless, although it is often suggested that the provision of better information will lead to better decisions, much depends on the way in which information, and in this case the weather forecast, is given to, and used by, the consumer.

The decision factors already mentioned are usually considered only in relationship to the short-range weather forecast, but they apply equally and probably more importantly to the longer-range forecasts. In addition the use of weather forecasts for decision-making in weather modification activities must not be underestimated. It could, of course, be argued that weather modification and weather forecasting are not associated with each other. But it is clear that modification of tomorrow's (or next month's) weather is dependent on the weather that is forecast for tomorrow (or next month), if decisions of whether to modify are going to be based on the predicted weather, as indeed they must be.

a. Weather-sensitive decisions

Any survey of the literature appropriate to the problem of weather-sensitive decisions in any economic enterprise leads into at least two subject areas, in addition to meteorology. These are the theory of decision-making, and the management of the particular economic enterprise. Literature concerning formal or mathematical theory of decision-making has increased markedly during the last 25 years and may be found in the journals of mathematics, statistics, operations research, engineering, economics and meteorology. A brief review by McQuigg (1965) indicates some typical work.

In the 1950's, two pioneering articles concerning decision-making appeared in the meteorological literature. First, Thompson (1952) pointed out the need to consider the operational deficiencies in categorical forecasts; second, Thompson and Brier (1955) presented an objective criterion which could be used to evaluate weather forecasts as a tool in decision-making. In both cases, the essential point was that one needs to know the magnitude of the cost/loss ratio for specific weather-sensitive decisions before choosing a course of action. Third, Gringorten (1959) developed mathematical models for making weather-sensitive decisions; Nelson and Winter (1960) discussed a variety of decision situations and provided some upper limits on the value of weather forecasts; and Miller (1962) provided another approach to the decision-making problem, that of multiple discriminate analysis. A few examples of applications of decision theory to agricultural problems include work by Kolb and Rapp (1962) on raisin drying, Williamson and Riley (1960) concerning the cotton harvest and McQuigg and Doll (1961) on grain drying.

The defining of the relationship between weather and decision-making is discussed in some detail by McQuigg (1965). Included is the problem of decisions with more than two discrete alternatives and more than two distinct types of weather, and the problem relating to decision in sequence. A particular problem arises when decisions are based on forecast weather, and McQuigg reports on a study by Borgman and Brooker (1961) involving the decision-making problem associated with haymaking in central Missouri.

In this example, the method of haymaking requires three consecutive favourable days for mowing, field curing, baling and hauling into a barn, a favourable day being defined as a day having all of the following characteristics: less than 0·10 inch of rain on that day, at least 70% of the possible sunshine, and less than 1·00 inch of rain on the previous day. The maximum feeding value of the hay is usually reached in early June, and yields two tons per acre, and if harvested under favourable weather at that time it has a gross value of $40.00 per acre and a net value (after harvest costs) of $26.00 per acre. If, on the other hand, the hay is cut, and rain of more than 0·10 inch falls while it is curing, the gross value declines, and an additional raking operation is required which costs $2.50 per acre. The exact loss in value of wet hay depends on the amount and intensity of the rain, and the length of time the hay lies on the ground before it is cured, but the authors suggested that it seems reasonable to set a minimum decline in value of $4.00 per acre, which added to the additional cost of raking brings the net value of the hay to $19.50 or less per acre. However, if the farmer decides to delay mowing, and the weather is actually favourable, the hay declines in value. McQuigg, in fact, notes that if weather information is to be useful in the decision-making process, the various 'alternative weather' combinations must be assigned increments of yield, labour, time or some other unit which is to be maximized (or minimized). Further, both the meteorologist and the farm manager must be aware of the definition of terms, and the 'structure' of decision problems that are economically important, and communication of the results of appropriate meteorological analyses must be in terms that are clearly defined.

b. Scale of weather-related decisions

The number of decision-makers interested in weather forecasts varies considerably in time and space, but Crow (1965) suggests that in the United States the number of 'general public' decision-makers interested in weather observations and forecasts might vary from a few million on a relatively unimportant weather day to over 100,000,000 on a day with considerable weather activity throughout the nation. By contrast, during the quiet winter days when field work is at a minimum, the number of farmers interested in the weather may be less than one million, this number increasing in the spring planting season to several million farmers in the United States. Aviation and marine interests are also catered for, and the number of these decision-makers in the United States may vary from probably 1,000 to over 100,000. Nevertheless, all decision-makers are not of equal economic importance, and the economic values of weather-related decisions in the United States are shown in Fig. VI.6. This shows that a very large proportion of the information prepared by the Weather Bureau and disseminated to the

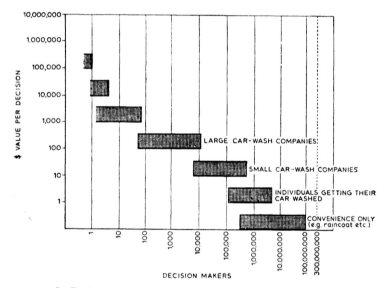

FIG. VI.6. *Daily range of the economic values of weather-related decisions in the United States (after Crow, 1965).*

public is used for *convenience only.* The decision as to whether or not to wear a sweater or raincoat is a typical example, and the daily fluctuation in uses of information for convenience only ranges from a few hundred thousand to many millions.

At the other end of the economic scale are weather-related operations numbering probably less than 10 in the United States each day, but with a $1,000 to probably over $100,000 price tag attached. The $1,000 (or more) weather-related decision-maker, Crow suggests, should use a professional weather consultant, and it is clear that when decisions either single or multiple reach $100,000 one or more staff positions in meteorology in the relevant companies are justified.

c. Weather sensitivity of user groups

The various user groups in any economy vary considerably in their sensitivity to weather. A study by the U.S. Weather Bureau (1964), for example, ranked 18 user groups in the United States (Table VI.6). To obtain a true relative measure of the 'economic benefit

TABLE VI.6: *Ranking of weather influence by user groups in the United States (after U.S. Weather Bureau, 1964)*

User group	(a) Weather sensitivity	(b) Decision latitude	Product of (a) and (b)	Rank	Projected 1972 national income* ($ × 10⁹)
General public	2·6	2·0	5·2	1	†
Fishing	2·3	2·3	5·3	2	0·26
Agriculture	2·0	3·0	6·0	3	28·10
General science	2·7	2·3	6·2	4	†
Air transportation	2·0	3·3	6·6	5	2·30
Forestry	2·3	3·3	7·6	6	1·68
Construction	2·6	3·0	7·8	7	39·30
Land transportation	2·6	3·3	8·6	8	12·90
Water transportation	3·0	3·0	9·0	9	2·05
General welfare	3·6	2·6	9·4	10	†
Energy production and distribution	3·3	3·0	9·9	11	30·00
Health and safety	2·9	3·6	10·4	12	†
Resources utilization	3·0	3·6	10·8	13	†
Merchandising	3·3	3·3	10·9	14	145·30
Water supply and control	4·2	3·0	12·6	15	4·28
Communications	3·6	3·6	13·0	16	22·70
Recreation	4·0	3·3	13·2	17	3·35
Manufacturing	4·2	4·3	18·1	18	262·40

Weather sensitivity	*Decision latitude*
1 = Great – involves life, very large money values	1 = Very flexible
2 = Important – much money and/or direct health effects	2 = Quite flexible
	3 = Some flexibility
3 = Significant – meaningful costs, secondary health effects	4 = Little flexibility
	5 = Rigid
4 = Modest	
5 = Small	
6 = None	

* Adapted from information supplied by National Planning Association. All data are in terms of 1962 dollars.

† Not assessed.

potential' that would result from satisfying the weather needs of the various user groups, however, it is necessary to include some weighted measure of the dollar value of each user category in the national economy. This is also shown in Table VI.6, which indicates the projected 1972 national income of the various user groups.

d. Role of decision-maker

The importance of the decision-maker in the use of weather forecasts has been assessed by Rapp and Huschke (1964), who suggest that the three main socio-economic components of a nation (people, governments and businesses) are also the three main weather-service users. They indicate that people are motivated to employ weather information in a very personal and individual way: to save lives, protect property, save time, maintain health and comfort, and have more fun. Furthermore, they are also highly flexible economic units, more so than units of either government or business, and can more often afford to 'wait and see' before deciding on a course of action. It has been suggested that people make and act on beneficial decisions either in response to weather-service information or in

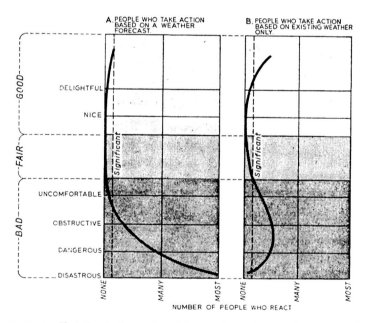

FIG. VI.7. *Variation in the number of people who react to weather as a function of weather extremes (after Rapp and Huschke, 1964).*

reaction to weather, according to a scale of potential weather effects (Fig. VI.7). This shows that 'most' (more than 80%) of the people will be able to and will in fact take some beneficial action based on a *forecast* of weather that threatens serious, negative ('disastrous') consequences, and that a much smaller 'significant' number of people makes some beneficial decision based on a *forecast* of 'obstructive' weather (weather that would cost them some loss of time and, possibly, minor property damage).

The number of people who take action based on *existing* weather only, is also shown in Fig. VI.7, and it indicates that when 'disastrous' weather (hurricane, flood, etc.) occurs, there are very few people who have not taken prior action. On the other hand, it appears that a more-than-significant portion of the public can await the arrival of 'dangerous' weather (heavy snow, ice storms, etc.), 'obstructive' weather (lighter snow, rains, etc.) or 'uncomfortable' weather and still choose a 'beneficial' action. The graph also suggests that the arrival of 'delightful' weather elicits favourable actions in a 'significant' number of people.

Although the arbitrary placement of 'significance' as used in Fig. VI.7 can and should be argued, and the 'activity' curves could be reshaped within limits, Rapp and Huschke believe that it represents reality quite well, for they say that people benefit from weather services when that service saves appreciable time, and helps them protect *lives* and *property*.

The value of weather forecasting to the second group of consumers, business, is seen in increased profits that are realized in a number of different ways: more sales, higher margins, reduced material losses, greater efficiency and lower costs. A few of these business decisions are amenable to the simple quantitative analysis already presented. In addition, Rapp and Huschke foresee an improved and more complex methodology that will make economic evaluations possible for many business problems. Nevertheless, even if rigorous analysis could be undertaken for all individual decision-makers, it would be impossible to estimate the value for an industry as a whole. For example, the value of rainfall forecasts to an individual raisin producer can be assessed; however, if all raisin producers utilized these forecasts to the utmost, raisin production would increase and the market for raisins would need to adjust to the increased supply. In addition, if all raisin producers tried to protect their product at the same time, the competition for labour might cause an increase in wages that would increase the cost of protection. Consequently, as the authors point out, 'we can be fairly sure that the value of improved weather forecasts to an industry is not necessarily the simple sum of the values to each decision-maker in that industry'.

e. Pay-off from weather decision-making

The 'pay-off' from weather decision-making is considered in a paper by Thompson (1963), who correctly points out that for any weather forecast two decisions are made: first, on the part of the meteorologist who provides the information, and second, on the part of the recipient who may be required to take action on the basis of the information he receives. Thompson describes a fictitious example of the weather decision problem, which he says involves principles both real and practical. This problem, essentially, was that the superintendent of an industrial organization had come to the conclusion that the weather forecast, instead of assisting him in his operation, might actually do his company more harm than good. The specific operation involved the cost of protecting equipment each day (about $300), and that damage which would be suffered each time the equipment was left unprotected during freezing weather ($2,000). Using the weather forecasts, protective measures would have been taken on 57 occasions when freezing or below conditions were forecast, but no protection would have been provided on the six occasions when freezing conditions occurred but were not forecast. The 'expense' of using weather forecasts, therefore, would have been $29,100 (57 × $300 plus 6 × $2,000), compared with an expense of only $27,000 (90 × $300) if forecasts were not considered, and protective measures were taken every night. Thus, the 'cost' for using the weather forecasts involved an additional $2,100.

Assistance and advice from a professional meteorologist indicated, however, that this 'cost' could be reduced, and the analysis of the decisions based on the forecaster's estimated probabilities is shown in Table VI.7. (For comparison the decisions based on conventional weather forecasts are also shown.) As shown, the 'expense' of using the forecaster's estimated probabilities was $23,000 (70 × $300 plus 1 × $2,000), a saving of $4,000 over taking protective measures every night, and a saving of $6,100 if decisions were based on conventional forecasts.

This, says Thompson, is the 'pay-off'. That is, for the expenditure of a few hours' time on the part of the meteorologist and the busi-

TABLE VI.7: *Analysis of decisions based on conventional weather forecasts and forecaster's estimated probabilities (after Thompson, 1963)*

Observed weather	Conventional forecasts			Forecaster's estimated probabilities		
	Protection	No protection	Total	Protection	No protection	Total
Above freezing	6	27	33	14	19	33
Freezing or below	51	6	57	56	1	57
Total	57	33	90	70	20	90

nessman working together, a saving of $4,000 over the three-month period of this example could be achieved. The example itself is obviously over-simplified, yet it illustrates an important principle concerning the meteorologist's service to the business and industrial community, notably that *both* the meteorologist and the user must provide information in order that efficient decisions can be achieved.

The decision-maker must, however, assess the particular nature of an operation in determining the appropriate strategy. This strategy is a complex problem, but Thompson (1966a) states that the operational 'utility' must be evaluated, 'a task which may involve considerations as concrete as the ability of the operator to survive a serious loss, or as intangible as an individual addiction toward gambling'. Thus, if the operator is working on a close economic margin, such that an initial loss or series of losses would preclude continuation of the operation, the strategy might appropriately be based on the minimum economic expectation. On the other hand, if the operator has sufficient capital to absorb potential initial losses, the mean, or even the maximum, economic expectation might be considered. The nature of the decision strategy may also, says Thompson, be influenced by the personal philosophy of the decision-maker, and the cautious operator may feel more comfortable with a decision based on the minimum expectation while the more adventurous entrepreneur may wish to gamble that the maximum expectation will be achieved.

f. Decision-making and long-range weather forecasts

The importance of the 'gambling' part of the weather forecast decision-making process should perhaps not be over-emphasized, but there are those who have predicted that the 1970's will be colder than the 1950's and 1960's. Lamb (1967), for example, says that although we cannot yet give a formal climatic forecast, we can already make some assessment of the probabilities of different climatic trends, and of the extra margins of variability that should be allowed for in any really long-range planning.

One associated aspect of meteorological decision-making is the complex kind of problem likely to arise by the late 1970's, when seasonal forecasts useful for decision-making one season ahead are expected to be available. If, for example, the U.S. Weather Bureau issues a forecast in June 1977 for the summer of 1978, and that such forecasts through previous successes can be relied upon, what would be the social and economic implications? Would, for example, farmers in the U.S. mid-west in June 1977 decide *not* to plant soybeans in the summer of 1978 because of an adverse weather forecast? In a similar way, if a forecast issued before the summer of 1977 indicated that weather conditions for tourists would be ideal in Europe in the summer of 1978, but not in 1977 or 1979, would European travel agents and associated tourist promoters have record incomes for the summer of 1978? And if these kinds of decisions are going to be made, what will be the psychological effects of knowing one season or one year in advance that the weather during your vacation time would be very bad? Both weather forecasters and psychologists should consider, therefore, the fact that there are many advantages in *not* knowing what the weather is going to be next month or next year, a fact which has received surprisingly little attention from those concerned with atmospheric resources. Perhaps gambling on the weather and the weather forecast has its rewards.

6. Benefits and costs of weather forecasting services

New technological developments, including high-speed electronic computers and meteorological satellites, have made it possible to provide significant improvements in international weather services. The benefits arising from these improvements include the 'value' derived from better and longer-range weather forecasts. The costs of providing such forecasts have, however, been questioned with the result that probably for the first time national meteorological services have done some serious 'weather forecast book-keeping'.

Few estimates of the total world cost and benefits of weather forecasts have been made, but Thompson (1966b), writing in the

World Weather Watch Planning Report No. 4, gives some data from a Stanford University (1966) report which indicates the annual potential benefit to the world economy from forecasting to be $16,900 million, a figure which is estimated to be about 50 times the cost. The data refer specifically only to the costs and benefits from the weather satellite system, which is only a part of the total programme of the World Weather Watch. Nevertheless, Thompson suggests that, at least in order of magnitude, large international benefits are to be obtained from significant improvements in weather services, including forecasting.

On a national basis benefit/cost ratios of weather forecasts are also difficult to assess. At a meeting of the W.M.O. (1968b) some attempts were made, however, to evaluate the relevant benefit/cost ratios of meteorological services in such countries as Australia, U.S.S.R. and the United Kingdom. Dr B. J. Mason, speaking for the United Kingdom, considered that although the benefits of a weather service to the community are difficult to evaluate in purely financial terms, it is important to attempt benefit–cost analyses even though, as in most economic problems, neither the data nor the assumptions can be very firm. Aviation is generally the largest customer for weather services in the United Kingdom, the value of such services being estimated to be at least £6·5 million with an overall benefit/cost of 10 to 1. Similarly, savings due to specialized services to agriculture are estimated to be £10 million, while the cost of providing such services is less than £0·1 million, thus a benefit/cost ratio of 100 to 1 is achieved.

The benefit/cost ratios for the United Kingdom, and for the other nations as reported by W.M.O., indicate that the actual benefit/cost ratios of national weather forecasting services are of the order of 20 to 1 to probably 50 or even 100 to 1, a ratio which to any business executive would presumably be very favourable. It seems very clear, therefore, that increased expenditure for weather forecasting services is an investment that will reap rich dividends.

7. Foreseeing the future: problems and prospects

The developments of the 1960's in the atmospheric sciences have brought about significant improvements in weather forecasting. Still greater improvements are foreseen, and many – although not all – weather forecasters, including Namiis (1968), suggest that by 1975 or 1980 economically valuable forecasts such as: detailed weather predictions out to one week; forecasts of weekly weather regimes out to one month; prediction of monthly means and other weather statistics with double the present skill; prediction of seasonal general weather characteristics; rough estimates of the character of a season one year ahead; and, estimates of average climatic deviations from normal for a five- or possible ten-year period, will be available.

Other aspects of the atmospheric sciences do offer challenging prospects, but weather forecasting still remains the 'show place of meteorology'. Moorman (1968) in commenting on this says that there are those who make a clear distinction between the *science* of meteorology and the *services* of meteorology and express the fear that attention to the services, with emphasis on operational problems, reduces the scientific vigour of meteorology. This condition has certainly existed in the past, but as Moorman says: 'If we agree that the task of meteorology is to describe, to understand, to predict, and perhaps to modify the atmosphere, it seems . . . that the solid advances of the last several years in *describing* and *understanding* atmospheric processes now logically require the application of that understanding to forecast problems.'

The difficulty is, however, that many meteorologists still regard weather forecasting as 'applied physics'. Sawyer (1967), in his survey of weather forecasting, indicates that the economic value of 10- to 20-day forecasts would be considerable, and that the possibility of producing such forecasts is one of the main incentives of World Weather Watch. Nevertheless, except for the above observation, Sawyer appears in this paper to consider the future of weather forecasting only in terms of a physical science. Very little, if any, mention is made of the wider interpretation of weather forecasting, especially the economic and social aspects, such as the effects of weather modification, consumers' changing attitudes to weather information, methods of presenting weather forecasts and whether in fact people really want to know what next week's weather is going to be.

Some of the problems of looking at weather forecasting have been mentioned. What should be done to solve these problems? First, the atmosphere and the forecast atmospheric conditions must be considered a resource that can and must be 'usefully' managed. In this regard, it is important to appreciate that the inputs of the atmospheric resources into any socio-economic model of an area are variable and not static. Second, research is required on the role of weather forecasts in decisions relating to adjustments to the weather. Who uses weather information and weather forecasts, and in what way? Does an increase in the amount of information or an improvement in the accuracy of weather forecasts necessarily lead to changes in production schedules or alterations in other human activities? To what extent is the meteorologist's view of the value of increased information borne out by the manner in which people actually use the information which he provides? Knowledge of these matters would be of considerable value to those involved in designing programmes of weather information and weather forecasting, as well as to those who are concerned with developing policies to encourage more efficient adjustment to weather variations and more efficient use of weather forecasts. To do these things, the atmospheric sciences must broaden in vision by actively encouraging research in the social and economic aspects of the resources of the atmosphere. This could be done by encouraging atmospheric scientists to look anew at some of these problems, and weather forecasters, because of their particular interests, may be the best people to do this. In addition, employment of suitably qualified and interested geographers, economists, sociologists and others in forecasting services should be given particular consideration, and sponsoring of appropriate research by social scientists should be undertaken. But, in particular, there must be a trend away from the notion that only mathematicians and physicists can make a contribution to weather forecasting. No one questions the valuable contribution of mathematicians and physicists but we would do well to remember that the atmospheric sciences, and particularly weather forecasting, embrace many things besides the physics of the atmosphere.

D. WORLD WEATHER WATCH

1. History and formulation

Meteorologists have long recognized the need for international co-operation. An outstanding example is the World Weather Watch (WWW), which arose out of a United Nations General Assembly recommendation to the World Meteorological Organization in the latter part of 1961. The WWW reached the operational stage when the first World Weather Centre in Washington, D.C., began functioning in January 1965, and since then two other World Weather Centres have been established, in Moscow and in Melbourne, as well as a large number of regional centres.

The impetus for the WWW programme came in part from the

launching of the first artificial satellite, Sputnik 1, on October 4, 1957. This achievement, with its implications for frequent and large-scale observations of the atmosphere, was followed by specialized satellites capable of observing and returning information on the world's weather. As a result, President Kennedy in speaking before the U.N. General Assembly in 1961, said: 'we shall propose further co-operative efforts between all nations in weather prediction and eventually weather control. We shall propose, finally, a global system of communications satellites linking the whole world in telegraph, telephone, radio and television' (Petterssen, 1966), These proposals received the unanimous support from the U.N. General Assembly and in December 1961 a recommendation was made to all Member States and the W.M.O. that they undertake the early and comprehensive study of means to advance the state of atmospheric science and technology. In so doing, it would provide a greater knowledge of the basic physical forces affecting climate, and the possibility of large-scale weather modification. This would enable Member States to develop their forecasting capabilities through an effective use of world and regional meteorological centres.

In response to this recommendation, the W.M.O. assembled in Geneva a group of experts from the United States and the Soviet Union who, along with assistance from various organizations including the International Council of Scientific Unions (I.C.U.S.), UNESCO and the International Union for Geodesy and Geophysics (I.U.G.G.), prepared a report on the implementations of WWW which was presented to the Fourth World Meteorological Congress in April 1963. Further studies were made, by the W.M.O. Secretariat, member countries and consultants, and resulted in the formulation of the WWW Plan, which was approved in its final form by the Fifth Congress held in Geneva in April 1967. This plan is published in full by the W.M.O. (*Weather Watch: The Plan and Implementation Programme*) and is admirably reported by O. M. Ashford in the October 1967 issue of the *W.M.O. Bulletin*. The following summary is an abstract of this W.M.O. publication (O.M.A., 1967).

The primary purpose of the WWW is to ensure that all members obtain the meteorological information they require, both for operational work and research. It is also intended to stimulate and facilitate research work necessary to improve weather forecasts. In particular, WWW is conceived as a world-wide system composed of the national facilities and services provided by individual members, co-ordinated and in some cases supported by W.M.O. and other international organizations.

The essential elements of the WWW are stated to be as follows:

(1) The observational networks and other observational facilities, called the *global observing system*;
(2) The meteorological centres and the arrangements for the processing of the observational data and for the storage and retrieval of data, called the *global data-processing system*;
(3) The telecommunication facilities and arrangements necessary for the rapid exchange of the observations themselves and of the processed data, called the *global telecommunication system*;
(4) The research programme; and
(5) The programme in education and training.

These elements are closely linked with each other, and WWW is designed to be a dynamic system, flexible enough to be adapted to changing conditions. Thus, new techniques of observation, telecommunication and data-processing will be introduced as soon as they have been proved to be sufficiently reliable and economical.

2. The World Weather Watch plan

a. *Global observing system*

The global observing system aims at remedying the major deficiencies in the existing networks of observing stations by making operational by 1971, 270 new surface stations, 180 radiosonde stations and 275 radiowind stations. The regional expansion implementation of surface land stations is shown in Table VI.8.

The ocean areas where there are no possibilities of island stations will have to be filled by ship stations. The plan calls for the retention of all existing ocean weather stations, and for between five and ten additional ocean weather stations some of which are to be located at key points in the Southern Hemisphere. It is also proposed that by the end of 1971 there should be at least 100 merchant ships taking upper-air observations and that the number of selected ships taking surface observations will be doubled.

Another of the components of the global system is the meteorological satellite programme. Improved satellites will be introduced during the period and they will provide data on cloud distribution during both day and night and certain other global atmospheric parameters for operational purposes.

In addition, studies in observing systems will relate to such

TABLE VI.8: *Implementation, further plans (1968–1971) and deficiencies for observations (8 times daily) from surface land stations of basic regional networks (after W.M.O., 1968a)*

W.M.O region	Number of observations requested	Implemented 1.1.1968		Implementation by 31.12.1971		Expected increase	Remaining deficiencies	
		No.	%	No.	%	%	No.	%
I Africa	5,428	3,700	68	4,533	84	16	895	16
II Asia	7,216	6,653	92	7,076	98	6	140	2
III South America	2,600	1,469	56	2,478	95	39	122	5
IV North and Central America	3,888	3,020	78	3,107	80	2	781	20
V South-west Pacific	2,760	2,058	75	2,139	78	3	621	22
VI Europe	6,800	6,722	98·8	6,745	99·2	0·4	55	0·8
Antartica	240	199	83	212	88	5	28	12
Global totals	28,932	23,821	82	26,290	91	9	2,642	9

questions as the optimum density of various types of stations, the development of automatic weather stations, constant-level balloons, satellite-borne remote sensors, dropsondes and meteorological rockets, as well as the improvement of the existing observing techniques and equipment.

b. Global data-processing system

The global data-processing system is designed to process data but it will also provide storage and retrieval facilities. There will be three levels of data-processing centres. The highest level (i.e. concerned with the greatest areal extent) consists of the World Meteorological Centres at Melbourne, Moscow and Washington, which will concentrate on collection, analysis and distribution of data on a global or at least hemispheric scale in both graphic and digital (computer-to-computer) form. In addition during the period 1968–1971, 21 regional meteorological centres will be established to prepare analyses and prognoses on a regional scale (e.g. Wellington, New Zealand for the South-west Pacific area) and to distribute this data to both world and national centres.

It is expected that further planning studies will be carried out, in collaboration with the relevant international organizations, to define the role of various segments of WWW with respect to aviation, maritime interests, hydrology, agriculture and the general public.

c. Global telecommunication system

Essential to the functioning of both the global observing system and global data-processing system is the existence of a high-speed, efficient and accurate communication network. It is the purpose of the global telecommunication system to provide such a network to collect and distribute raw observational data to national, regional and world meteorological centres and subsequently to distribute the resulting processed information to other such centres. The system is to be organized on a three-level basis, the main trunk circuit, the regional telecommunication networks and the national telecommunication networks (Fig. VI.8).

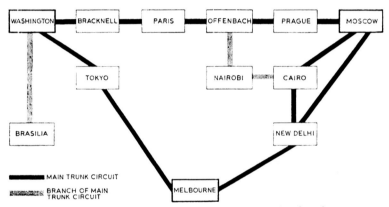

FIG. VI.8. *Routing of the main trunk circuit and its branches (after W.M.O., 1968).*

The national networks will collect the observations and distribute the data within the country concerned. The regional networks are responsible for the collection of observations in a given part of the world and for their transmission to meteorological centres and telecommunication hubs on the main trunk circuit. They will also provide for the distribution of the observational data and output products of world and regional meteorological centres to satisfy the needs of national meteorological centres. Finally, the main trunk circuit will interconnect world meteorological centres and appropriate regional telecommunication hubs.

d. Research and education

The programmes for research, education and training are as important to the success of the WWW as the global observing, data-processing and telecommunications systems. In fact, the real benefits of WWW are found in the utilization of weather forecasts in aiding various human activities.

Major improvements in the accuracy and range of operations forecasting and serious investigation of the feasibility of large-scale weather modification and control can, however, come only from a greatly increased understanding of the behaviour of the atmosphere. A vigorous programme of research on the general circulation of the atmosphere is therefore necessary, and the Global Atmospheric Research Programme has been drawn up as the research arm of WWW. Garcia (1967) gives details of this programme.

3. Potential benefits of World Weather Watch

In recent years more attention has been focused on the impact of weather on human activities in the hope of helping decision-makers gain greater efficiency and reduce losses in conducting their operations. Although these studies have been concentrated in the developed nations of the world, the results have consistently indicated substantial benefits, either realized or possible, relative to the costs of providing the existing or the improved weather services.

The costs of building weather ships, setting up computer centres and constructing telecommunication facilities are fairly easily identified and expressed in monetary terms. On the other hand, the benefits of the services which they can provide are much more elusive and frequently inexpressible in dollars and cents. Yet, if the WWW is to be more than a costly adventure in international co-operation, it must be able to demonstrate positive advantages and benefits to encourage nations to participate. It is also necessary that some attempt be made not only to *identify* possible benefits but also to *assess* them. This will help to assure that the limited resources available to WWW are invested in the optimal aspects of its plan and/or regions of concern. For example, it may be very useful for meteorological science in general to acquire new information and knowledge about the upper boundaries of the atmosphere, yet, as stated by the W.M.O. (1967): '. . . in newly developing regions of the world, the cost/benefit ratio associated with the establishment of simple temperature and precipitation stations may exceed that of the meteorological satellite by several orders of magnitude'.

While it is presently impossible to calculate the full potential benefit of the WWW, a selection of studies summarized by Thompson (1966b) provides some idea of the possibilities. This review suggested that '. . . a saving of perhaps 15% to 20% of protectable losses for prediction of weather for several days in advance' is possible. While these are ideal figures depending upon 'perfect' weather forecasts and optimal operational use of the meteorological advice, they do suggest the probable potential benefits to be realized from WWW. For example, the estimated total crop losses in the United States during 1961 due to various weather elements such as drought, hail and wind were $1,197 million. Therefore, a 10% saving arising from better weather forecasting and use of weather advice would reduce losses by $120 million. In the same manner, the current warning system of natural disasters in the United States saves as many as 2,610 lives each year and with an improved system it is expected that 461 more lives could be saved. Further, in Australia where the total national weather service costs about $5 million annually, prediction of weather conditions associated with bush and timber fires is estimated to reduce fire damage by approximately $34 million.

On a world scale the potential annual benefit to the economy from an improved satellite weather observing and forecasting system has been calculated by Stanford University (1966) to be $16,900 million, the relative benefits being: agriculture $6,000 m; construction $3,000 m; fuel, $300 m; utilities, $500 m; forestry, $100 m; government, $5,000 m; and marine transportation, $2,000 million. This is 50 times the estimated annual cost of the system.

If benefits of the magnitude indicated in the examples cited above can be realized in developed nations (where it might be assumed that the application of modern science and technology would have significantly reduced the harmful effects of weather elements), those to be expected in the lesser developed nations can hardly be less. Although the techniques required to take full advantage of meteorological services are largely lacking in the underdeveloped countries, at present, so are the weather forecasts themselves. The WWW will therefore not only encourage many of these nations to expand their own meteorological services, but will also make available some of the capital and knowledge required (Akingbehin, 1966). In addition, by reducing repetition and overlap of weather services between countries, the WWW will free personnel to concentrate more on national needs, an important consideration for both developed and underdeveloped nations. As one observer has pointed out (Langlo, 1966), even without any new discoveries in meteorological knowledge, the wider observational coverage, together with the faster collection and exchange of raw and processed data which the WWW is designed to provide, will result in better weather services to the world's customers. And, at the same time that the World Weather Watch is advancing and expanding the world's meteorological services, it will also be advancing and increasing world understanding and peaceful co-operation.

E. THE MODIFIED ATMOSPHERE

1. Weather knowledge and modification

In any assessment of the value of weather, it is necessary to take into account those aspects of the atmosphere which result from direct or indirect activities of man. These activities during the last few decades have resulted in two areas of concern to the applied meteorologist: first, the deliberate alteration of the atmosphere to produce changes in the 'natural' atmospheric conditions such as in the clearance of cold fog from airports; and second, the inadvertent alteration of the atmosphere, such as industrial or automobile pollution. It is clear, however, that some if not all pollutants have some effect on atmospheric conditions. Consequently, any study of the benefits and costs of weather knowledge must consider the 'contribution' of the modified atmospheric resources* whether they be a result of weather modification or atmosphere pollution.

2. Economic and social aspects

Some of the economic aspects of the modified atmosphere have already been discussed (see Chapter II, Section C) but it is worth repeating that the financial cost of air pollution *is* immense. Property damage in the United States due to airborne pollutants, for example, is estimated to be more than $10 billion a year, this representing a cost per person in the U.S. of over $50. In the cities the cost is much higher and it has been estimated that in New York City air pollution adds several hundred dollars per year in washing, cleaning, repairing and repainting bills to the budget of a family with two or three children. The benefits on the other hand are presumably gained by industry, and employment which 'causes' the atmospheric pollution

---

* A critical appraisal of how man is using these resources is available in *Pollution: what it is . . . what it does . . . what can be done about it . . .* (Maunder, 1969b), a publication resulting from a series of lectures given at the University of Victoria, B.C. Some of the comments which follow, concerning air pollution, are in part, from two chapters of this publication, Chapter 1: 'What is Pollution' (Maunder, 1969c), and Chapter 3: 'Weather, Weather Modification, and Air Pollution' (Maunder, 1969d).

in the first place, a factor which may imply that the coexistence of a modern industrial society and a polluted environment is almost inevitable.

With respect to 'true' weather modification, knowledge of the benefits and costs is no easier to assess, especially in some respects such as the effect of weather modification on agriculture land price differentials where Johnson and Haigh (1969) have shown that there are many complications. Nevertheless, the costs of most weather modification experiments are usually relatively small, and are normally believed to be only a fraction of the benefits obtained through weather modification. For example, the budget for the 1965 and 1966 fiscal year for the Federal Weather Modification programme in the United States totals less than $13 million (Table VI.9), an expenditure which is small compared with the probable

TABLE VI.9: *U.S. Federal weather modification programmes in 1966* (after National Science Foundation, 1966)*

Department or agency	Fog & cloud diss.	Precip. modif.	Hail supp.	Light-ning modif.	Severe storm modif.	Other	Total
N.S.F.	0·03	1·22	0·35	0·10	0·10	0·20	2·00
Commerce	—	0·035	0·035	0·13	0·32	0·13	0·65
Army	0·16	—	—	0·09	—	—	0·25
Navy	0·71	—	—	—	0·20	—	0·91
Air force	0·26	—	—	—	—	—	0·26
Interior	—	2·98	—	—	—	—	2·98
Agriculture	—	—	—	0·14	—	—	0·14
Totals	1·16	4·235	0·385	0·46	0·62	0·33	7·19

* Millions of dollars.

benefits. The National Science Foundation (1966) gives details of some of those benefits. A public utility on the Pacific Coast concluded, for example, that in the drainage area of one of its reservoirs an increase of less than 27% in annual precipitation would clearly justify cloud seeding, and that an increase of 10% for a large watershed might be worth $200,000. Similarly, an airline estimated that the immediate benefits in reduction of operating expenses from fog dispersal in an intermountain area were at least five times the seeding costs.

On another scale, it is obvious that if a farmer thinks he may increase his per acre wheat yield from seven to eight bushels by rainfall induction or hail suppression at a cost of a few cents per acre, he will be strongly inclined to take the risk of the expenditure even though the results are in doubt. It has also been estimated that the mean losses of $250,000,000 from hurricanes might be reduced by as much as one-third if only modest reductions in storm intensity or slight changes in storm paths could be achieved. Accordingly, benefits of probably $50 to $100 million per season from hurricane modification in the United States could probably be achieved for an expenditure of probably less than $20 million.

It is, nevertheless, evident that if *all* the benefits and *all* the costs are taken into account, the benefit/cost ratio may not be as high. For example, as the National Science Foundation (1966) states: 'The evaluation of fog dissipation requires not only the measurement of benefits and costs at the airport and in airline operations, but the assessment of the benefits and costs from installing equipment which could land aircraft notwithstanding fog or from re-routing traffic on the ground and in the air.' In the same manner, it is not at all clear that artificially induced precipitation in one area will automatically provide benefits to the surrounding areas. It is equally appropriate to ask what 'right' has an airline company to disperse cold fog from an airport so that it can make a profit at the expense of the restaurants, and news-stands in the airport terminal, or the bus, railroad and taxi companies. Some of these questions involve the problem of legal ownership of the atmosphere, and are discussed in Chapter VII.

3. Engineering aspects

The engineering aspects of artificially modifying the atmosphere are described in a number of papers (e.g. Hicks, 1967; Battan, 1968; Biswas *et al.*, 1967). However, the engineering control of atmospheric pollution is of more direct concern.

Air pollution can, according to Benline (1965), generally be controlled in three ways. The first is by preventing or minimizing its formation or process control. We can, for example, conduct an operation in such a manner or by using such materials as to eliminate objectionable emissions. Second, since the formation of all pollution cannot be prevented, removal of the pollutants from the discharge can be done. Many devices have been designed and built to accomplish this task, some of these being: gravity settling chambers, scrubbers, baghouses, electric precipitators, absorbers, and blow-by and engine devices, the latter being used to reduce pollutants emitted from cars. With regard to the last device, it may be noted that the U.S. Federal Government now requires automobile manufacturers to place air-pollution devices on new cars. However, not all pollution can be controlled by process change, nor is any control equipment economically 100% efficient. Consequently, the third method of control must be to discharge the residual pollution at such a height or in such a location that it may be quickly and readily diffused.

Most of the arguments advanced for one form of control over another stress the economic aspects, and although the dollars and cents of pollution abatement are politically foremost, it should also be remembered (and taken into account) that social, health, recreational and planning considerations are also important and to many people may far outweigh any argument in dollars and cents. Pollution can and must be controlled by engineering methods, but this does not mean that only the economic-engineering aspects of the problem should be taken into account.

4. Medical aspects

In addition to the economic and engineering aspects of the benefits and costs of atmospheric modification, is the very important question of air pollution and health. When it comes to assessing the effects of unclean air on public health, however, considerable debate ensues, and often completely opposite viewpoints arise. But whether pollution is to be endured as a nuisance or suppressed by vigorous governmental action, it must be considered as an unpleasant and expensive consequence of twentieth-century living.

In discussing air pollution and public health, McDermott (1961) says that if something is happening to the public health from the widespread pollution of the air, it must presumably be happening to large numbers of people. Yet, since there is no real public outcry, it must be something that goes on undramatically in its individual manifestations, otherwise it would attract public notice as an 'epidemic'.

Most of the pollutants in the atmosphere are a result of burning. In many cities the garbage is burned, sometimes twice, in the backyard and on the city dump. Automobiles are highly mobile 'burners' throughout their active lifetime, and when they are outmoded, they too often end up on the pyre. In short, much of the energy of civilization is supplied by burning, and most of its debris is likewise burned. Two of the final products of a completed combustion are water and carbon dioxide, which in the amounts involved in urban life would be entirely harmless. In many cases, however, fuel and debris are only partially burned, and a wide variety of chemical substances are given off into the air. Some of this material is visible smoke, made up of particulate matter, such as fly-ash or soot; but some of the material entering the air is composed of complex invisible chemicals, and it is these substances that seem to represent the greater menace to health. In particular, the aggravation of chronic bronchitis–emphysema by air pollution has been most drastically demonstrated in a few epidemics of acute illness attributed to air pollution. In London, in 1952, for example, severe air pollution was associated with an 'excess mortality' of over 4,000 persons during one week, and *Time* (Copyright Time Inc. 1953) (November 9, 1953) in reporting on the event stated:

> London's fogs, which once romantically shrouded the nocturnal prowlings of Sherlock Holmes' Professor Moriarty . . . now veil an even grimmer killer. . . . In one smog-bound week last December, 4,000 Londoners died from trying to breathe the noxious combination of smoke and fog that choked the city.

A similar situation occurred in Donora, Pennsylvania, in 1948, and McDermott (1961) in commenting on this case said:*

> The air of Donora, on a bend of the Mongahela River with high hills on all sides, must take up the smoke and fume of blast furnaces, steel mills, sulfuric acid mills and slag-processing plants. In October 1948, a thermal inversion occurred over most of the U.S. including the Donora basin. There the usual smog, instead of lifting each day at noon as was its custom, remained unabated. By the third day of the constant smog, 5,910 persons were reported ill. More than 60 per cent of the inhabitants 65 and older were affected, and almost half of these were seriously ill. In all, 20 persons died, 17 of the deaths occurring on the third day of unremitting smog.

The smog in these two situations was of the sulphur type, but there appears to be no reason to doubt that a hydrocarbon smog could have the same effect. The most significant finding, says McDermott, is that no single smog component in either disaster was present in a higher concentration than usual, this finding pointing therefore to the ugly conclusion that the same smog breathed by everyone a day or two at a time without immediate or apparent ill effect may be highly injurious to substantial numbers of people when it is breathed continuously for only a few days more.

It should be noted, however, that in some smog-prone areas recent

* From Walsh McDermott, *Air Pollution and Public Health.* Copyright 1961 by Scientific American Inc. All rights reserved.

legislation has considerably reduced the incidence of lethal smogs. In London, for example, during the 1960's, there has been a marked reduction in the incidence of 'pea-soup' fogs, a reduction chiefly attributable to the establishment of smokeless zones under the Clean Air Act of 1956.

5. Controlling the urban environment

The increasing use of controlled environments in urban areas is perhaps an indication that man is becoming less dependent on the weather than he was. Nevertheless, the ever-increasing cost of weather, particularly in disasters such as hurricanes and tornadoes, or in flooding or heavy snowfalls, points clearly to the fact that man controls only a relatively small part of his atmospheric environment.

The possibility of controlled modification of the urban environment is nevertheless a natural sequel to the inadvertent modification that has already gone on. Two possibilities emerge immediately, according to Hilst (1967): first, the generation of atmospheric wind and turbulence during critical periods to assure adequate ventilation; second, enclosure and air conditioning of the city itself. Sustained modification of wind and weather over areas of hundreds of square miles are presently the subject of theoretical arguments but, the Houston Astrodome notwithstanding, no one has yet enclosed a sizeable city. But even though these deliberate controls are denied us until we have much larger and cheaper energy sources, we can anticipate the time when more subtle control techniques are available. We do know, for example, that incipient atmospheric phenomena are critically balanced and, at least in theory, could be modified with more modest energy expenditures.

The need for totally enclosed artificial environments for work bases on the Moon and planets other than Earth has stimulated conceptual design of enclosures for earthbound cities. In principle, these enclosures could be for temperature and humidity and for the maintenance of air quality. Hilst (1967), for example, considers the problems associated with a hemispherical dome one mile in radius – a feasible dimension structurally and economically. The external environmental stresses on this structure will depend upon where it is located, but as the major source of protection of the inhabitants and enclosed property against the elements, it can be assumed that the structure must be capable of withstanding wind forces associated with wind speeds over 100 mph. Further, a one-inch rainfall in two hours would lead to a steady flow of water around the base of the dome of 24 million gallons/day, and a 10-inch snowfall that accumulates uniformly would produce a load of 420 million pounds. These, therefore, are a few of the external environmental stresses with which the envelope over the city must contend.

Whether or not controlled environments on a large scale will become a reality, remains for the future, but in considering the benefits and costs of atmospheric modification in the broadest sense it is important to consider *all* the costs and *all* the benefits. Employment opportunities, for example, are closely associated with industry, and many industries under present economic conditions could be said to necessitate some air pollution. But whatever method is used for evaluation, it is clear that we must *first* find out the value of atmospheric resources at the present time, polluted or unpolluted, for then and only then will we be able to decide whether they are worth controlling, and in what way they should be controlled.

BIBLIOGRAPHY

AKINGBEHIN, N. A., 1966: World Weather Watch: a means of accelerating development. *W.M.O. Bull.*, 15: 120–31.

ALLEN, L., 1964: How to beat the weather. *Virginia Municipal Review*, Sept., p. 300.

BATTAN, L. J., 1968: Some problems in changing the weather. *Weatherwise*, 21: 102–6.

BENLINE, A. J., 1965: Air pollution control problems in the city of New York. *Trans. N.Y. Acad. Sci.*, 27: 916–22.

BISWAS, K. R., *et al.*, 1697: Cloud seeding experiment using common salt. *Jour. App. Met*, 6: 914–23.

BORGMAN, E. and BROOKER, D. B., 1961: *The Weather and Haymaking in Missouri*. Univ. of Mo., Coll. of Agr., Agr. Exp. Sta. Bull. B 777, 7 pp.

BORGMAN, L. E., 1960: Weather-forecast profitability from a client's viewpoint. *Bull. Amer. Met. Soc.*, 41: 348.

BRIER, G. W., 1950: Verification of forecasts expressed in terms of probability. *Monthly Weather Rev.*, 78: 1–3.

BRIER, G. W. and ALLEN, R. A., 1951: Verification of weather forecasts. In: MALONE, T. F., ed., *Compendium of Meteorology*. Amer. Met. Soc., Boston, pp. 841–8.

BROSS, I. D. J., 1953: *Design for Decision*. Macmillan Company, New York, 276 pp.

BRYAN, J. G. and ENGER, I., 1967: Use of probability forecasts to maximize various skill scores. *Jour. App. Met.*, 6: 762–9.

CHARNEY, J. G., *et al.*, 1966: *The Feasibility of a Global Observation and Analysis Experiment*. Publ. 1290, Nat. Acad. Sci., Nat. Research Council, Washington, D.C., 172 pp.

CONTROLLER, CANADIAN METEOROLOGICAL SERVICE, 1949: University of Toronto poll of students on weather terminology. *Bull. Amer. Met. Soc.*, 30: 61–2.

CROW, L. W., 1965: *Economic Values of Weather-related Decisions.* U.S. Dept. of Commerce, E.S.S.A., U.S. Weather Bureau, 4 pp.

CURTISS, J. H., 1968: An elementary mathematical model for the interpretation of precipitation probability forecasts. *Jour. App. Met.*, 7: 3–17.

DEMSETZ, H., 1962: *Economic Gains from Storm Warnings: Two Florida Case Studies.* Rand Corp. RM 3168 N.A.S.A., 43 pp.

EPSTEIN, E. S. and MURPHY, A. H., 1965: A note on the attributes of probabilistic predictions and the probability score. *Jour. App. Met.*, 4: 297–9.

FREEMAN, M. H., 1966: The accuracy of long-range forecasts issued by the Meteorological Office. *Met. Mag.*, 95(1132): 321–5.

GARCIA, R. V., 1967: The global atmospheric research programme. *W.M.O. Bull.*, 16: 212–18.

GLEESON, T. A., 1967: On theoretical limits of predictability. *Jour. App. Met.*, 6: 213–15.

GRINGORTEN, I. I., 1951: The verification and scoring of weather forecasts. *Jour. Amer. Stat. Assn.*, 46: 279–96.

GRINGORTEN, I. I., 1959: Probability estimates in relation to operational decisions. *Jour. Met.*, 16: 663–71.

GRINGORTEN, I. I., 1965: A measure of skill in forecasting a continuous variable. *Jour. App. Met.*, 4: 47–53.

GRINGORTEN, I. I., 1967: Verification to determine and measure forecasting skill. *Jour. App. Met.*, 6: 742–7.

HALLANGER, N. L., 1963: The business of weather: its potentials and uses. *Bull. Amer. Met. Soc.*, 44: 63–7.

HIBBS, J. R., 1966: Evaluation of weather and climate by socio-economic sensitivity indices. In: SEWELL, W. R. D., ed., *Human Dimensions of Weather Modification.* University of Chicago, Dept. of Geography, Research Paper No. 105, pp. 91–109.

HICKS, J. R., 1967: Improving visibility near airports during periods of fog. *Jour. App. Met.*, 6: 39–42.

HILST, G. R., 1967: What can we do to clear the air? *Bull. Amer. Met. Soc.*, 48: 710–12.

HOLLOWAY, J. L. and WOODBURY, M. A., 1955: *Application of Information Theory and Discriminant Function Analysis to Weather Forecasting and Forecast Verification.* University of Pennsylvania, Institute of Cooperative Research, Contract No. 551(07), Tech. Rept. 1, 85 pp.

JOHNSON, L. B., 1965: Presidential messages: On weather modification. *Bull. Amer. Met. Soc.*, 46: 399–402.

JOHNSON, S. R. and HAIGH, P. A., 1969: Agricultural land price differentials and their relationship to potentially modifiable aspects of the climate. *Rev. Econ. Stat.* (in press).

KOLB, L. L. and RAPP, R. R., 1962: The utility of weather forecasts to the raisin industry. *Jour. App. Met.*, 1: 8–12.

LAMB, H. H., 1967: Britain's changing climate. *Geog. Jour.*, 133: 445–68.

LANGLO, K., 1966: A new look in meteorology: the World Weather Watch. *Impact of Science on Society*, 16: 65–92.

LORENZ, E. N., 1966: Atmosphere predictability. In: *Advances in Numerical Weather Prediction*. 1965/66 Seminar Series, The Travellers Research Center, Inc., Hartford, Conn., pp. 34-9.

McDERMOTT, W., 1961: Air pollution and public health. *Scientific American*, 205(10): 49-57.

McQUIGG, J. D., 1964: *The Economic Value of Weather Information*. Ph.D. Dissertation, University of Missouri, Columbia, Missouri.

McQUIGG, J. D., 1965: Forecasts and decisions. In: D: Foreseeing the future. *Met. Monographs*, 6(28): 181-8.

McQUIGG, J. D. and DOLL, J., 1961: *Weather Variability and Economic Analysis*. Univ. of Mo., Coll. of Agr., Agr. Exp. Sta. Res. Bull. 771, 40 pp.

McQUIGG, J. D. and THOMPSON, R. G., 1966: Economic value of improved methods of translating weather information into operational costs. *Monthly Weather Rev.*, 94: 83-7.

MASON, B. J., 1966: The role of meteorology in the National Economy. *Weather*, 21: 382-93.

MAUNDER, W. J., 1969a: The consumer and the weather forecast. *Atmosphere*, 7: 15-22.

MAUNDER, W. J. (Editor), 1969b: *Pollution: What it is . . . What it does . . . What can be done about it . . .* A publication of the Evening Division, University of Victoria, B.C., 115 pp.

MAUNDER, W. J., 1969c: What is pollution? In: MAUNDER, W. J., ed., 1969b: *Pollution: What it is . . . What it does . . . What can be done about it . . .* Univ. of Victoria, B.C., pp. 1-10.

MAUNDER, W. J., 1969d: Weather, weather modification and air pollution. In: MAUNDER, W. J., ed., 1969b: *Pollution: What it is . . . What it does . . . What can be done about it . . .* Univ. of Victoria, B.C., pp. 21-30.

MEGLIS, A. J., 1960: Annotated bibliography on forecast verification. *Met. Geoastrophys. Abstr.*, 11: 1129-74.

MILLER, R. G., 1962: Statistical prediction by discriminant analysis. *Met. Monographs*, 4(25): 54 pp.

MOORMAN, T., JR, 1968: The forecasting problem. *Bull. Amer. Met. Soc.*, 49: 12-15.

MURPHY, A. H., 1960: Meteorology and heating load requirements. *Met. Monographs*, 4(22): 65-8.

MURPHY, A. H. and EPSTEIN, E. S., 1967a: Verification of probabilistic predictions: a brief review. *Jour. App. Met.*, 6: 748-55.

MURPHY, A. H. and EPSTEIN, E. S., 1967b: A note on probability forecasts and 'hedging'. *Jour. App. Met.*, 6: 1002-4.

NAMIAS, J., 1968: Long range weather forecasting – history, current status and outlook. *Bull Amer. Met. Soc.*, 49: 438-70.

NATIONAL SCIENCE FOUNDATION, 1966: *Weather and Climate Modification*. Report of the Special Commission on Weather Modification, National Science Foundation Report No. 66-3, Washington, D.C.

NELSON, R. R. and WINTER, S. G., JR, 1960: *Weather Information and Economic Decisions, A Preliminary Report*. Rand Corp., RM-2620-N.A.S.A.

NELSON, R. R. and WINTER, S. G., JR, 1963: *A Case Study in the Economics of Information and Coordination: The Weather Forecasting System.* Rand Corp., Santa Monica, Calif., pp. 11–34.

ODDIE, B. C., 1964: The language of forecasts. *Weather,* 19(5): 138–43.

O.M.A., 1967: The World Weather Watch Plan. *W.M.O.* Bull., 16: 195–200.

PANEL ON INTERNATIONAL METEOROLOGICAL COOPERATION, 1966: *The Feasibility of a Global Observation and Analysis Experiment.* National Academy of Science – National Research Council, Publ. 1290, 172 pp.

PETTERSSEN, S., 1966: Recent demographic trends and future meteorological services. *Bull. Amer. Met. Soc.,* 47: 950–62.

RAPP, R. R. and HUSCHKE, R. E., 1964: *Weather Information: Its Uses, Actual and Potential.* The Rand Corporation, Santa Monica, Calif., Rand Memo. RM-4083-USWB (U.S. Weather Bureau), 126 pp.

REICHELDERFER, F. W., 1964: Using weather services in your business. In: *External Sources of Help.* U.S. Government Printing Office, p. 86.

RICHARDSON, L. F., 1922: *Weather Prediction by Numerical Process.* Cambridge University Press, Cambridge, 236 pp.

ROBINSON, G. D., 1967: Some current projects for global meteorological observation and experiment. *Quart. Jour. Roy. Met. Soc.,* 93: 409–18.

ROOT, H. E., 1962: Probability statements in weather forecasting. *Jour. App. Met.,* 1: 163–8.

SANDERS, F., 1958: *The Evaluation of Subjective Probability Forecasts.* Massachusetts Institute of Technology, Department of Meteorology, Contract AF 19(604)–1305, Sci. Rept. 5, 60 pp.

SAWYER, J. S., 1967: Weather forecasting and its future. Part I – Forecasting from 1850 to the present. *Weather,* 22: 350–60; Part II – Forecasting in the years to come. *Weather,* 22: 400–6.

SEWELL, W. R. D., 1966: Humanity and the weather. *Chicago Today,* 3(2): 24–9.

SHAK, G. W., et al., 1966: Temperature terminology for public forecasts. *Weatherwise,* 16: 161–3.

SHANNON, C. E. and WEAVER, W., 1949: *The Mathematical Theory of Communication.* University of Illinois Press, Urbana, 117 pp.

SHERROD, J. and NEUBERGER, H., 1958: Understanding forecast terms – results of a survey. *Bull. Amer. Met. Soc.,* 39(1): 24–6.

SHORR, B., 1966: The cost/loss utility ratio. *Jour. App. Met.,* 5: 801–3.

SHUMAN, F. G. and HOVERMALE, J. B., 1968: An operational six-layer primitive-equation model. *Jour. App. Met.,* 7: 525–47.

STANFORD UNIVERSITY, 1966: SPINMAP: *Stanford Proposal for an International Network for Meteorological Analysis and Prediction – Summary Report.* Stanford University, Palo Alto, Calif., May 31.

TENENBAUM, O., 1966: A new look at the forecast story. *Bull. Amer. Met. Soc.,* 47: 25–7.

THOMPSON, J. C., 1952: On the operational deficiencies in categorical weather forecasts. *Bull. Amer. Met. Soc.,* 33: 223–6.

THOMPSON, J. C., 1962: Economic gains from scientific advances; operational improvements in meteorological prediction. *Jour. App. Met.,* 1: 13–17.

THOMPSON, J. C., 1963: Weather decision making – the payoff. *Bull. Amer. Met. Soc.*, 44: 75–8.

THOMPSON, J. C., 1966a: A note on meteorological decision making. *Jour. App. Met.*, 5: 532–3.

THOMPSON, J. C., 1966b: The potential economic and associated values of the World Weather Watch. *World Weather Watch Planning Report*, No. 4. W.M.O., Geneva, 35 pp.

THOMPSON, J. C. and BRIER, G. W., 1955: The economic utility of weather forecasts. *Monthly Weather Rev.*, 83: 249–54.

THOMPSON, J. W., 1964: Prediction in physics and the social sciences. *General Systems*, 9: 15–23.

TOBIN, R. L., 1966: Weather or not. *Saturday Rev.*, 49c: Nov. 12, pp. 95–6.

U.S. DEPARTMENT OF COMMERCE, WEATHER BUREAU, 1964: *The National Research Effort on Improved Weather Description and Prediction for Social and Economic Purposes*. Federal Council for Science and Technology, Interdepartmental Committee on Atmospheric Sciences, 84 pp.

WHITE, R. M., 1964: New weather discoveries will serve you. *Nation's Business*, Nov.

WHITE, R. M., 1966a: A perspective on the present status of weather prediction. In: *Advances in Numerical Weather Prediction*. 1965/66 Seminar Series, The Travellers Research Center, Inc., Hartford, Conn., pp. 1–5.

WHITE, R. M., 1966b: Broadcasting the weather. *Bull. Amer. Met. Soc.*, 47: 21–4.

WILLIAMSON, E. and RILEY, J., 1960: *The Inter-related Effects of Defoliation, Weather and Mechanical Packing in Cotton Quality*. Paper presented at Winter Meeting of Amer. Soc. Agr. Engr., Memphis, Tenn.

WINKLER, R. L. and MURPHY, A. H., 1968: 'Good' probability assessors. *Jour. App. Met.*, 7: 751–8.

W.M.O., 1967: Assessing the economic value of a National Meteorological Service. *World Weather Watch Planning Report*, No. 17. W.M.O., Geneva, 14 pp.

W.M.O., 1968a: The World Weather Watch – a progress report. *W.M.O. Bull.*, 17: 172–81.

W.M.O., 1968b: Economic benefits of meteorology. *W.M.O. Bull.*, 17: 181–6.

——, 1953: Smoggles. *Time*, Nov. 9, 1953, p. 28.

ADDITIONAL REFERENCES

AD HOC WEATHER WORKING GROUP, 1966: Biological aspects of weather modification. *Bull. Ecological Soc. of America*, 47: 39–78.

ARAKAWA, H., 1966: Usefulness of weather forecasting and storm warnings. *Weather*, 21: 46–7.

BATES, F. C., 1962: Severe local storm forecasts and warnings and the general public. *Bull. Amer. Met. Soc.*, 43: 288–91.

CRUTCHFIELD, J. A., 1968: Economics of weather modification. *Proc. 1st. Natl. Conf. Weather Modification*, Albany, N.Y., pp. 181–9.

EPSTEIN, E. S., 1969: The role of initial uncertainties in prediction. *Jour. App. Met.*, 8: 190–8.

GARCÍA, R., 1969: The global atmospheric research programme. *W.M.O. Bull.*, 18: 186–8.

GENTRY, R. C., 1969: Project STORMFURY. *W.M.O. Bull.*, 18: 146–54.

GLAHN, H. R., 1964: The use of decision theory in meteorology – with an application to aviation meteorology. *Monthly Weather Rev.*, 92: 383–8.

GOSLINE, C. P., 1949: The role of the meteorologist in a large company. *Bull. Amer. Met. Soc.*, 30: 119–23.

HALE, D., 1957: Eliminate weather worries with far-reaching forecasts and long-range planning. *Amer. Gas Jour.*, 184: 40–1.

HENDRICK, R. L., 1968: Evaluating the desirability of weather modification. *Proc. 1st Natl. Conf. Weather Modification*, Albany, N.Y., pp. 241–50.

HOLFORD, INGRID, 1964: The public and the weather. *Weather*, 19: 12–14.

KRICK, I. P., 1955: The value of weather modification operation to the pulp and paper industries. *Pulp and Paper Mag. of Canada*, 56(3): 263–74.

LEIGHTON, P. A., 1966: Geographical aspects of air pollution. *Geog. Rev.*, 66: 151–74.

LESSING, L., 1968: Doing something about the weather, in a big way. *Fortune*, 77: 132–7.

LIEURANCE, N. A., 1969: Aviation weather service and requirements: 1970–80. *W.M.O. Bull.*, 18: 91–3.

LODGE, J. P., JR, 1969: An atmospheric scientist views environmental pollution. *Bull. Amer. Met. Soc.*, 50: 530–3.

LORENZ, E. N., 1969: Three approaches to atmospheric predictability. *Bull. Amer. Met. Soc.*, 50: 345–9.

McCORMICK, R. A., 1969: Meteorology and urban air pollution. *W.M.O. Bull.*, 18: 155–65.

MACHATTIE, L. B., 1969: Modern meteorological services and forest fire control. *Pulp and Paper Mag. of Canada*, 70(16): 87–9.

McQUIGG, J. D., 1961: The weather bureau state climatologist and his service to agriculture. *Weatherwise*, 14: 154–7.

McQUIGG, J. D., 1968: A review of problems, progress and opportunities in the use of weather information in agricultural management. *UNESCO Symposium on Agroclimatological Methods*. Reading, England, pp. 175–85.

MALONE, T. F., 1968: New dimensions of international cooperation in weather analysis and prediction. *Bull. Amer. Met. Soc.*, 49: 1134–40.

MIYAKODA, K., SMAGORINSKY, J., STRICKLER, R. F. and HEMBREE, G. D., 1969: Experimental extended predictions with a nine-level hemispheric model. *Monthly Weather Rev.*, 97: 1–76.

MORGAN, J. J., 1961: Use of weather factors in short-run forecasts of crop yields. *Jour. Farm. Econ.*, 43: 1172–8.

MURPHY, A. H., 1969: Measures of the utility of probabilistic predictions in cost–loss ratio decision situations in which knowledge of the cost–loss ratios is incomplete. *Jour. App. Met.*, 8: 863–74.

NEMOTO, J., 1958: The economic value of weather forecasts. *Jour. Met. Res.* (Tokyo), 10: 823–5.

NOFFSINGER, T. L., MEANS, L. L. and STOFFER, D. R., 1964: A weather service programme for agriculture (United States). *Agric. Met.*, 1: 93–106.

M

PALMER, M., 1962: Extra weather data system pays operating dividends. *Power Engineering*, 66: 48–9.

PINK, J., 1965: Weather warnings produce profits. *Gas*, 41: 43–5 (Dec.).

REEDY, W. W. and HURLEY, P. A., 1968: Making weather modification a working partner in water resources development. *Proc. 1st Natl. Conf. Weather Modification*, Albany, N.Y., pp. 65–70.

RIDKER, R. G. and HENNING, J. A., 1967: The determinants of residential property values with special reference to air pollution. *Rev. Econ. Stat.*, 49: 246–57.

ROTH, R. J., 1963: Further application of weather information. *Bull. Amer. Met. Soc.*, 44: 72–4.

RYDELL, C. P. and SCHWARZ, G., 1968: Air pollution and urban form: a review of current literature. *Jour. Amer. Inst. Planners*, 34: 115–20.

STARK, D. W., 1957: Weather modification. Water – three cents per acre foot. *California Law Review*, 45: 698–711.

SUTCLIFFE, R. C., 1966: *Weather and Climate*. Weidenfeld & Nicholson, London, 206 pp.

TEFERTILLER, K. R. and HILDRETH, R. J., 1961: Importance of weather variability on management decisions. *Jour. Farm Econ.*, 43: 1163–9.

THEIL, H., 1966: Miscellaneous forecasts concerning the weather and the construction sector. In: THEIL, H., 1966, *Applied Economic Forecasting*. North Holland Pub. Co., Amsterdam, pp. 283–301.

THOMPSON, P. D., 1963: An essay on the technical and economic aspects of the network problem. *W.M.O. Bull.*, 12: 2–6.

WASHINGTON, W., 1968: Computer simulation of the earth's atmosphere. *Science Journal*, 4(11): 36–41.

WHITE, R. M., 1969: Geophysical data management – why? *Bull. Amer. Met. Soc.*, 50: 143–6.

W.M.O., 1967: *Activities and Plans of the World Meteorological Centres*. Weather Watch Publication, World Meteorological Organization, Geneva, 104 pp.

W.M.O., 1968: The economic benefits of national meteorological services. *World Weather Watch Planning Report*, No. 27, W.M.O., Geneva.

VII Political, planning and legal aspects

A. THE SETTING

Several recent trends in the *use* of the atmosphere on the one hand, and the actual *variations* in the atmosphere on the other, have focused attention on the political, planning and legal aspects of weather and climate. These growing trends include: the pollution of the atmosphere, man's ability to either modify and/or forecast the weather with greater precision, and the environmental control that man is exerting in urban areas. Of these trends, variations in the 'natural' atmosphere usually involve planning and at times political decisions, whereas modification of the 'natural' atmosphere initiates many legal and political problems.

Although climates have always been changing, it is only in recent decades that man has considered the implications of these changes on his economy. This awareness of climatic change has been brought about largely by the general warming of most parts of the earth (particularly some areas of the Northern Hemisphere) in the period 1900–1940, and the substantial cooling in these same areas in the decades of the 1950's and 1960's. Such 'climatic variations' emphasize the need for appropriate long-term planning, for it is clear that planning must be involved if we are to live within the limit of our atmospheric resources. In particular, the cooling in the last two decades is important, for if it continues, agricultural production could be reduced considerably, particularly in the 'temperate' areas of the world. Plans are therefore required to produce food reserves to meet emergencies if widespread crop failure should occur. Associated research, involving the development of crops and animals able to withstand harsher conditions, as well as more economical methods of heating buildings are also urgently required. If this is to be accomplished, however, the politician and planner must become more weather-orientated, for only then will the climatic resources of the 1970's and 1980's be used to the best advantage.

Legal concern with atmospheric resources is usually restricted to

331

those aspects concerned with their modification, whether they be intentional such as rainmaking and fog dispersal, or non-intentional such as atmospheric pollution. Concern has arisen, chiefly because weather modification is not uniformly beneficial, for some people may sustain large losses as a result of weather modification activities. Increased rainfall, for example, may benefit the wheat farmer but only at the expense of the fruit farmer. Similarly, increased snowfall may increase the profits of the hydro-electric power company, but the same 'man-induced' snow may cause large stock losses and great inconvenience to the travelling public. Such 'losses' can result in court cases, and a considerable amount of evidence has already accumulated in the United States regarding 'losses' resulting from weather modification 'errors'. It is, of course, extremely difficult (if not impossible) to prove that 'man-made' rain did cause a loss, for there is no way of distinguishing man-made rain from natural rain. In this regard, however, court cases have resulted and will presumably continue to occur in the future.

The political and administrative aspects of the atmospheric resources are also pertinent, particularly in regard to legislation relating to weather modification and air pollution. For example, the 'simple' question of 'who owns the atmosphere' has still to be resolved, and it seems clear that some form of international agreement on the use and control of atmospheric resources must be formulated in the near future.

B. POLITICS AND POLICIES

1. Politics and the atmosphere

Politicians and political scientists have for the most part taken little interest in the human use of the atmosphere. Generally they have accepted the atmosphere as 'given', and have directed little thought to the effects of weather or weather modification on human behaviour. The relationship between technology and the political system has not, of course, been ignored. Mann (1968), for example, discusses a publication by Price (1965) on the role of the scientist in political decision-making and the status of science in the policy process. Political scientists are certainly aware of the complex interplay between technological developments and social and political conditions, because strategy in international relations is affected. The political significance of technological breakthroughs in engineering and economic analysis is also recognized by political scientists, as illustrated by their extensive work on water resources. Nevertheless, the problems associated with weather, weather modification and weather forecasting have, in general, not been considered.

The political and economic consequences of weather modification have many associated problems, and some of these are discussed in a review by Mann (1968). He asks, for example, if the U.S. Federal Aviation Agency undertakes fog suppression, should those benefited by the service, notably the airlines, be required to support such services financially? Furthermore, if rainfall is induced in one location, will it not result in a long-time decline in precipitation in the downwind area, thus damaging the interests of those receiving less rain? And, if this does occur, what public mechanism can balance the benefits and the costs between those receiving the additional rainfall and those receiving less?

In addition to the problems already mentioned, there are those concerned with the public investment aspect of weather modification. Mann, for example, says:

> . . . weather modification has tremendous implications for public investment at all levels of government. Increased certainty with regard to water supplies may reduce the need and affect the design criteria for such major water installations as dams, levees, and urban storm drainage systems. Reduced spending for these facilities may then provide opportunities for increased investment in other areas. Such changes . . . will create new needs to re-examine and revise public policy.

In a similar manner, weather modification also has a potential as a weapon for conflicts among nations, since the ability to modify weather and to use this capacity to affect crops, river systems and urban water supplies may give a nation strategic advantages. Further, it is possible that weather control could be used in more subtle ways to induce behaviour changes over time.

a. Research needs

The research needs of some of the problems already mentioned are examined in some detail by Mann. In their studies of public policy and decision-making processes political scientists have used many approaches including traditional institutional analyses, case studies, enquiries into the attitudes and perceptions of decision-makers, and investigations of the relationships between given environmental conditions and policy outcomes.

All of these techniques, suggests Mann, are of potential value in examining and understanding human reactions to atmospheric change. First, institutional studies would reveal some of the consequences of alternative methods of organizing regulatory programmes in weather modification, such as what should be the formal and informal relationships between official weather modification programmes and private groups concerned with such activities, and at what point is technical judgment required, and at what point administrative judgment? Second, case studies of decision-making in weather modification would provide some insights into how weather modification problems arise and are resolved. Such studies would reveal how various individuals and groups, both private and public, reach decisions about weather modification activities. In addition, Mann suggests that more needs to be learned about the behaviour of legislators, bureaucrats and private parties in reaching decisions about weather modification activities. Third, little is known at present about the attitudes and perceptions of decision-makers in weather modification activities, but as these practices become more widespread, public awareness will increase, and possibly more definable attitudes and views will emerge. For example, Mann asks:

> . . . If the snow pack is increased in a given area, what effect will this have on investments in ski areas and on individual decisions to ski? If water supplies are made more dependable, and destructive storms less likely, how will cities adjust their investment budgets and policies? What changes will farmers make in water management practices, types of crops grown, and soil erosion controls? How will public policy change when traditional programs (dams, irrigation systems) are no longer necessary? How will vacation patterns be affected by planned weather changes?

The long-term effects of weather modification activities from a political and decision-making viewpoint must also be considered, and Mann in his paper asks how decision-makers will operate within different time parameters. In contrast with short-term programmes, for example, the effects of long-term programmes will be felt only gradually. Nevertheless, these effects will have to be anticipated and appropriate decision needs considered, such as the decisions relating to the design and construction of dams, farm operations, forestry practices, storm drainage systems, recreational patterns, and numerous other social and economic activities, if modification of the atmosphere is to be long-term.

b. Natural and modified weather

It is clear from the writing of Mann, in his survey entitled 'Human dimensions of the atmosphere from the perspective of a political scientist', that a considerable amount of research is required if we are to understand fully all the political aspects of the modified as well as the natural atmosphere. Indeed, it is perhaps noteworthy to emphasize that most if not all studies so far have been concerned with the modified atmosphere, or with the 'side effects' of modifying atmospheric conditions, and few if any studies have concerned the more basic problems of the political aspects of the natural atmosphere.

The need for such research is emphasized by Ostrom (1968), who poses several major questions. These include: how will human welfare be affected by efforts to modify atmospheric conditions; what are the essential characteristics of the atmosphere as a resource system which must be taken into account in the organization of an appropriate enterprise and management system; and, what enterprise–management systems are appropriate for the development of atmospheric resources?

The first question relates to the fact that the natural atmosphere behaves as, to quote Ostrom: 'a dynamic system which sustains a flow of both goods and bads when measured by the criteria of human evaluation'. Thus, hail, lightning, tornadoes, hurricanes and other atmospheric phenomena can be the source of major destruction to human values, whereas on the credit side the atmosphere performs the essential work of distilling and transporting nearly all of the drinkable water supplies found upon the land masses of the earth. A major political question is, therefore, whether the developing technology of weather modification can be used to modify atmospheric conditions in order to increase the yield of precipitation under certain conditions, and to decrease the magnitude of turbulence and the yield of 'bads' under other conditions. In this regard, Ostrom points out that an economic advance in human welfare can occur wherever human intervention increases the yield of goods in excess of aggregate costs or decreases the yield of injuries or 'bads' with the same net result. The difficulty is, of course, that a 'good' at one time and place may create injuries or losses at another time and place. In the same way, increased precipitation at the time of harvest may be as detrimental to agriculture as decreased precipitation during the early growing season. Further, although men may be willing to take a calculated risk and tolerate the losses suffered by an 'act of God', they may not be as tolerant of losses suffered as a consequence of the actions of other men.

A further aspect of this first question is the need to know something of the pattern of interaction among the different potential uses of the atmosphere and the effect that these patterns have upon the general resource economy of the atmosphere. For example, what effect will any weather modification attempt to reduce storm damage have on the atmosphere's yield of water in the form of precipitation, its use for air transportation and its capacity to discharge diverse waste loads? Conversely, asks Ostrom, how do efforts to increase the atmospheric yield of precipitation affect storm patterns and intensities and other patterns of use which form a part of the total configuration of demands made upon the atmosphere?

The second question concerns the appreciation of the fact that the atmosphere as a natural resource appears to meet all of the criteria of a common-pool, flow resource system. It should be noted, however, that the resource is not easily isolated except within the controlled air-space of a shelter, and as a consequence it cannot be easily bounded or contained, and in most of its aspects it cannot be subject to a high level of control. Ostrom also suggests that the atmosphere as a common-pool, flow resource has some of the most perplexing characteristics of any of the common-pool resources, at least as perplexing as those of the oceans. One particular problem is that unlike water resource systems, the atmosphere has no distinct and definite boundaries for distinguishing sub-units. There are, for example, no boundaries for easily characterizing microclimates, airsheds and sub-systems in the atmosphere. The task of specifying boundary conditions appropriate to the field of effect produced by one form of action in the use and development of atmospheric resources may, therefore, be quite independent of the boundary conditions appropriate to another pattern of use and development. As a result, says Ostrom, different scales of organization and different forms of enterprise may be appropriate to different types of development.

The third question asks what enterprise–management systems are appropriate for the development of atmospheric resources which involve the concept of a common property resource. This conception usually includes reference both to the individual interests in the common-pool or flow resource and to the community of interests which are shared in such a resource, one having reference to the elements and relationships comprising the subsets of interests, the other to the universal set of interests associated with a common-pool or flow resource.

Many complex problems result, therefore, from the common-pool aspects of the atmosphere, and although the field of water resource development provides the most extensive experience in the use of mixed private and public institutional arrangements for the development of a common-pool, flow resource, this experience, according to Ostrom, is not directly applicable to problems concerning the development of atmospheric resources. Nevertheless, many of the basic concepts can be used, subject to appropriate modification, to deal with the distinctive circumstances associated with the development of atmospheric resources.

2. Policies towards weather modification

During the 1960's a number of reports were published in the United States about the success of various weather modification activities in that country. As already pointed out in Chapter II, it now seems clear that in certain areas, at certain times and with certain kinds of atmospheric conditions, weather modification is a reality. The 'success' of such atmospheric modification has focused attention on the public policy towards weather modification, not only in the United States, but in Canada and other countries as well.

In the United States, for example, commercial weather modification firms are finding a growing market for their services and a number of federal government agencies are accelerating their efforts to develop operational programmes. Sewell (1968) reports, for instance, that the U.S. Bureau of Reclamation is spending more than $3 million a year on basic and applied research to provide the capability to augment natural precipitation in regions in which some of its irrigation and hydro-electric power projects are located. The additional streamflow which this precipitation would provide is expected to result in significant improvements in productivity of the Bureau's irrigation and power schemes. In addition, the National Science Foundation and the Departments of Agriculture, Commerce and Defense have expanding programmes on research and development relating to weather modification.

a. Government programming in United States

The place of government programming in weather modification in the United States has been discussed in some detail by Kahan

(1968), who used statistical information published by the Library of Congress (1966) and the National Science Foundation (1967). A table in the latter publication shows that in the fiscal year 1966 the Departments of Agriculture, Commerce, Defence and the Interior joined with the National Aeronautics and Space Administration and the National Science Foundation in spending \$7,034,382 on 84 identifiable projects, under the leadership of 81 identified principal investigators. Twenty-three of the investigators were Federal government employees, one was a non-university state employee, 42 were employed by universities and 15 were private research groups. The pattern of participation by Federal Agencies in areas of weather modification in 1966 is shown in Table VII.1.

TABLE VII. 1 : *Participation of U.S. Federal Agencies in weather modification, 1966* (from Kahan, 1968; after Library of Congress, 1966)*

	Cloud physics	Precipi-tation	Hail	Light-ning	Cold fog	Warm fog	Hurri-canes	Torna-does	Miscel-laneous	Total
Agriculture	—	—	—	140	—	—	—	—	—	140
Commerce	—	35	35	130	—	—	320	—	130	650
D.O.D. – Army	95	—	—	95	30	—	—	—	30	250
D.O.D. – Navy	196	40	—	—	—	196	67	—	125	624
D.O.D. – Air force	60	—	—	—	—	60	—	—	60	180
Interior	—	1,465	—	—	—	—	—	—	103	1,568
N.A.S.A.	—	—	—	—	—	69	—	—	—	69
N.S.F.	314	581	393	182	—	25	49	57	459	2,060
Total	665	2,121	428	547	30	350	436	57	907	5,541

* Thousands of dollars.

The absence of major duplication and the presence of some co-operation are important aspects of the governmental scene in the United States, says Kahan. Nevertheless, gaps exist between programmes, and in addition there are under-emphasized areas that deserve attention and funding support.

One area of concern, for example, is that although the scientific controversy surrounding efforts such as precipitation management is being replaced by factual evidence concerning the conditions under which cloud seeding produces the desired result, this knowledge will be of debatable value until the necessary economic and social analyses required for effective application have been made. As Kahan states:

The governmental efforts in these areas to date have been sincere but inadequate. Limitation of funding is not the whole reason.

There has been a great preoccupation with answering the scientific and engineering questions involved. We have not communicated effectively the urgent need for participation of the social scientists. There have been governmentally sponsored efforts to stimulate their interest, but there is a need for a much greater effort in this direction. It may take a long time to solve the 'people problems' the rapidly developing technologies for modifying the weather will create.

Some of these problems may be solved by drawing them to the attention of social scientists and it is clear that their participation in the design of operating systems should contribute to identification of those circumstances in which the greatest net benefits are likely to be obtained. Some studies have in fact been made in this regard, such as those by Lyden and Shipman (1966) and Sewell and Day (1966). The first paper mentioned is concerned with the public policy issues raised by weather modification, and a most relevant question, posed by Lyden and Shipman, is in defining the function of weather modification:

> Is weather modification a public function, analogous to military protection, postal services, or highway construction? Or is it a private, non-governmental function? If the latter, is it a private function vested with a public interest, as had been determined to be the case, for example, in rail or air transportation? . . . Or is weather modification an ordinary, lawful business, subject only to reasonable regulation, as in the manufacture and sale of clothing? This view raises questions about property rights, i.e. who owns the clouds?

At least in the United States, however, weather modification as a public function does not simplify matters, for a considerable amount of private capital has already been invested in cloud seeding enterprises, and any proposal of a governmental monopoly would undoubtedly meet with strong opposition from such investors. Even if these objections could be overcome, additional questions arise, such as should the programme be a Federal governmental operation, or one administered by the states with co-ordination, and perhaps partial financing, by the Federal government.

b. Government jurisdiction

To overcome the uncertainty of policy development proceeding from government arbitration in cases of conflicting activities by private firms, a government may resort to more positive forms of action, but the question of governmental jurisdiction then arises.

Lyden and Shipman ask if weather modification activity efforts can be undertaken by the Federal government on the basis of its powers enumerated in the U.S. Constitution. It appears that the interstate commerce and war powers could be interpreted to encompass weather modification. Any such activities having international ramifications would, of course, be potentially subject to federal control, but Lyden and Shipman point out that the Federal government is not obliged to pre-exempt this whole field for itself. Where it does not act and does not preclude activity by other governmental jurisdictions, the states under their residual powers could presumably operate in the field, and the authors suggest that one may argue that if control is left to the individual states, the resulting governmental activities will better reflect the country's particularized regional needs, both physical and cultural. On the other hand, in the United States 50 sets of State Laws, or in Canada 10 sets of Provincial Laws, could cause innumerable administrative problems. The alternative of the concurrent jurisdiction of the Federal and state or provincial governments, could minimize many of these problems if the respective activity efforts of each was properly designed to interrelate with and support the other. The problems that would still arise in many areas would, of course, be those associated with international boundaries, and it seems reasonable to suggest that some form of international agreement must be formulated in the near future.

Regardless of how the jurisdictional problem is handled, there is a range of possible positive actions that can be undertaken. Several categories are suggested by Lyden and Shipman. The first concerns the various uses of compulsory power, that is, the use of restraints or positive requirement behind which stands the threat of enforcement through the imposition of penalties or disabilities. This category carries the designation coercive, to emphasize the central point that legal standards are established and enforced as an expression of the public interest. However, because of the crudity of criminal law enforcement as a means for the expression of public policy, it is most frequently used, according to Lyden and Shipman, as a means of supporting other more conclusive 'determinative' types of intervention. These fall into two general categories, 'prior restraints' and 'corrective intervention'. When prior restraint is employed, licences or permits are used to enable qualified persons to exercise a designated privilege, subject to the limitations applying to the authorized field of activity, and several U.S. states employ prior restraint in the regulation of weather modification activities. Accordingly, in most instances licences or permits to conduct cloud seeding are issued only to those applicants who can demonstrate that they have certain technical qualifications and the financial means, usually through insurance, to meet any claims for damages incurred. However, the authors state that this approach is of questionable effectiveness at this stage in the development of weather modification, because the potentialities and limitations of weather modification are not yet clear enough to be translated into meaningful qualifications for issuing licences or permits. Further, since restraint operates most effectively in areas where public policy is quite well defined, and technical considerations well understood, then prior restraint can be used in weather modification only in the rather limited areas where clearly defined criteria have been established.

The second form of coercive action, corrective intervention, stems from a statutory standard, often supplemented and interpreted by formal rule-making, applicable to all persons who may enter the field. Private initiative is tolerated as long as it does not transgress the applicable standards, but if in the judgment of the administering agency a violation occurs, the offending party is placed on defence and is required to show cause why he should not be restrained from further activity of the type complained of. Lyden and Shipman state that the advantages of this type of action derive from selective intervention in a field of private action. Such intervention can be used to influence a desirable level of conduct in a flexible, mobile fashion, and it is especially applicable where there is need to explore towards a realistic balance between toleration and restraint, and one that reflects a workable accommodation of private initiative and general public interest. For this reason, the authors suggest that it would appear to merit much more consideration than it has thus far been given for the regulation of weather modification activities, for it could be applied when weather modifiers had misrepresented their potentialities or their accomplishments, and used in addition to control the scope or range of operations of such entrepreneurs.

In addition to the coercive forms of action already discussed, there are those non-coercive forms of service, assistance and proprietary action. The differences among them can be seen in the product supplied: service provides a specific product such as protection, defence, research, education, etc.; assistance ordinarily results in a change in the capacity of persons to meet their needs; proprietary action results in a purchasable service or commodity.

The United States, for example, has Federal, state and local governments engaged in a variety of non-coercive actions in weather modification. Local governments, in particular, frequently undertake cloud seeding activities to increase their water supply, and at the federal level many departments and agencies are performing cloud seeding and conducting research in cloud physics, typical examples being Project Skyfire (Forest Service) and Project Stormfury (Navy). These ventures can be classified as non-coercive service actions. The U.S. Federal government and the states are also involved in non-coercive assistance actions. Congress, for example, has directed the National Science Foundation 'to initiate and support a program of study, research, and evaluation in the field of weather modification', and towards this end, it has embarked upon a course of providing financial assistance to encourage the sponsoring of research efforts in cloud physics and related areas of investigation.

In their survey of public policy issues, Lyden and Shipman assess the forms of action most appropriate to weather modification. The types of activity that a government can undertake are classified as government-performed operations (i.e. government activities designed to modify the weather for itself and others), research, control over privately performed operations, and risk protection. In addition, the forms of action most appropriate for each type of activity are indicated for those areas in which policy is emerging, those in which public policy is well defined and all areas regardless of the state of policy definition.

In summarizing the position, the authors suggest that in the United States the wise choice of weather modification is probably a strategy falling between the extremes of emphasis, and it is important to note that non-governmental activity is already active in the field, whereas government experience is limited, research and development work appearing to be the government's most significant effort. As pointed out earlier, the international implications need to be clarified, and under these circumstances the use of either a governmental monopoly or the toleration of open-ended private initiative is questionable.

c. The role of government in Canada

The position of government policy in other countries is not, of course, the same as in the United States, and it is perhaps appropriate briefly to contrast the position in Canada. Sewell (1968) states that weather modification in Canada to date has generally met with broad public acceptance and few major conflicts of interest have appeared. Nevertheless, it seems probable that as the intensity and geographical incidence of weather modification expand, the number of potential conflicts of interest will multiply and court cases will arise.

Governments in Canada have in general played a very passive role in the field of weather modification. The Federal government, for example, has confined its activities to the funding of research, mainly on cloud physics, and the provincial governments have shown little or no interest in weather modification potentialities or problems. In the field of cloud seeding, government agencies have hired firms in an effort to augment precipitation to raise the levels of hydro-electric power reservoirs, to supply water for agricultural purposes and to reduce forest fire hazards, but little information is collected as to the nature and location of weather modification attempts and their claimed successes. A further noteworthy point is that there are presently no regulations relating to weather modification in Canada, and almost anyone who wishes to modify the weather can do so. Sewell indicates, however, that increasing activity in weather modification in Canada will inevitably lead to governments taking a much more active role than they have in the past in controlling operations, and in determining circumstances under which weather modification is the appropriate means of dealing with problems created by variations in the weather.

The management of Canada's atmospheric resources, at least up to 1969, appears to have been assumed by the Federal government. The Meteorological Branch of the Department of Transport is the principal federal government agency involved in atmospheric resources, but other departments including the Department of National Health and Welfare and the Department of Agriculture are both concerned with the 'use' of atmospheric resources regarding air pollution and agriculture. By contrast, and with few exceptions, there are no organized programmes at the provincial level of administration in Canada that deal entirely with the atmosphere.

Sewell considers that the question is open, whether the present administrative framework in Canada is adequate to deal with the problems that are likely to emerge from increasing use of the atmosphere, and from an expansion of weather modification. This depends upon the resolution of the question as to which level of government has jurisdiction over the resources of the atmosphere, particularly in view of the fact that the ownership and control of these resources is not dealt with specifically by the British North American Act. It is possible that the provinces may claim jurisdiction over the atmospheric resources of 'their' territory, as appears to be the case with off-shore mineral rights, but the problems created by the dynamic nature and common property characteristics of atmospheric resources seem to make it almost certain that the Federal government will assume the major control over their use.

d. Perception and attitudes of public policy

The perception of the possibilities of weather modification and the attitudes of people towards government involvement in weather modification are questions of considerable merit, whatever the 'political' attitude towards atmospheric resources is. A review of the associate problems of perception and attitudes is given by Sewell and Day (1966), who comment:

> . . . not everyone is convinced that the weather can be altered. Some people believe that increased rainfall does result from cloud seeding, others believe that the rain would have fallen in any case. A number suggest that weather modification is a fraud. There are also differences of opinion as to whether modification is beneficial. Some people feel that it can bring considerable economic benefits: others feel that the overall results are negative. A few go as far as to say that it is immoral since it involves 'tampering with Nature'.

Such difference of opinion is clearly of importance to the formulation of public policy, and in 1965 Sewell and Day administered a questionnaire to 140 people in several parts of the United States and Canada. The two main groups involved were farmers (those who were actively engaged in farming on a full-time basis) and non-farmers (all others), the non-farmers being drawn from a wide variety of occupations, such as forestry, aviation, shipping, meteorology and engineering. In brief, five main questions were asked: who believes that attempts to modify weather are effective; to what extent is the belief in the feasibility of weather modification influenced by the degree of aridity, and by various socio-economic characteristics; which groups feel that the government should be involved in weather modification; what kinds of government activity related to weather modification appear to have support and which kinds appear to be opposed; and, how do attitudes towards government involvement vary occupationally and regionally?

The results, according to Sewell and Day, revealed that there were serious doubts on the part of many people (at least in those areas of the U.S. and Canada studied) as to whether weather modification is effective, and even if it is, whether its results are generally beneficial. In addition, the study revealed what appeared to be overwhelming support for government involvement in weather modification. Among the additional comments most frequently encountered were those relating to the potential 'external effects' of weather modification, many of the respondents fearing that cloud seeding in one state might rob another of the precipitation it would otherwise receive, and several feared that seeding on one side of the international boundary would affect people on the other side. The setting up of some kind of international agreement in weather modification has already been suggested, and some of the respondents in the questionnaire of Sewell and Day in fact suggested that immediate attention should be given to the establishment of an appropriate administrative framework to deal with the interstate and international disputes which will inevitably arise.

C. PLANNING ASPECTS

1. The importance of atmospheric resources

The importance of assessing the value and utilization of atmospheric resources in any planning programme has already been emphasized. Planning, of course, can have a very wide meaning, including planning for future sales and for industrial and urban design facilities, buildings, towns and cities. All of these aspects of planning involve looking into the future, and perhaps a fundamental question in any programme is whether or not the economic, social and physical variable will be the same as at present, or will be different. For example, demographic specialists assist by predicting the likely characteristics of population, economists predict trends in interest rates, gold prices, international trade and volume of transport, and geographers, sociologists and psychologists assist the planner in predicting possible trends in the 'social' characteristics of communities. All of these predictions are considered by the planner, in one way or another, irrespective of whether he is planning shopping centres, sewage systems, new towns, new industries, new products or new transportation facilities. In most cases, however, the physical environment is considered to be static, and is usually not considered (at least as a variable) in any planning. For example, ocean currents are usually expected to remain constant in any ocean sewage disposal system; landforms, rivers and lakes are usually considered static features in any town, industrial or transportation planning; and in a similar manner the atmospheric resources are normally considered 'non-variable'. It is, nevertheless, clear from the evidence of the twentieth century that the resources of the atmosphere are anything but 'non-variable'. In any one year, for example, there are wide fluctuations in the weather in any area, and the climatological records are constantly being broken. Such variations are well known and many – although by no means all – short-term planners do take these variations of the atmosphere into account. Nevertheless, most planners do not appear to consider the fact that such variations in atmospheric resources, if repeated in the same area, can have significant economical and social effects, and even more important, if a trend towards drier, wetter, colder or warmer conditions continues – as indeed has happened – then plans made now, on the assumption that the climate is *not* changing, may well be totally inadequate. A proper appreciation of the fact that the atmosphere *is* a variable is therefore an essential aspect in any short- or long-term planning.

The need for research on planning and the related decision-making involved in the human uses of the atmosphere are reviewed by Hufschmidt (1968):

> Viewed most broadly, the planning and decision-making questions involving use of the atmosphere as a resource should be of the same type as questions arising from the use of any other resource. . . . In fact, the notion of the atmosphere as a natural resource is a very powerful one: it allows one to apply the theories, methods of insight of natural resource planning (including water-resource planning) to the problems of human use of the atmosphere. . . . In economic terms, its beneficial effects – rain, sunshine, clouds and winds in the proper proportions of time, place and intensity – have been considered free goods, its malevolent effects – floods, droughts, heat waves and cold waves – have been considered unavoidable costs.

In considering planning and decision-making for atmospheric resources, as analogous to that for water resources, Hufschmidt gives several examples, but most are concerned with the effects of weather modification and short-range forecasts, and little mention is made of the problems involving long-term planning using atmospheric resources. The importance of systems research is emphasized, however, and Hufschmidt suggests that the social and economic effects of atmospheric management can be thought of as outputs of a physical–technologic–economic system. Weather modification, for example, may change rainfall inputs to a given hydrologic system, which may or may not be controlled or disturbed by the works of man. Consequently, using a systems analysis such as simulation, the effect of rainfall input can be traced through a water resources system to output in the form of water supplies of a particular quantity and quality, or in the form of controlled high and low flows, electricity and energy produced, and recreation services provided.

The systems approach in such cases is undoubtedly most appropriate, and with modification could presumably be considered in the use of the very long-range forecast (i.e. for a decade or more) for long-term planning for the 1980's and 1990's. However, such forecasts are not yet available, and according to Namias (1968) will not be available until the late 1970's. In the meantime, the only knowledge of the input of the atmospheric resources into the world economy of the 1970's comes from the short-range forecast (month to season), and the 'climatic trend'. But, as most if not all atmospheric scientists readily agree, the 'climatic trend' is extremely difficult to predict; it would indeed be a bold planner who would base his planning ideas on a forecast of the climatic trend. Nevertheless, there are some indications, as has already been pointed out, that the climate of the 1970's and 1980's is unlikely to be as warm as the 1930's and 1940's, and if, as some climatologists are predicting, the temperature trend is downward, then it is perhaps not unreasonable for a planner to take these predictions into account.

2. Man the modifier

One of the problems of forecasting any long-term climatic trend is man's influence through weather modification and atmospheric pollution on the 'natural' climatic trend. Indeed, it seems likely that the climatic trend produced by man could become more important than that produced by nature, and it is not clear whether the man-made trend will be any more predictable than the natural trend.

In a study of urbanization and the new weather technology, Meier (1966) has some very pertinent comments, and says that since weather is part of a geophysical world-system, repeated attempts at its modification, even though only local effects are intended, can have world-wide repercussions. Further, he comments that it may be safely concluded that either the weather modification programme must be limited to a single isolated political system for which uniform legal provisions can be legislated (e.g. Australia, New Zealand or Iceland), or it must have a multi-billion dollar net profit, part of which can readily be applied to the equalization of benefits. Meier says that, to be successful, weather modification must have major immediate economic effects of a highly predictable sort; accordingly this argument rules out virtually all agricultural applications, for even in market crops a more regular supply of water will tend to reduce unit value of the crop, leading to debatable improvements on welfare. But Meier emphasizes that cities would benefit, since they have developed at least half the property value in the world, and accumulated a much larger share of the human capital.

In the operation of contemporary cities, Meier says that only a small part of the losses attributed to weather are likely to be controllable. Nevertheless, truly large gains and losses may be attributed to growth potentials. Where huge populations need to be urbanized, for example, so that economic development may proceed, and weather is a limiting factor, then weather becomes a crucial variable and the values involved become very large. The evidence for this, suggests Meier, is the problem of urban water supply. In some cases, such as Tokyo and Hong Kong, the problem is perennial, and periodic typhoons of destructive magnitude are needed to fill the reservoirs and prevent a worse disaster. By contrast, in most other vulnerable metropolitan areas, such as New York, it is the onset of an occasional drought, or a series of subnormal years, that causes a crisis.

The effect of long-term climatic changes on cities is also commented upon by Meier, who suggests that some present-day cities are rather finely adjusted to their climatic environment. Indeed, even cities facing a water shortage would be likely to experience significant losses, at least over a five- to ten-year span, if the rainfall were to be increased substantially, since natural vegetation would change, production cycles starting from agricultural crops would need to be worked out all over again, the costs of floods would be raised, and unexpected insect pests would appear for which there would be no already existing defences. Thus, many of the adjustments to the change are likely to result in over-compensations which create a new chain of losses. Meier considers that the talk about helping an already existing city by significantly changing its climate is ill-considered and he suggests that it would be in the interest of the cities to object to projects for bringing about permanent changes in climate.

On the other hand, the prospective improvements in weather forecasting should enable the city to adapt more readily to normal changes, and the present trend towards automation of production, marketing and transportation scheduling should enable automatic adjustments to potential weather-caused interruptions to be readily introduced into the system. For example, Meier says that the effect of the computerization and distribution is to produce more sales per unit of inventory. The effect of better weather forecasting would therefore mean that the products used by people are more likely to be on hand when asked for, and less likely to be in stock when not needed. Improvements in the system attributable to improved weather data may amount to only 1 or 2%, says Meier, but in view of the fact that a $1,000 billion American economy is anticipated for 1975 a rather large weather data collection system can be afforded.

In association with, and possibly in addition to, improved weather forecasting, various programmes of weather control are possible. These include a review of the vulnerability of cities to extreme deviations in weather which will suggest economical proposals for warding off any weather catastrophe, and the redirection of very severe storms into relatively unpopulated areas, or their dissipation. In addition, programmes to reduce the likelihood of weather extremes would reduce the stringency of engineering and operational specifications. Such a programme, however, involves extremes which are closely interrelated in the atmospheric system; hence a programme undertaken to reduce the range of variation would probably have to be national or international in scope and probably must await a much more thorough accumulation of data about the urban economy and the atmospheric variables.

The problems that are, and will be, encountered in any planning based on a forecast of atmospheric conditions are therefore considerable, but it is clear that a much greater understanding of the economic and social aspects of the environment – particularly of the urban environment – will be required before planning on either climatological predictions or long-term weather forecasts can be accomplished. Nevertheless, the recognition that the weather and climate during the last three decades of the twentieth century will be different from that already experienced is essential if the best uses of the planner's talents are to be achieved.

D. LEGAL ASPECTS

1. The law and the atmosphere

In the past little concern has been expressed over the legal aspects of the atmosphere. This was due in part to man's inability to control the atmosphere, and because man, although causing pollution of the atmosphere, was not aware of the consequences. The decades of the 1950's and the 1960's have demonstrated, however, that the ownership of the resources of the atmosphere is important, since man is now modifying the atmosphere in various ways to alter the weather in certain areas, and he is also polluting the atmosphere to such an extent that most people are aware that this problem exists.

Questions such as who 'owns' the air, and what are the legal boundaries of the air, are therefore of considerable importance. However, except in a few cases few decisions have been made or agreed to by either provinces, states or nations as to its ownership. Perhaps the most pressing question to be answered is the legal ownership of the atmosphere, for the emission of 'pollutants' (whether these be directly or indirectly associated with any weather modification activity), is at present controllable only at their source, and once these 'pollutants' move across boundaries (local, government, provincial, state or federal), a breakdown in jurisdictional control results. In effect, control is exercised over air *space*, but as yet few if any controls have been agreed on as to the control of the atmosphere *per se*.

2. The law and weather modification

a. Status of atmospheric resources

Although the legal ownership of atmospheric resources raise problems of considerable complexity, a few lawyers have expressed an interest in the problem, including Morris (1965, 1966), Johnson (1968) and Taubenfeld (1968), and some of their findings are briefly commented upon. One of the early papers on 'The Law and Weather Modification' was by Morris (1965), who commented on the sharp increase in the number of proposed and passed laws designed to curtail, control or outlaw weather modification activities, and he indicated that even counties (in the U.S.) are initiating ordinances to prevent or severely impede the progress of weather modification. Many involved closely with weather modification, he said, feel that regulation at the present time is premature and that a scientific breakthrough might be lost by too closely harnessing scientists in their research. However, as Morris points out, that argument overlooks the fact that the courts and the legislatures in the U.S. have already been forced to act, and will be further 'prodded by the public to continue to enjoin cloud seeding and to hastily enact laws concerning weather modification'.

In general, lawyers appraise past rules and precedents when asked to advise on entirely new and experimental matters, and Morris indicates that most of the present legal literature on the subject of weather modification refers to rules pertaining to the flow of ground and surface waters, and is in many cases the only guide to weather modification regulation. Yet, the knowledge of the scientist may indicate to the legal profession that the ancient laws pertaining to rivers should not be applied to weather modification. Consider, for example, the following legal and social problems which may arise from hurricane steering (Fig. VII.1), as presented by Morris.

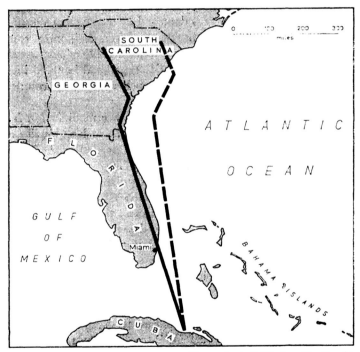

FIG. VII.1. *Natural track of Hurricane Cleo, 1964 (solid line), with altered track (dashed line) if 10° direction change were initiated 24 hours prior to landfall (after Morris, 1965; after Hendrick, 1965).*

. . . let us assume that a means is found, perhaps by seeding one sector of a hurricane, of getting it to wobble off with a ten-degree direction change. Consider . . . the course of Hurricane Cleo in 1964. . . . What agency is going to give the power to decide which way to divert a hurricane? Who will pay the damages to homes in South Carolina if the decision is made to steer it from Florida? Or, who will pay the damages to homes in Florida if the agency decides not to steer it away from Florida? Will insurance companies continue to write coverage in South Carolina when they learn that every hurricane headed for Florida will be diverted to there? Will there be some type of agreement between the states ahead of time for a bonus payment from the State of Florida to the people of South Carolina? Will Florida refuse to pay if the scientists report that the hurricane would have hit South Carolina even if there had been no seeding?

It would seem in this case that the Federal government should handle these problems, but if the United States government steps in and makes the payments, would the government then be 'forced' to pay the one and one-half billion dollars each year for everyone in the United States whose property is damaged by any type of storm?

Such are some of the problems envisaged in a decision to steer a hurricane away from its natural path, and probably of even greater significance from the point of view of international law and international relationships, is the effect of such steering on countries outside of the United States.

The necessity for legal regulation of weather modification is of course not limited to catastrophic type storms, such as hurricanes. For example, Morris quotes the case of the airports in the United States which are engaging in supercooled fog removal by silver iodide and dry ice seeding. So far, he says, the only serious side effects appear to be occasional snow flurries in the communities downwind of the airports, and in one case, the airport manager agreed to pay for snow removal in the downwind city in order to be allowed to continue with the seeding programme. However, the problems will become more numerous when the removal of the warmer fog is possible, and it seems clear that if fog dispersal is allowed to go on unchecked at airports, legal recourse by the bus lines, taxi companies, and the airport news-stands and restaurants, etc. to such 'interference' will probably be forthcoming. A significant point in such an argument, it seems, is the question as to whether a company or individual has the legal right to alter the 'natural' weather, or whether the advantages that accrue to some economic enterprises through adverse weather conditions should be permitted to continue.

b. The courts and weather modification

In a later paper, Morris (1966) stresses that the law is lagging too far behind the advance of science and technology, and asks: '. . . on what principles should these laws and regulations be based?'

During the last two decades, in the United States, several cases have gone to the courts, and Morris says that although no general body of principles has been developed, a few general principles have been suggested such as 'the greatest social utility' or 'negligence of the defendant'. For example, the rule providing 'the greatest social utility' was suggested in *Slutsky v. City of New York* ([1950] 97 NYS 2d 238), in which a resort owner sought an injunction to prevent the City of New York from engaging in cloud seeding. The court, in denying the injunction, stated:

> This court must balance the conflicting interests between a remote possibility of inconvenience to plaintiffs' resort and its guests with the problem of maintaining and supplying the inhabitants of the City of New York and surrounding areas, with a population of about 10 million inhabitants, with an adequate supply of pure and wholesome water. The relief which the plaintiffs ask is opposed to the general welfare and public good; and the dangers which plaintiffs apprehend are purely speculative. This court will not protect a possible private injury at the expense of a positive public advantage.

The New York ruling represented a broad-minded and liberal attitude, according to Morris, but not all courts concur. For example, in *Southwest Weather Research, Inc. v. Rounsaville*, ([1958] 327 SW 2d 417), a Texas court enjoined all future cloud seeding activities above the plaintiff's land.

At this point it should be noted that in most lawsuits for damages the plaintiff must *prove* that the defendant was negligent or careless. Such proof is, of course, difficult in most, if not all, weather modification activities, for there is no known way of distinguishing 'man-made rain' from 'natural rain'. Further, even if the modified could be distinguished from the natural element, and modified rain falls on say farmer Jones's property, this does not necessarily prove that this rain was caused by a specific rainmaker.

Up to 1966, eight lawsuits involving weather modification had been filed in the United States; one of these was abandoned before the court had reached a decision on its merits, two were under consideration, and of the remaining five, one was decided in favour of persons desiring to enjoin weather modification activities, and four were decided in favour of the defendant weather modifiers. Three of the latter four decisions held that no damage was attributable to the attempted cloud seeding; accordingly, we do not know what those courts would have done had it been established that the seeding did somehow modify the weather in question. Two of the three cases cited by Morris are as follows:

> . . . *Samples v. Irving P. Krick, Inc.,* (Civil Nos. 6212, 6223 and 6224, Western District of Oklahoma, 1954). This case arose out of cloud seeding sponsored by Oklahoma City in 1953. The plaintiff, a landowner, sued for property damages incurred in a cloudburst and flood which were coincident with the cloud seeding operations. The plaintiff failed to prove to the satisfaction of the jury that the seeding could have influenced the storm, and their verdict was for the defendant.
>
> . . . *Auvil Orchard Co., Inc., et al. v. Weather Modification Inc., et al.* (Case No. 19268, Superior Court, Chelan County, Washington, 1956), involved cloud seeding for the prevention of hail. Flash floods had occurred on farms adjacent to the hail prevention target area. The court therefore granted a *temporary* order banning hail suppression attempts for one season. At a later date, however, after hearing expert meteorological testimony, the court refused to grant a permanent injunction. It was not convinced that cloud seeding had brought about the exceptional rainfall which caused the floods.

In these cases the plaintiffs failed to *prove* that cloud seeding caused injury to the persons suing for the redress, although it is important to appreciate that even though proof of injury might not be made, a temporary restraining order can effectively stop an entire season's weather modification activities as happened in the above-mentioned Auvil Orchard case. In the two other cases, discussed by Morris, one (*Slutsky v. City of New York,* already discussed) resulted in a decision in which cloud seeding was held to be permissible even though the plaintiff might be injured thereby; in the other, the Supreme Court of Texas barred cloud seeding undertaken by a group of private farmers. Morris described the Texas case as follows:

> . . . *Southwest Weather Research Inc. v. Rounsaville,* (*1958*) 327 SW 2d 417. The court found that the purpose of the cloud seeding was to disperse gathering hail storms which might injure certain crops of the farmers. However, ranchers in the area desired moisture in any form, including hail, and sued to stop cloud seeding which might interfere with gathering storms. Conflicting expert and lay testimony was presented by the parties, but the trial court reached the conclusion that the cloud seeding attempts were effective and did deprive the ranchers of moisture. The Texas Supreme Court reviewed the case and found that it was proper for the trial court to issue an order in these circumstances, banning further cloud seeding until additional evidence could be produced

showing that hail prevention activities would not reduce the amount of precipitation on the ranchers' lands. . . .

Morris suggests that these cases illustrate that research guidance is needed by the courts. For similar reasons, guidance is also needed by the legislators, and Morris sounds a note of warning that (at least in the U.S.) powerful lay groups of special interest, with no real scientific background, are exerting pressure on legislators. For example, Maryland in 1965 made it a crime, subject to three years' imprisonment, to engage in any form of weather modification over the state (Maryland Senate Bill 348, March 30, 1965), and the State of Pennsylvania, in November 1965, created a law which gives to each county the option to outlaw weather modification, this Act specifically prohibiting weather modification whenever as Morris quotes 'The County Commissioners shall adopt a resolution stating that such action is detrimental to the welfare of the country.'

c. Problems in legal research

Another viewpoint of weather modification and the law is given by Johnson (1968), who outlines the state of existing legal research with regard to weather modification, particularly as it applies to the U.S. National Science Foundation and other U.S. government agencies concerned with research on legal problems raised by weather modification. Five major questions in the field of law and law-policy are suggested as deserving early study. First, study of the existing federal regulatory agencies to determine the relevance of their competence and experience to atmospheric control problems; second, study of the relevance of the federal-review-board concept as expressed in the 1949 Hoover Commission Report and subsequent studies of federal organization related to water projects; third, the study of why and to what extent the Federal government, rather than the states, should control human uses of the atmosphere; fourth, the study of why and to what extent the court-centred, common law system should control the human use of the atmosphere, and why and to what extent administrative agencies should exercise control; and fifth, the reason why and to what extent the Federal government should provide indemnity for damage caused by weather modification activities.

During the 1950's a considerable amount of literature appeared on the subject of the law of weather modification, but in the 1960's, except for certain government-sponsored research, little interest has been shown by legal scholars in weather modification problems.

Exceptions include the work of Morris already discussed, and that of Taubenfeld and Oppenheimer. For example, Oppenheimer (1965) deals with the activities of various federal agencies in the area of weather modification, and state legislation in respect to weather modification, and the advisability of conducting federal research projects in conformity with state statutes. Also discussed is the recovery of damages under the Federal Tort Claims Act and other federal laws.

In the early stages of legal research on weather modification most if not all problems were concerned primarily with rainmaking. During the 1960's, however, a much wider range of weather modification activities have emerged, such as hurricane modification, cold fog dispersal, warm fog dispersal, hail suppression, snow diversion, lightning suppression and regional climate modification. In addition, the more accurate weather predictions of the 1960's and the promise of even more accurate and longer-ranged forecasts in the 1970's has focused attention on a far wider range of legal problems, for, as Johnson points out, the problems raised by cold fog dispersal over airports are vastly different from those raised by hurricane modification.

In view of these 'new' techniques of weather modification, Johnson suggests several areas of research that justify further consideration by legal experts.

First, the use of the 'traditional legal research'; that is, using the 'traditional' materials of the appellate court decisions. In the United States all state supreme court decisions, most lower state appellate court decisions and all Federal Appellate Court decisions are printed and published in book form and are kept in numerous law libraries throughout the nation. Accordingly, 'library research' is possible, but has certain limitations, for even if one can accurately predict what the courts of one state will hold in a given situation, this does not necessarily imply what the state courts in the other 49 states, or federal courts, may decide in the same or similar situation. To date, however, only a few weather modification decisions have been rendered in all the state and federal courts combined; accordingly, the traditional case law scholar must search further afield to find analogies in cases involving air and water pollution.

Second, the concept of 'reasonableness' as expressed in the law of nuisance, and the law of riparian rights, should be thoroughly examined for possible applications to weather modification problems. For example, Johnson suggests that a study of how the concept of 'reasonableness' as expressed in the law of nuisance, which has been applied in cases involving use of the atmosphere for smoke disposal, would be useful. In addition, an examination of how 'reasonableness' as expressed in the law of riparian rights, which has been applied in cases involving use of streams for waste disposal, would provide valuable insights in problems associated with atmospheric resources. However, it should be noted that, so far, the doctrine of nuisance has been explored only lightly for its possible applications to the use of the atmosphere for weather modification. Nevertheless, there is a similarity between the use of the atmosphere for waste discharge and for intentional weather modification, since both involve the use by individuals of an unowned but readily available natural resource, and both involve intentional modifications of that resource for the 'benefit' of the modifier's own land. Because of these similarities, Johnson suggests that it would seem that a careful appraisal of the law of nuisance in smoke damage cases might be helpful in demonstrating how the courts are likely to treat weather modification problems in the future, and in a similar way, the concept of 'reasonableness' as applied in the riparian rights law pertaining to waste disposal in rivers and streams, could provide some interesting analogies. This law says that a man can use a river or stream flowing past his property for waste disposal so long as his use is 'reasonable' in relation to the uses of others below him. There are therefore some obvious similarities between the 'pollution' situation and that posed by weather modification, as well as presumably the situation involving the 'use' of the natural atmosphere as it flows past a state's or man's property.

The third area of research suggested by Johnson involves the relevance or otherwise in weather modification of various legal concepts such as 'strict liability' (which applies in cases of an activity considered 'ultra-hazardous'), 'trespass' (relating to the unauthorized entry of a person or thing upon land in the possession of another) and 'negligence', which is defined as conduct which falls below the standard established by the law for the protection of others against unreasonable risk of harm. With respect to 'trespass', Johnson states that causing rain to fall on another's land has been held to be a trespass, as has flooding another's land, as well as on occasion the casting of smoke into the atmosphere which then is carried by the wind over another's land. Thus, in similar circumstances, if cloud seeding is performed negligently, and no harm would have resulted *except* for that negligence, then liability may be imposed on the cloud seeder. However, before this concept can be applied effectively, a standard of performance against which to measure the conduct of the weather modifier must be established.

The fourth point involves one of the most critical questions in court actions for damages or injunctions resulting from weather modification; that is, the proof of cause and effect. Before the courts award damages or injunctions for weather modification activities, according to Johnson, they must be convinced under rules long established, that the modification attempt did in fact alter the weather, and that as a result of this alteration the plaintiff was damaged (and would *not* have been damaged otherwise). It is obvious, however, that because of the nature of the weather modification problem, cause and effect will ordinarily be difficult to prove in that it is extremely difficult if not impossible to prove that the modification attempt did in fact alter the weather. However, as Johnson points out, although it will probably not be possible to prove that on a certain day one inch more rain fell than would have fallen but for the weather modification attempt, and that this additional rainfall was the cause of farmer Brown's damaged crops, it may be possible to prove that over a given three-month or six-month period, one inch of rain was added to what normally would have fallen on a given area. Consequently, if this can be proved, it means that when the rainfall of an area is changed, certain benefits and costs will be 'shared' among the residents there. A problem remains, however, that the courts may be relatively powerless to allocate the costs so that they are also shared by those receiving the benefits, since it will be difficult to prove who received the benefits.

The fifth area for further research, suggested by Johnson, is assessing the role of the courts and the Federal government in the field of weather modification. This is a difficult area requiring legal research, and Johnson indicates five inadequacies of the court-centred system as applied to weather modification problems. They are the inability: to handle the more sophisticated problems of cause and effect, to control the air polluter or weather modifier whose effect is so small on other users that these users find the expense of a lawsuit too great in relation to the recoverable damages, to protect the public interest in a given condition of air or weather, to develop uniform standards for resource quality, and to develop long-range plans for the development and use of atmospheric resources.

d. National and international aspects

The legal problems associated with hurricane modification, as has already been pointed out, will no doubt lead to international questions, and in fact the position of international law and the use, misuse and alteration of atmospheric resources has already received attention. The position of the international lawyers and weather modification, for example, has been reviewed by Taubenfeld (1968), who poses five questions. First, what are the questions lawyers ought to be asking themselves about the impact of weather modification activities and the need for legal direction and control; second, what potential areas of international friction can be foreseen; third, what mechanisms exist, or might be necessary, to prevent and resolve such disputes; fourth, what efforts to date have been made to ask and attempt to answer these questions; and fifth, what potentially are the most fruitful lines for future investigation and action?

According to Taubenfeld, answering these questions involves an understanding and appreciation of the role of atmospheric resources in several areas. For example, consider questions such as: Can weather 'belong' to a person, to a county, to a state or to a country? Are there rights in 'man-produced' weather? If such rights can be acquired, is it by capture and/or by first user? Further, can a person 'buy' someone else's weather? Assuming that 'harm' (which may mean no more than 'change') occurs, and that responsibility can be traced, is the award of damages the 'best' result – or is the best remedy prohibition of 'harm' or 'change'? Are there other remedies one could pursue? Should there be 'immunities', or absolute liabilities, or some compromise restraints? All of these questions do of course apply equally well to 'domestic' weather modification. Nevertheless, questions of international concern include: does a state 'own' *its* weather; what measures can it take to protect itself against a deprivation or a change of its natural weather; can another state obtain a 'right' to experiment by offering compensation; if the losses are small, what arrangements exist or can be created to adjust them amicably; if they are large, will the same procedures work; and, if losses are major and threaten another country's survival, or its survival in comfort, what can be done? Taubenfeld further asks what is the role of the United Nations Charter in regulating state action and reaction in this field? For example, does Article 51 of the U.N. Charter, which limits the use of force in response to an 'armed' attack, legally bar the use of force to prevent another country from destroying the weather *status quo*?

Several international agencies, including the World Meteorological Organization, are of course directly concerned with international meteorology, but their role, including that of World Weather Watch already discussed in detail, has generally been concerned with the non-legal aspects of the atmosphere. Indeed, it is clear that one of the prime aims of the W.M.O. is to stimulate international co-operation, and few meteorologists would consider that the atmosphere over a W.M.O. member's territory had a 'status' in any way different from the atmosphere over the territory of any other member country.

The question of whether existing international treaties can serve as guides for weather modification activities is posed by Taubenfeld. For example, does the prohibition of nuclear explosion established in the Antarctic Treaty of 1959, and the Test Ban Treaty of 1963, have implications for possible controls of weather modification activities? Second, what treaty regime (air law, international river controls, outer space provisions) offers useful precedents for dealing with potentially harmful experiments?

As will readily be appreciated, the law as it is applied to the atmosphere, whether it be the natural or modified atmosphere, raises problems of great complexity, and many of these problems are interdisciplinary in their nature. Thus, the atmospheric scientist's view on the magnitude and time-scale of potential changes are essential, as are the views of political scientists, economists and sociologists.

The problems associated with the ownership of the resources of the atmosphere *are* part of the economic and social environment of the 1970's. If these problems are to be solved, a considerable amount of research will be required by those social scientists interested and trained in law and meteorology. If this research is done, it is clear that the best use of the world's atmospheric resources will be achieved.

BIBLIOGRAPHY

HENDRICK, R. L., 1965: *Potential Impact of Weather Modification on the Insurance Industry.* Travellers Research Center, Hartford, Conn., 21 pp.
HUFSCHMIDT, M. M., 1968: Needs for research on planning and decision-making aspects of human uses of the atmosphere. In: SEWELL,

w. r. d., *et al.*, 1968: *Human Dimensions of the Atmosphere.* National Science Foundation, Washington, D.C., pp. 21–8.

JOHNSON, R. W., 1968: Weather modification and legal research. In: SEWELL, W. R. D., *et al.*, 1968: *Human Dimensions of the Atmosphere.* National Science Foundation, Washington, D.C., pp. 87–98.

KAHAN, A. M., 1968: The place of government programs in weather modification. *Bull. Amer. Met. Soc.*, 49: 242–6.

LIBRARY OF CONGRESS, 1966: *Weather Modification and Control.* Legislative Reference Service, Library of Congress, Rep. No. 1139, 89th Congress, 2nd Session, Washington, D.C.

LYDEN, F. J. and SHIPMAN, G. A., 1966: Public policy issues raised by weather modification: possible alternative strategies for government action. In: SEWELL, W. R. D., ed., 1966: *Human Dimensions of Weather Modification.* University of Chicago, Dept. of Geography, Research Paper No. 105, pp. 289–303.

MANN, D. E., 1968: Human dimensions of the atmosphere from the perspective of a political scientist. In: SEWELL, W. R. D., *et al.*, 1968: *Human Dimensions of the Atmosphere.* National Science Foundation, Washington, D.C., pp. 81–6.

MEIER, R. L., 1966: Urbanization and the new weather technology: problems, possibilities and institutional implications. In: SEWELL, W. R. D., ed., 1966: *Human Dimensions of Weather Modification.* University of Chicago, Dept. of Geography, Research Paper No. 105, pp. 261–76.

MORRIS, E. A., 1965: The law and weather modification. *Bull. Amer. Met. Soc.*, 46: 618–22.

MORRIS, E. A., 1966: Institutional adjustment to an emerging technology: legal aspects of weather modification. In: SEWELL, W. R. D., ed., 1966: *Human Dimensions of Weather Modification.* University of Chicago, Dept. of Geography, Research Paper No. 105, pp. 279–88.

NAMIAS, J., 1968: Long range weather forecasting – history, current status and outlook. *Bull. Amer. Met. Soc.*, 49: 438–70.

NATIONAL SCIENCE FOUNDATION, 1966: *Weather Modification*, Eighth Annual Report, National Science Foundation, Rep. NSF 67–9, Washington, D.C.

OPPENHEIMER, J. C., 1965: Legal aspects of weather modification. In: *Weather and Climate Modification.* Report to Chief, U.S. Weather Bureau, U.S. Department of Commerce, Washington, D.C., July 10, 6 pp.

OSTROM, V., 1968: Needs for research on the political aspects of the human use of the atmosphere. In: SEWELL, W. R. D., *et al.*, 1968: *Human Dimensions of the Atmosphere.* National Science Foundation, Washington, D.C., pp. 71–9.

PRICE, D. K., 1965: *The Scientific Estate.* Harvard University Press, Cambridge, Mass., 323 pp.

SEWELL, W. R. D., 1968: Weather modification and Canadian public policy, *Queen's Quart.*, 75(2): 289–301.

SEWELL, W. R. D. and DAY, J. C., 1966: Perception of possibilities of weather modification and attitudes toward government involvement.

In: SEWELL, W. R. D., ed., 1966: *Human Dimensions of Weather Modification.* University of Chicago, Dept. of Geography, Research Paper No. 105, pp. 329–44.

TAUBENFELD, H. J., 1968: The international lawyer and weather modification. In: SEWELL, W. R. D., *et al.*, 1968: *Human Dimensions of the Atmosphere.* National Science Foundation, Washington, D.C., pp. 99–102.

ADDITIONAL REFERENCES

BAUM, W. A., 1968: Congressional action on weather modification. *Bull. Amer. Met. Soc.*, 49: 234–7.

CHANGNON, S. A., JR, 1969: Recent studies of urban effects on precipitation in the United States. *Bull. Amer. Met. Soc.*, 50: 411–21.

CLEVELAND, H., 1966: The politics of outer space. *Bull. Amer. Met. Soc.*, 47: 105–10.

DAVIS, R. J., 1968: Special problems of liability and water resources law. In: TAUBENFELD, H. J., ed., 1968: *Weather Modification and the Law.* Oceana Publications, Dobbs Ferry, New York, pp. 103–40.

DUCKWORTH, F. S., 1961: The meteorologist as an expert witness. *Bull. Amer. Met. Soc.*, 42: 447–51.

GABITES, J. F., 1965: Air pollution and the siting of industry. *Proc. 4th New Zealand Geog. Conf.*, Dunedin, pp. 161–5.

GREENWOOD, P. G. and HILL, R. R., 1968: Buildings and climate in Singapore. *Jour. Trop. Geog.*, 26: 37–47.

HAAS, J. E., 1968: Social and political aspects of planned weather modification. *Proc. 1st Natl. Conf. Weather Modification*, Albany, N.Y., pp. 202–9.

HOWELL, W. E., 1965: Cloud seeding and the law in the Blue Ridge area. *Bull. Amer. Met. Soc.*, 46: 328–32.

JOHNSON, R. W., 1968: Legal implications of weather modification. In: TAUBENFELD, H. J., ed., 1968: *Weather Modification and the Law.* Oceana Publications, Dobbs Ferry, N.Y., pp. 76–102.

LAWRENCE, E. N., 1954: Microclimatology and town planning. *Weather*, 9: 227–32.

MANN, D. E., 1968: The Yuba City flood: a case study of weather modification litigation. *Bull. Amer. Met. Soc.*, 49: 690–714.

MERCER, A. G., 1968: International law and the French nuclear weapons test. *Pacific Viewpoint*, 9: 51–68.

MORRIS, E. A., 1968: Preparation and trial of weather modification litigation. In: TAUBENFELD, H. J., ed., 1968: *Weather Modification and the Law.* Oceana Publications, Dobbs Ferry, N.Y., pp. 163–84.

OSOKIN, I. M., 1962: The study of the climate of cities as an urgent present day need and a task for university geographers. *Soviet Geog.*, 3: 55–8.

SCHWARTZ, L. E., 1965: *International Relations and Weather Modification.* A report to the Special Commission on Weather Modification. Operations and Policy Research, Inc., Washington, D.C., 100 pp.

SEALY, K. R., 1967: The siting and development of British airports. *Geog. Jour.*, 133: 148–71.

TAUBENFELD, H. J. (Editor), 1968: *Weather Modification and the Law.* Oceana Publications, Dobbs Ferry, N.Y., 228 pp.

WALLINGTON, C. E., 1965: Weather in court. *Weather,* 20: 290–3.

WOHLERS, H. C., PURDOM, P. W., JACKSON, W. E. and SCHOENBERGER, R., 1969: Can air pollution be controlled by legislation? *Scientia,* 104(681–2): 58–64.

VIII Conclusion

It is clear that greater ingenuity and human effort, as well as an increasing portion of national incomes, must be devoted to the atmospheric sciences. Advances in technology have, for instance, created new demands for weather information, but they have also generated new atmospheric hazards. For example, man by creating new compounds and 'dust' in the atmosphere causes new kinds of pollution, some of which give rise to difficult problems of analysis, prevention and dispersal. To help him cope with these problems, man quite naturally looks to the atmospheric sciences.

The evaluation of the atmospheric environment, however, transcends the borders of customary disciplines, and as Petterssen (1966) has noted, meteorology may therefore be called upon to perform an even greater number of services to meet the needs of a crowded world. Many of the services required are types with which meteorologists are already familiar: disaster warnings, services to aviation, tourism, shipping and the general public. Others, including longer-range forecasts for agriculture, fishing, water-resource management, air pollution control, construction works and certain industrial and retail operations, will provide new avenues for the 'weatherman'. But how will the atmospheric scientist cope with these hazards, and more importantly, what guidance can we expect in the future in order that all people may be able to live in harmony with their atmospheric resources?

Fortunately, developments in the 1960's advanced the concept of atmospheric resources, the international projects of the World Weather Watch (W.W.W.) and the Global Atmospheric Research Programme (G.A.R.P.) being especially noteworthy. The World Weather Watch, in particular, will provide a significant advance over the existing international weather system, and is expected to lead to a marked improvement in operational weather forecasting over the entire earth for the protection of life and property and for economic growth. Obviously, this improvement will benefit not only the emerging nations, whose economies are primarily agricultural and whose weather services are poorly developed, but also the advanced nations, for the effect of meteorological hazards and long-term climatic trends is probably even more important in those areas where long-term planning and development are well established.

The Global Atmospheric Research Programme should also provide a research resource that may stimulate global weather modification processes so that in the future we may know just how far we can go deliberately, or may go inadvertently, in changing the weather and climate. However, to assure that large-scale environmental modification, if it can be accomplished, is used only under co-operative international control, and solely for the benefit of the people of the world, remains a problem for the politicians to solve.

In this book, an attempt has been made to bring together the most significant and pertinent associations between man's economic and social activities, and the variations in his atmospheric environment. Basically, this has involved an understanding of, and appreciation for, the resources of our atmospheric environment. To provide maximum benefit, however, these resources must be recognized, and their importance correctly evaluated.

BIBLIOGRAPHY

PETTERSSEN, S., 1966: Recent demographic trends and future meteorological services. *Bull. Amer. Met. Soc.*, 47: 950–63.

ADDITIONAL REFERENCES

BATTAN, L. J., 1969: Weather modification in the U.S.S.R. – 1969. *Bull. Amer. Met. Soc.*, 50: 925–45.

CRESSMAN, G. P., 1969: Killer storms. *Bull. Amer. Met. Soc.*, 50: 850–5.

DAY, H. J., BUGLIARELLO, G., HO, P. H. P. and HOUGHTON, V. T., 1969: Evaluation of benefits of a flood warning system. *Water Resources Res.*, 5: 937–46.

DE ANGELIS, R. M., 1969: Enter Camille. *Weatherwise*, 22: 173–9.

FLOWERS, E. C., McCORMICK, R. A. and KURFIS, K. R., 1969: Atmospheric turbidity over the United States, 1961–66. *Jour. App. Met.*, 8: 955–62.

GENTRY, R. C., 1969: Project STORMFURY. *Bull. Amer. Met. Soc.*, 50: 404–9.

HAHN, LE ROY, 1969: Predicted versus measured production differences using summer air conditioning for lactating dairy cows. *Jour. Dairy Sci.*, 52: 800–2.

HERBERT, G. A., HASS, W. A. and ANGELL, J. K., 1969: A preliminary study of atmospheric effects on the sonic boom. *Jour. App. Met.*, 8: 618–26.

JULIAN, P. R., KATES, R. W. and SEWELL, W. R. D., 1969: Estimating probabilities of research success in the atmospheric sciences: results of a pilot investigation. *Water Resources Res.*, 5: 215–27.

KLEIN, W. H., 1969: The computer's role in weather forecasting. *Weatherwise*, 22: 195–201.

KREITZBERG, C. W., 1969: Prospective applications of models to forecasting. *Bull. Amer. Met. Soc.*, 50: 947–55.

KUETTNER, J. P. and HOLLAND, J., 1969: The BOMEX project. *Bull. Amer. Met. Soc.*, 50: 394–402.

LALLY, V. E. and LICHFIELD, E. W., 1969: Summary of status and plans for the GHOST balloon project. *Bull. Amer. Met. Soc.*, 50: 867–74.

LAMB, H. H., 1969: The new look of climatology. *Nature*, 223 (5212): 1209–15.

McKAY, G. A. and THOMPSON, H. A., 1969: Estimating the hazard of ice accretion in Canada from climatological data. *Jour. App. Met.*, 8: 927–35.

McQUIGG, J. D., 1969: *The Economic Impact of Weather and Weather Information.* Invited Lecture, W.M.O. Commission for Climatology, Geneva, October, 28 pp.

MAINE, R., 1969: A real-time interactive control system for meteorological operations. *Jour. App. Met.*, 8: 845–53.

MAUNDER, W. J. and WHITMORE, A. D., 1969: The value of weather: challenge of assessment. *Aust. Geog.*, 11: 22–8.

MYRUP, L. O., 1969: A numeral model of an urban heat island. *Jour. App. Met.*, 8: 908–18.

RASOOL, S. I. and HOGAN, J. S., 1969: Ocean circulation and climatic changes. *Bull. Amer. Met. Soc.*, 50: 130–4.

ROBERTS, W. O., 1967: Climate control. *Physics Today.* 20(8): 30–6.

SEWELL, W. R. D., 1968: Emerging problems in the management of atmospheric resources: the role of social science research. *Bull. Amer. Met. Soc.*, 49: 326–36.

SEWELL, W. R. D. and McMEIKEN, J. ELIZABETH, 1967: Emerging problems in the management of atmospheric resources in Canada. *Atmosphere*, 5(4): 34–8.

SEWELL, W. R. D., et al., 1968: *Human Dimensions of the Atmosphere.* National Science Foundation, Washington, D.C., 174 pp.

SMAGORINSKY, J., 1969: Problems and promises of deterministic extended range forecasting. *Bull. Amer. Met. Soc.*, 50: 286–311.

SPAR, J., 1969: An experiment in localized numeral weather prediction. *Jour. App. Met.*, 8: 854–62.

STAFF, THE RAND CORPORATION, 1969: Weather modification progress and the need for interactive research. *Bull. Amer. Met. Soc.*, 50: 216–46.

SUGDEN, L., 1969: Provision of meteorological services for supersonic transport operations. *Quart. Jour. Roy. Met. Soc.*, 95: 789–94.

SUTTON, O. G., SINGER, S. F. and DAVIES, D. A., 1964: Weather and climate in 1984. *New Scientist*, 22(388): 208–12.

THOMPSON, J. C., 1969: The value of weather forecasts. *Science Journal*, 5(12): 62–6.

THOMPSON, L. M., 1968: Impact of world food needs on American agriculture. *Jour. Soil & Water Cons.*, 23(1): 3–9.

THOMPSON, L. M., 1969: Weather and technology in the production of corn in the U.S. Corn Belt. *Agron. Jour.*, 61: 453–6.

U.S. DEPARTMENT OF COMMERCE, DEFENSE, INTERIOR, STATE, TRANSPORTATION, ATOMIC ENERGY COMMISSION, NATIONAL AERONAUTICS AND SPACE ADMINISTRATION, NATIONAL SCIENCE FOUNDATION, 1969: World Weather Program plan for fiscal year 1970 (Report to the U.S. Congress). *Bull. Amer. Met. Soc.*, 50: 658–86.

WILLIAMS, G. D. V., 1969: Weather and prairie wheat production. *Canad. Jour. Agric. Econ.*, 17: 99–109.

— 1969: Taming the hurricane. *Newsweek*, Sept. 1, pp. 47–8.

— 1969: A killer named Camille and her toll. *Newsweek*, Sept. 1, pp. 16–19.

Subject Index

Geographical Index

Organization Index

Author Index

Including bibliographies and additional references

Morgan, J. J., 329
Morris, E. A., 350, 351, 352, 353, 354, 355, 356, 361, 362
Moule, G. R., 80, 178
Munn, R. E., 38, 49
Murphy, A. H., 262, 285, 325, 326, 328, 329
Murphy, F. C., 19, 49, 286, 287
Musgrave, J. C., 183
Musgrave, W. F., 61, 173
Myrup, L. O., 366

Namias, J., 269, 270, 271, 272, 309, 326, 347, 361
Nash, A. H., 87, 179
Naya, A., 84, 178
Neill, J. C., 70, 172
Nelms, W. P., Jr, 107, 178
Nelson, R. R., 230, 255, 282, 284, 300, 326
Nemoto, J., 329
Neuberger, H., 280, 327
Nicodemus, M. L., 70, 131, 178, 234, 255
Noffsinger, T. L., 329
Nordman, D., 183
Nuttonson, M. Y., 66, 178
Nye, R. H., 128, 178

Oddie, B. C., 278, 327
Odell, R. T., 69, 178, 179
Ogden, D. C., 35, 50, 228, 255
Oppenheimer, J. C., 356, 361
Orazem, F., 231, 255
Osmun, W. G., 178
Osokin, I. M., 362
Ostrom, V., 335, 336, 337, 361
Oury, B., 257

Palleson, J. E., 66, 173
Palmer, M., 330
Parker, E. M., 172
Parr, A. E., 185, 215
Passel, C. F., 198, 215
Paterson, J., 224, 255
Patterson, R., 78, 178
Paul, D. E., 183
Pengra, R. F., 66, 178
Penman, H. L., 63, 179

Perry, R. A., 25, 50
Peterson, G. A., 231, 255
Peterson, J. T., 89, 179
Petterssen, S., 312, 327, 364, 365
Petty, M. T., 150, 179
Phillips, L. E., 3
Philpott, B. P., 78, 179
Pink, J., 330
Plank, V. G., 183
Plant, E., 183
Popkin, R., 3
Portsmouth, G. B., 71, 179
Poulter, R. M., 198, 215
Prest, A. R., 228, 255
Price, D. K., 332, 361
Prohaska, F., 210, 215
Pruter, A. T., 92, 171
Purdom, R. W., 363

Rackliff, P. G., 198, 215
Ragland, J. L., 171
Rainey, V. G., 183
Rapp, R. R., 73, 176, 262, 300, 304, 305, 325, 327
Rasool, S. I., 366
Rayner, J. N., 52
Reedy, W. W., 330
Reichelderfer, F. W., 266, 281, 327
Revelle, R., 249, 255
Rice, R. W., 216
Richard, O. E., 126, 179
Richardson, L. F., 268, 327
Ridker, R. G., 228, 256, 330
Rigg, T., 73, 179
Riley, J., 300, 328
Roberts, W. O., 37, 50, 366
Robertson, G. W., 184
Robertson, N. G., 178
Robinson, A. H. O., 121, 179
Robinson, F. E., 182
Robinson, G. D., 269, 270, 327
Rogerson, T. L., 87, 179
Romanoff, E., 224, 254
Rooney, J. F., Jr, 52, 121, 179
Root, H. E., 295, 327
Rose, J. K., 69, 179, 227, 256
Rosenwaike, I., 209, 215
Ross, D. A., 79, 80, 180
Roth, R. J., 330